Practical
Ultrasound

Practical Ultrasound

Edited by
R.A.Lerski

Medical Physics Department,
Ninewells Hospital and Medical School,
Dundee DD1 9SY, UK

IRL PRESS
OXFORD · WASHINGTON DC

IRL Press
Eynsham
Oxford
England

First Published 1988

British Library Cataloguing in Publication Data

Practical ultrasound.
   1. Man. Diagnosis. Ultrasonography
   I. Lerski,R.A.
   616.07′543

Library of Congress Cataloging in Publication Data

Practical ultrasound.
   Includes index.
   1. Diagnosis, Ultrasonic. I. Lerski, Richard A.
RC78.7.U4P73 1988 616.07′543 88-13313

ISBN 1 85221 068 0 (hardbound)
ISBN 1 85221 157 1 (softbound)

Typeset by Infotype and printed by Information Printing Ltd, Oxford, England

Contents

Contents

Contents

List of Contributors

R.Blackwell
Department of Medical Physics and Bio-Engineering,
University College Hospital,
1st Floor, Shropshire House,
11−20 Capper Street, London WC1E 6JA, UK

M.C.Bossi
Ospedale San Paulo,
Milan, Italy

P.D.Clark
Department of Clinical Physics and Bio-Engineering,
West of Scotland Health Boards,
11 West Graham Street, Glasgow G4 9LF, UK

D.Cooke
Senior Registrar in Radiology,
St James University Hospital,
Beckett Street,
Leeds LS9 7TF, UK

D.O.Cosgrove
X-ray Department, Royal Marsden Hospital,
Downs Road, Sutton, Surrey SM2 5PT, UK

P.A.Dubbins
Consultant in Charge,
Plymouth General Hospital,
Freedom Fields, Plymouth, UK

D.H.Evans
District Department of Medical Physics and
Clinical Engineering,
Leicester Royal Infirmary, Leicester LE1 5WW, UK

J.A.Evans
Department of Medical Physics,
The General Infirmary,
Leeds LS1 3EX, UK

A.Hollman
Department of Diagnostic Radiology,
Ultrasonic Unit,
Western Infirmary,
Glasgow G11 6NT, UK

H.C.Irving
St James University Hospital,
Beckett Street,
Leeds LS9 7TF, UK

R.A.Lerski
Director of Medical Physics,
Medical Physics Department,
Ninewells Hospital and Medical School,
Dundee DD1 9SY, UK

M.B.McNay
Department of Midwifery,
University of Glasgow,
Queen Mother's Hospital,
Glasgow G3 8SH, UK

P.Morley
Department of Diagnostic Radiology,
Ultrasonic Unit,
Western Infirmary,
Glasgow G11 6NT, UK

J.P.Neilson
Department of Obstetrics and Gynaecology,
University of Edinburgh,
Centre for Reproductive Biology,
37 Chalmers Street,
Edinburgh EH3 9EW, UK

M.Restori
Senior Physicist,
Department of Ultrasound,
Moorfields Eye Hospital,
City Road,
London EC1V 2PD, UK

J.C.Rodger
Monklands District General Hospital,
Monkscourt Avenue,
Airdrie ML6 0JS, UK

C.D.Sheldon
Gartnavel General Hospital,
1053 Great Western Road,
Glasgow G12 0YN, UK

Preface

In the thirty years since the initial introduction of ultrasound into obstetrics and gynaecology by Ian Donald, its spread into virtually all areas of medical diagnosis has been relentless. These new applications have constantly proved invaluable in the non-invasive and hazard-free investigation of disease. The pace of technological advance, starting with the earliest bistable B-scanning followed by grey-scale methods and progressing through real-time equipment and digital electronics to Doppler applications, has been intense. At present, it is not clear where the next breakthrough in equipment design will come, except perhaps in the more widespread availability of low-cost software controlled digital systems, and perhaps in colour Doppler mapping. It seems likely that ultrasound will maintain its significant cost advantages over other medical imaging techniques [e.g. X-ray CT and Magnetic Resonance Imaging (MRI)]. As a consequence its impact on health care will continue to be great, particularly in view of the very high costs and low availability of MRI.

This text has been constructed from chapters contributed by experts in their individual specialist applications of ultrasound with the aim of providing a broad coverage of virtually all the present uses of the technique. Also included are chapters on the physics and technology of ultrasound. The book will be of value to trainees seeking a concise text covering all aspects of ultrasound and to users requiring additional information on applications that they wish to start in their own departments.

R.A.Lerski

Abbreviations

AIPP	axial imaged pulse profile
AIUM	American Institute of Ultrasound in Medicine
BPD	biparietal diameter
CT	computerized tomography
CTS	Cardiff Test System
CW	continuous wave
GTC	grey scale transfer curve
GTO	grey scale test object
hCG	human chorionic gonadotrophin
HCP	high contrast penetration
HSG	hysterosalpingography
IAR	imaged acoustic range
I(SATA)	spatial average temporal average intensity
ISB	intrinsic spectral broadening
I(SPTA)	spatial peak temporal average intensity
I(SPTP)	spatial peak temporal peak intensity
IUCD	intrauterine contraceptive device
IVC	inferior vena cava
LCP	low contrast penetration
LDPE	low density polyethylene
LIBP	lateral imaged beam profile
NTD	neural tube defect
PI	pulsatility index
PID	pelvic inflammatory disease
POD	polycistronic ovarian disease
PPI	plan position indicator
PRF	pulse repetition frequency
PVD	posterior vitreous detachment
PW	pulsed wave
RAM	random access memory
RF	radiofrequency
RTO	resolution test object
SCE	sister chromatid exchange
SFD	small-for-dates
ST	slice thickness
TGC	time gain compensation
TIUV	total intrauterine volume
TMM	tissue mimicking material
VHR	ventricular/hemisphere ratio
VTR	video tape recorder

Chapter 1

Physics—the nature of ultrasound

J.A.Evans

1. Introduction

1.1 *Ultrasound as sound*

For those of us with normal hearing, sounds and noises are part of everyday existence, and indeed sound-proof rooms have been suggested as a form of torture. The word 'ultrasound' is much less familiar and can conjure up images of scientific mystery and inaccessible technology. Ultrasound is however, merely sound of a particular kind and its nature is best understood by considering it as a special form of sound.

All sound is simply a mechanical vibration produced at one location and travelling through some medium to another. The banging of a drum causes the drumskin to vibrate. This vibration is passed on to nearby air molecules which in turn pass it on further. Eventually, this wave of vibration may travel far enough to enter someone's ear drum where it causes the membrane to vibrate giving us the sensation of hearing.

Sounds are distinguished from each other not only by their loudness but also by their pitch, or frequency. The frequency of a sound is the number of vibrations which occur per second. For example, the musical note middle C has a pitch of roughly 256 vibrations per second. If such vibrations are well ordered, they are termed cycles and a frequency of one cycle per second has a unit which is termed the *Hertz* and given the symbol Hz. Thus middle C is at a frequency of 256 Hz. It turns out that increasing the pitch by one musical octave represents a doubling of frequency. The C above middle C therefore has a frequency of 512 Hz (*Figure 1*). Six octaves above middle C brings us to a frequency of 16 640 Hz and it is difficult for the average human ear to perceive notes above this frequency. By convention, the limit of human hearing is normally taken to be 20 000 Hz or 20 kHz (kilo-

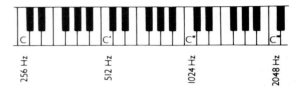

Figure 1. Typical piano keyboard showing middle C and three octaves.

hertz) and vibrations with frequencies higher than this are said to be ultrasonic. 'Ultra' in this context means beyond and so the name implies that these vibrations are 'beyond sound' whereas really they are 'beyond hearing'.

Ultrasound is therefore defined as mechanical vibration with a frequency above 20 kHz.

It is important to realize that this distinction between audible sound and ultrasound is arbitrary. Indeed, there are many mammals such as dogs, bats and dolphins whose hearing range extends well beyond 20 kHz.

In fact, medical ultrasound uses frequencies which are very much higher than anything occurring in nature. Typically the range 2−10 million cycles per second, i.e. 2−10 MHz (megahertz) is used for ultrasonic scanners. The reasons for this choice of frequency will be explained later in this chapter.

1.2 *Ultrasound as waves*

The sort of vibrations produced by a musical instrument are relatively well ordered and this distinguishes them from the chaotic, random vibrations we call noise. Scientifically, a vibration can only be said to be at a certain frequency if it is completely regular and vibrates at exactly the same rate continuously. This would be called a pure tone, and the movement of the vibrating body and the nearby air molecules can then

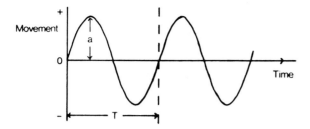

Figure 2. Sine wave movement of a body emitting a pure tone.

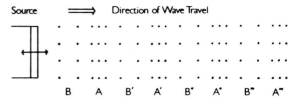

Figure 3. Regions of compression and rarefaction due to a piston-like source.

be described as a sine wave (*Figure 2*).

We can see that the body is initially at point 0 and moves firstly in the positive direction and then back to 0 before moving in the negative direction. The time, T, which it takes to execute one complete cycle is called the *period*. For a pure sine wave this is always exactly the same from one cycle to the next. The height of the curve indicated as 'a' in the diagram is known as the *amplitude*.

If the frequency of such a wave is f, then it must be executing f complete cycles per second. Therefore the period T of each one will be given by

$$T = \frac{1}{f} \qquad (1)$$

In the case of ultrasound the vibrating source is normally called the *transducer* (see Section 2.2) and in medical use it will normally be in contact with gel or the patient's skin so the molecules surrounding the source will be gel or tissue molecules rather than air molecules. Nevertheless the principle is the same.

It is often useful to think of the transducer as a vibrating piston. As it moves backwards and forwards it displaces particles of the adjoining medium in the same direction. This means that both the pressure and the density in that region of the medium will fluctuate in a similar way. Of course pressure and density will also have steady state values to which they will return when the wave has passed. However, the temporary changes in pressure and density produced by the source in the medium are passed to neighbouring molecules over a period of time (*Figure 3*).

Thus at any time there will be some regions such as A, A′, A″ etc. where the pressure is higher than usual, and the medium has undergone a *compression* and also there will be other regions such as B, B′, B″ etc., where the pressure is lower than usual and the medium is said to have undergone a *rarefaction*.

At some time later, the pattern will have changed

and regions A, A′ etc. will themselves be in *rarefaction*. In this way the disturbance or *displacement* introduced by the source is passed on through the medium as a series of compressions and rarefactions. It is important to realize that any one individual particle in the medium is merely oscillating backwards and forwards even though the wave itself is actually moving through the medium away from the transducer.

This form of wave in which the particles of the medium vibrate in the same direction as the wave motion is known as a *longitudinal* wave. It is also possible in some media (particularly solids) for waves to travel or propagate at right angles to the direction of particle displacement. These are known as *transverse* waves. Transverse waves, also known as *shear* waves, do not propagate well in soft tissue or gas and are therefore of very little significance in normal medical ultrasound practice. They can however, travel appreciable distances in bone.

1.3 *Speed of propagation*

The speed at which the wave propagates through the medium depends largely upon its *compressibility* (normally given the Greek letter 'kappa', $\varkappa$, as a symbol). This is a measure of how easy it is to compress or deform the medium. Thus a pressure wave propagating through a very compressible material will produce large deformations, i.e. relatively large particle displacements. This normally results in slower wave propagation.

If the speed of propagation of the wave through the medium is c, then

$$c = \frac{1}{\varrho \varkappa} \qquad (2)$$

where ϱ (Greek letter 'rho') is the density of the medium and $\varkappa$ is the compressibility, as above.

Normally solids, as relatively incompressible or 'stiff' materials, have low values of $\varkappa$ and therefore

Table 1. Values of compressibility, density and speed of sound in biological materials.

Material	Compressibility, $\varkappa$ 10^9 m s^2 kg^{-1}	Density, ϱ 10^3 kg m^{-3}	Speed of sound, c m s^{-1}
Aluminium	0.009	2.70	6400
Perspex	0.12	1.19	2680
Bone (skull)	0.08−0.05	1.38−1.81	3050−3500
Liver	0.38	1.06	1570
Kidney	0.40	1.04	1560
Blood	0.38	1.06	1570
Fat	0.51	0.92	1460
Lung	5.92	0.40	650
Air (NTP)[a]	7650	1.2×10^{-3}	330

Data from McDicken (1981), Hussey (1985) and Wells (1978).
[a]NTP, normal temperature and pressure.

high values of c. Gases being very compressible have high $\varkappa$ and low c and liquids and soft tissues are in between (see *Table 1*).

It can be seen from *Table 1*, that the values for the speed of sound in different soft tissues are very similar. On modern scanners, it is assumed that the value is 1540 m s^{-1} which is a reasonable approximation in most cases but can cause difficulties in specific instances where bone or eye tissue is involved or when there is a significant quantity of fat present.

It is extremely important to know the speed of sound assumed by a scanning machine for the following reasons:

(i) The principle of pulse−echo imaging depends upon an accurate knowledge of the speed of echo propagation (see Chapter 2).

(ii) The accuracy of most ultrasonic measurements relies upon the correctness of the 'c' value (see Chapter 2).

(iii) The interpretation of reflection coefficients at major interfaces relies upon knowledge of c along with the density (see later this chapter).

(iv) Interpretation of Doppler effect data is only possible quantitatively if c is accurately known (see Chapter 4).

(v) Focusing of real-time scanners relies upon accurate information on speed of sound values (see Chapter 2).

1.4 *Wavelength*

If a plot is made (*Figure 3*) showing particle displacement, pressure or density along the direction of

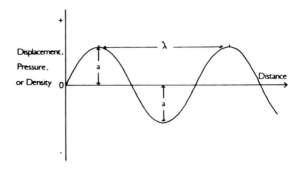

Figure 4. Graph showing displacement, pressure or density variations with distance. The wavelength λ and amplitude a are indicated.

propagation frozen at some point in time, it also takes the form of a sine wave (*Figure 4*). The vertical axis can represent particle displacement pressure or density with 0 being the equilibrium values. The distance measured in the medium between nearest points with identical values of one of those quantities is known as the *wavelength*, λ (Greek symbol 'lambda').

If the frequency is f, then that means that f cycles per second pass through any point in space. These f cycles have a length λ each and therefore occupy a total length of $f\lambda$ which must be the total distance travelled by the wave in one second, which is the speed c. Therefore

$$c = f\lambda \qquad (3)$$

As we have seen, medical ultrasound frequencies are normally in the range 2−10 MHz, which assuming c to have the value of 1540 m s^{-1}, implies that wavelengths in tissue are in the range 0.77−0.154 mm. Note that higher frequencies mean shorter wavelengths.

2. Generation and detection of ultrasound

2.1 *Piezoelectric effect*

Since ultrasound is a form of rapid mechanical vibration it is necessary to find a source which is capable of such a high frequency oscillation. This is not altogether trivial since there are no naturally occurring generators of ultrasound. However, materials do exist which exhibit the phenomenon of piezoelectricity or the *piezoelectric effect*. This means

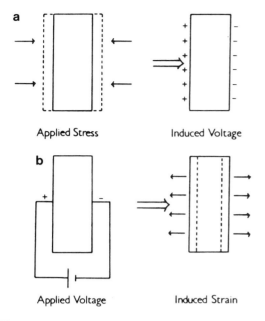

Figure 5. (a) Piezoelectric effect and (b) inverse piezoelectric effect.

that when a pressure is applied to the surface of such a material, an electric voltage appears on its surface. Further, this phenomenon also works in reverse, i.e. if a voltage is applied to the surface, the material will change shape normally by expanding or contracting. This is called the *inverse piezoelectric effect* (*Figure 5*). A housing incorporating a slab of piezoelectric material is called a *transducer*.

2.2 Frequency responses of ultrasonic transducers

If the frequency of the electric signal applied to the transducer is varied, its response will also vary. There will be several frequencies at which its response, i.e. the magnitude of the displacement produced, will be very large and others at which it will be quite small. The large responses are known as *resonances* and normally the largest resonance is found at the lowest resonant frequency f_R (*Figure 6*).

This corresponds to a wave inside the crystal travelling from one side to the other and back in exactly the *period* of the applied signal. If this happens this internally reflected wave reinforces the next cycle and so on. This happens when the distance travelled by the wave inside the crystal, i.e. twice its thickness, is equal to the wavelength (*Figure 7*). Therefore for

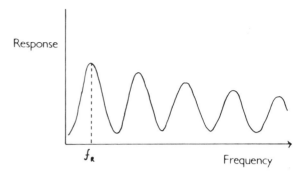

Figure 6. Frequency response of ultrasonic transducers showing multiple resonances. The frequency of the first peak is f_R.

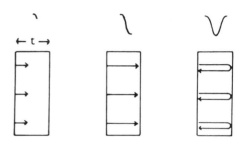

Figure 7. Half-wave resonance in an ultrasonic transducer. The wave travels from left to right arriving at the other face after travelling half a wavelength. After reflection, the journey back completes one whole wavelength.

the major resonance, the wavelength λ_R has to be equal to $2t$ where t is the thickness of the transducer crystal. If the speed of sound in the crystal material is c_T then the resonant frequency f_R will be given by

$$f_R = \frac{c_T}{\lambda_R} = \frac{c_T}{2t} \qquad (4)$$

and the device is said to be *half wave resonant*.

Frequently, only the lowest resonant frequency carries significant energy and so the device is insensitive away from this frequency. The term *bandwidth* is used to describe the shape of the resonant peak (*Figure 8*).

If the displacement produced at the resonant frequency f_R is made equal to 1, then the width of peak Δf is measured where the displacement falls to roughly 0.7 (-3 dB point, see Section 3.5). This value Δf is called the *transducer bandwidth*.

Most frequently scanners operate using short pulses of ultrasound, rather than continuous sine waves at a

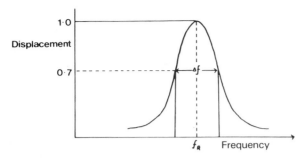

Figure 8. Transducer bandwidth Δf, shown as the width of the resonance peak at 0.7 maximum amplitude.

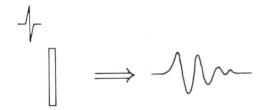

Figure 9. Short pulse at resonance frequency by a shock-excited transducer.

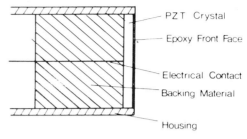

Figure 10. Construction of typical medical ultrasound transducer for pulse−echo imaging.

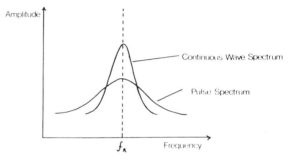

Figure 11. Comparison of bandwidth of a pulse and continuous wave transducer with the same resonant frequency.

single frequency. These short pulses are produced by applying a voltage step to 'shock' the crystal into oscillation. As is the case for any resonant system such as a spring or pendulum, the transducer responds to a step shock by vibrating at, or near to, its resonant frequency f_R (*Figure 9*).

Once shocked into oscillation, a transducer without energy loss would continue to ring like a bell, indefinitely. In order to produce very short pulses it is therefore necessary to cause the energy to be lost quickly and this is done by providing a suitable backing material to absorb sound travelling in the reverse direction. A diagram of a complete transducer is given in *Figure 10*.

The result of an effective backing is to shorten the response of a typical transducer to $3-4$ cycles, producing a pulse as in *Figure 9*. This obviously differs from a continuous wave output since there are now long periods in between pulses when no output is being produced and this is equivalent to adding some lower frequency components into the signal. Also, when the pulse starts, it does so very rapidly which is equivalent to adding some higher frequency components. The net result is that a well backed transducer of this sort, producing very short pulse outputs has

a much broader bandwidth than a continuous wave transducer (*Figure 11*).

There is, however, a penalty to be paid for short pulse production by this method. Much of the original energy is absorbed by the backing material and hence wasted. The sensitivity and efficiency of such transducers are therefore relatively poor. Nevertheless, the technique remains, far and away, the commonest in use in medical ultrasound today.

2.3 *Ultrasonic fields from transducers*

Once a suitable transducer has produced the required short pulse, it is important to study how that pressure wave spreads out from its source, as it travels through the medium in contact with the transducer. The shape of the ultrasound beam emitted from a transducer is governed by a process known as *diffraction*. Diffraction can be found whenever a wave approaches an obstacle or aperture which has dimensions comparable to a wavelength. It applies equally to light waves in optics although light wavelengths are many times smaller. The ultrasonic transducer can be con-

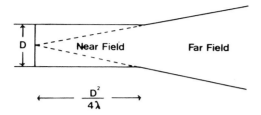

Figure 12. Idealized form of intensity distribution of an ultrasound transducer showing near field and far field.

sidered to represent such an aperture and therefore its dimensions strongly influence the beam shape.

For example, as we have seen, a circular transducer operating at 3 MHz produces ultrasound with a wavelength of roughly 0.5 mm in soft tissue. Such a transducer might have a diameter of 20 mm which is then just 40 wavelengths. Diffraction effects can be clearly identified in such situations. The main manifestation is that the region in front of the transducer has two distinguishable zones (*Figure 12*) which are called the *near field* and the *far field*.

Figure 12 shows an idealized theoretical prediction of the beam shape of a transducer producing continuous waves (CW). The prediction is that the *near field* will extend to a distance $D^2/4\lambda$ from the transducer where D is the diameter and λ the wavelength. In the near field the width of the beam will be roughly equal to the transducer diameter. Beyond $D^2/4\lambda$ the beam diverges in the *far field* which theoretically can be traced back to the transducer centre. In other words, in the *near field* the system behaves as though the transducer were a large 'torch-like' source sending out a pencil beam whereas in the *far field* it behaves as though the transducer were a very small point source.

This rather naive view has many limitations. In particular it does not strictly apply to pulsed or focused transducers. Even CW transducers frequently deviate in practice from this ideal behaviour. Beam profiles of real transducers show much more complex patterns. Nevertheless, the concept is valid to some extent most of the time and is therefore of value. It is important to realize that the aim of most scanner designs is to perform all or most of the imaging in the near field where the beamwidth is small and therefore transverse resolution (see Chapter 2) is good. As we will see later, the ultrasound can only penetrate a limited depth into tissue and so the objective normally is to design the system so that the end of the near field is roughly at the limit of penetration.

This diffraction model of near and far fields is a good example of one of the many compromises in medical ultrasound. Increasing the transducer frequency f would lead to a decrease in the wavelength λ, (since $f = c/\lambda$) and therefore an increase in near-field length. This is desirable since it increases the depth range over which good transverse resolution is achievable and also because higher frequencies should lead to better axial resolution (see Chapter 2). However, higher frequencies penetrate shorter distances in the body and so there is a severe limitation on what can be achieved. Similarly, smaller transducer diameters produce narrow near fields and hence improved resolution. Unfortunately, since the near field length is given by $D^2/4\lambda$, smaller diameters also produce shorter near fields and hence smaller useful depth ranges. To make matters worse, small diameters also result in more rapid divergence in the far field. To paraphrase the well-known rhyme 'When they are good, they are very very good, but when they are bad they are horrid!'.

2.4 Pressure and intensity

It was shown earlier in this chapter that an ultrasonic wave propagates as an oscillatory pressure disturbance with associated changes in particle displacement, particle velocity and density. Because of the oscillatory nature of these changes their average value over a long time period is zero, i.e. the positive parts of the wave are matched by the negative parts. In other words the positive *amplitudes* are the same as the negative *amplitudes*.

At first sight it might appear that the end result of these oscillations is zero. However, it is clear that the wave itself does travel in a specific direction and therefore some energy is transferred from one point in the medium to another. The energy travelling through the medium is confined to the beam and we can therefore define an *intensity* which is the amount of energy passing per second through unit area at right angles to the direction of wave propagation (*Figure 13*).

If the total energy passing through area A is E Joules, and the time taken for the energy to pass through is t s, then the intensity I is given by

$$I = \frac{E}{tA} \tag{5}$$

The units of I are therefore Joules/s/sq. metre ($J\ s^{-1}\ m^{-2}$). However the quantity E/t is the energy flow per second which is power and has units of Watts.

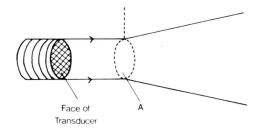

Figure 13. Definition of intensity. Beam cross-sectional area is A.

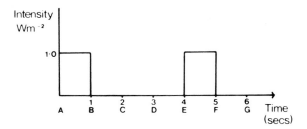

Figure 14. Hypothetical pulsing regime for a transducer.

Therefore I is normally given in W m^{-2} or mW cm^{-2}.

It can also be shown that if the pressure amplitude of a continuous wave is P_o, then

$$I = \frac{1}{2} \frac{P_o^2}{\varrho c} \qquad (6)$$

Thus the intensity is proportional to the square of the pressure. Since negative pressure values give positive numbers when squared, the average value of P_o^2 over a long time is *not* zero, and this helps us to understand the apparent contradiction referred to above. In practice, intensity values are not normally measured directly. Calibrated precision transducers known as hydrophones are used to measure pressure amplitudes and the corresponding intensity values are calculated from the above equation.

For continuous wave fields, provided the measurement is carried out over many cycles, the intensity will always have the same value. However for pulsed fields, this is not so. As an example imagine a simple case in which the ultrasound is turned on at an intensity of 1 W m^{-2} for 1 s and then off for 3 s (*Figure 14*) and this pattern is repeated indefinitely.

If the intensity is measured for 1 s between times A and B in the diagram, the average value measured will be 1 W m^{-2}, and the same answer will be obtained in the period E to F. However, if the measurement is made between B and C, or C to D, or D to E or F to G, the result will be zero.

On the other hand, the average value from time A to time C will be 0.5 W m^{-2} since it is one for half the measurement time and zero for the other half. Similarly, if the measurement is made between A and E, the average intensity value will be 0.25 W m^{-2}.

The situation for intensity measurement in pulsed fields therefore needs to be approached with care. In practice, intensity figures are normally quoted in one of three ways: temporal average, pulsed average or temporal peak. The temporal average intensity I_{TA} is the measurement obtained after averaging over many cycles. It is implicit that increasing the measuring time further would not give a different value for I_{TA}. The pulse average intensity I_{PA}, is the value obtained by averaging only during the duration of a pulse and not during the 'off-time'. The temporal peak intensity I_{TP} is the maximum instantaneous value measurable and corresponds to the peak value of the pulse.

The sampling in space must also be considered carefully. Within the beam there will be 'hot' spots and 'cold' spots. The regions of maximum intensity might for example be at the focus of a focused system, but this 'hot' spot may extend for only a few mm in any direction. Thus if the intensity is measured over that small region, the value will be high. On the other hand if it is averaged over the whole beam cross-section a much smaller value will result. It has therefore become conventional to refer to *spatial peak* (SP) and *spatial average* (SA) intensities. The I_{SP} value is the maximum intensity sampled over a very small distance found anywhere in the beam. The I_{SA} value is the average value across the beam at some distance from the transducer.

Combining the above concepts we get a whole range of intensity parameters, but it is probably the I_{SPTA} (Spatial Peak Temporal Average) value which is most frequently reported.

3. Propagation of ultrasound in tissue

As the ultrasound wave propagates through tissue, its intensity changes due to the diffraction effects described in the previous section but also because of several other processes. The most important processes in medical ultrasound are:

(i)   Reflection
(ii) Scattering
(iii) Refraction
(iv) Absorption

These will be considered individually.

3.1 Reflection

This is a process which is intuitively straightforward because we can make an analogy with an optical mirror. Under certain circumstances an ultrasound wave can be re-directed at a surface in a manner identical to that of light striking a reflecting surface. The laws of optical reflection are then applicable to the ultrasound case i.e. (*Figure 15*) and the process is called *specular reflection*.

The laws which characterize specular reflection are:

(i) The angle of incidence *i* equals the angle of reflection *r*.
(ii) The incident beam, reflected beam and normal all lie in the same plane.

The conditions under which *specular reflection* occurs at an interface are that the size of the interface must be large compared to a wavelength and that the roughness of the interface must be small compared to a wavelength. In clinical terms there are many situations where this is the case. Notable examples are: the midline of the fetal brain (thereby facilitating the BPD measurement), large blood vessel walls (e.g. aorta and vena cava) and organ boundaries (e.g. edge of liver, kidney, etc.).

One very important consequence of specular reflection is that the reflected beam, in general, does not come back along the path of the incident beam. Thus if the same transducer which has produced the incident beam is also acting as a receiver of echoes, then reflections from specular reflectors will not normally be seen. Only when the incident beam approaches the surface at 90° (or very close to 90°) will the reflected beam be detected by the transmitting transducer. This is crucial for the biparietal diameter (BPD) measurement since non-visualization of the midline therefore implies that the transducer is not orientated at 90° to it and thus an incorrect anatomical section has been chosen (see Chapter 6).

The other question which must be considered is how much of the energy impinging upon a specular reflector is reflected according to the above laws and how much is transmitted, unaffected by the interface.

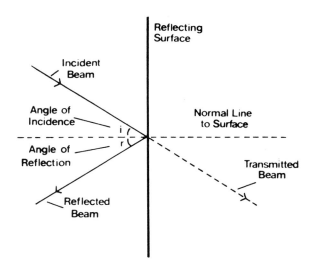

Figure 15. Specular reflection. Note that the angle *i* is equal to the angle *r*.

This depends critically upon a quantity known as the *acoustic impedance* (Z). Strictly Z for any given medium is defined by

$$Z = \frac{p}{v} \qquad (7)$$

where *p* is the instantaneous excess pressure associated with the ultrasonic wave and *v* is the instantaneous particle velocity.

The quantities *p* and *v* can be difficult to measure and it can be shown that in most cases, Z is also given by

$$Z = \varrho c \qquad (8)$$

where *ϱ* is the density of material and *c* is the speed of sound in the material.

The fraction of the incoming energy which is reflected at an interface depends upon the change in Z across that interface. The greater the change in Z, the greater the fraction of energy reflected will be. The emphasis is on the *change* in Z across the boundary. Having either a high or low value on one side will not in itself predict how much energy will be reflected.

The fraction of energy reflected at an interface at which specular reflection takes place is called the *reflection coefficient, R*.

$$R = \frac{\text{Reflected intensity}}{\text{Incident intensity}} \qquad (9)$$

R obviously cannot have a value greater than one. It is also possible to define a *transmission coefficient* T, describing the fraction of energy transmitted across an interface.

$$T = \frac{\text{Transmitted intensity}}{\text{Incident intensity}} \qquad (10)$$

Since any energy not reflected is assumed to be transmitted

$$T + R = 1 \qquad (11)$$

As stated above a special case for specular reflection occurs when the angle of incidence is 90°. Under these conditions echoes return to the transmitting transducer and are detected and visualized. The 90° incidence case is also special in that a simple calculation for the value of R becomes possible. Thus for 90° incidence (called 'normal' incidence)

$$R = \left[\frac{Z_1 - Z_2}{Z_1 + Z_2} \right]^2 \qquad (12)$$

Where Z_1 and Z_2 are values of Z on either side of the boundary (*Figure 16*).

Some typical values of Z are shown in *Table 2*. The important point to note is that most soft tissues have similar values of Z around 1.6×10^{-6} kg m^{-2} s^{-1}. This is naturally the case since $Z = \varrho c$ (Equation 8) and both ϱ and c are similar for all soft tissues (*Table 1*). It is important to notice the very low value for air which is typical of gases and the high value for bone which is typical of solids. Thus from equation 12, small changes in Z would be expected at interfaces between adjacent soft tissue, e.g. muscle/liver or kidney/blood vessel, but larger changes at interfaces between soft tissue and either bone or air. The reflection coefficients R between some of these combinations are calculated in *Table 3* for the case of normal incidence using equation 12. It is because the fraction of energy reflected at soft tissue/bone and soft tissue/air boundaries is so large that ultrasonic 'shadowing' (see Chapter 2) occurs. Particularly in the latter case it is difficult to get any information from beyond the boundary since it acts as a near-perfect mirror allowing almost no energy to pass through.

3.2 Scattering

The preceding section dealt with the situation which arises when an ultrasonic beam strikes a large, flat, smooth target. In other circumstances targets may be

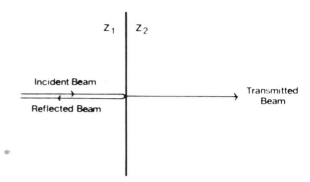

Figure 16. Specular reflection with normal incidence.

Table 2. Values of acoustic impedance (Z) for biological materials.

Material	Acoustic impedance 10^6 kg m^{-2} s^{-1}
Aluminium	17.28
Perspex	3.19
Bone	7.80
Liver	1.65
Kidney	1.62
Blood	1.61
Fat	1.38
Lung	0.26
Air (NTP)[a]	0.00004

Data from McDicken (1981).
[a]NTP, normal temperature and pressure.

Table 3. Values of intensity reflection coefficient (R) for biological interfaces.

Reflecting interface	Intensity reflection coefficient (R)
Muscle/blood	0.0009
Fat/kidney	0.006
Fat/muscle	0.01
Bone/muscle	0.41
Bone/fat	0.48
Soft tissue/air	0.99

small or rough on a scale comparable with the wavelength. This is particularly true within regions of soft tissue. Thus images from liver, kidney and thyroid parenchyma are composed largely of scattered echoes.

The two main characteristics of scattering are:

(i) Most of the energy passes beyond the single scatterer without change.

Figure 17. Diagram showing scattering in all directions from a small point target.

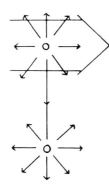

Figure 18. Multiple scattering. Some of the energy scattered by the small target in the main beam goes on to interact with a second scatterer outside the beam.

(ii) The energy which does interact with the target is redirected (scattered) in almost all directions (*Figure 17*).

Note that some of the scattered energy is redirected back towards the direction of the source. This is called the *back-scattered* signal. Also because of the multi-directional nature of the interaction the angle of approach to the target is far less critical than for specular reflection.

The relative contribution which scattering makes to the total loss of signal in the propagating wave depends critically upon the size of the scattering targets relative to the wavelength. Thus either an increase in scatter size or frequency (corresponding to a shorter wavelength) will cause a rapid increase in the scattering contribution.

It is also possible for a wave scattered in some direction to go on to hit another scatterer and undergo a second scattering process and so on. The closer the scatterers are, the more likely this process known as *multiple scattering*, becomes (see *Figure 18*). This is very common in biological situations and is of particular importance in blood. Here individual red blood cells form closely packed scattering targets.

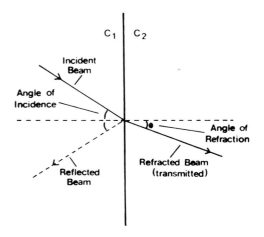

Figure 19. Diagram of refraction or 'beam bending' at an interface with velocity c_1 on one side and c_2 on the other.

3.3 *Refraction*

In Section 3.1 the interaction of an ultrasonic wave with a large flat interface was examined. The interface consisted of the boundary between two materials of differing acoustic impedance, Z. Another process which occurs at boundaries is 'beam-bending' or *refraction*. In this case, it is not the change in acoustic impedance which matters but a change or difference in sound velocity.

The amount of bending is described by Snell's Law

$$\frac{\text{Sine } i}{\text{Sine } \theta} = \frac{c_1}{c_2} \quad (13)$$

where i and θ are the angles of incidence and refraction, respectively (*Figure 19*). Note that if c_1 is *greater* than c_2 then i will be greater than θ. The bending will therefore be *towards* the normal. Conversely if c_1 is *less* than c_2, the bending will be away from the normal. It should also be noted that if the angle 'i' is zero then so too is 'θ' and no bending takes place. In other words, if the incidence is along the normal, then the beam passes straight through.

In practice, refraction is normally insignificant in medical ultrasound. A relatively large smooth boundary is required and thus conditions are those for specular reflection (q.v.). Since relatively large changes in the velocity c are needed for significant beam bending, this means that it is likely to be important where there are also large changes in the acoustic impedance Z ($Z = \varrho c$). Thus strongly refrac-

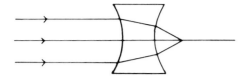

Figure 20. Focusing action of an ultrasonic lens with a sound velocity higher than the surrounding medium.

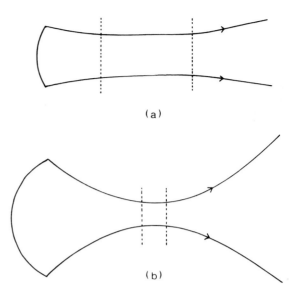

(a)

(b)

Figure 21. Focusing by ultrasonic lenses. (**a**) Weak focusing with a slightly curved small aperture lens leads to a long depth of focus with relatively large beamwidth. (**b**) Strong focusing from a wide aperture, steeply curved lens gives a narrow beamwidth over a small focal depth.

ting surfaces tend to be strongly reflecting surfaces and the refracted (transmitted) beam is of relatively low intensity. The other important factor is that materials such as gas and bone which provide dramatic changes in Z and c also absorb ultrasound rapidly (see Section 3.4) and thus any energy which is transmitted at the boundary is then rapidly absorbed. The one situation where this is not so is in the eye, where the interface between aqueous and vitreous humour provides significant refraction.

Another important aspect of refraction in medical ultrasound is in the use of lenses to achieve beam focusing. In this context the beam is deliberately bent by a calculated amount in order to redirect it to the desired focus.

Assuming that the surrounding medium is water or soft tissue, the diagram in *Figure 20* shows how focusing might be achieved using solid material such as perspex or polystyrene, with a higher sound velocity. Since the wave entering the lens is travelling from a medium with a lower speed of sound to a medium with a higher one, refraction bends the beam *away* from the normal. On exit from the lens the reverse is true and bending towards the normal is found. Thus the required shape is biconcave which is the opposite of that required in optics to achieve focusing. This is because in the optical case light travels faster in air than glass.

The position of the focus is governed by the curvature of the lens. The depth of focus depends upon the relative diameter or *aperture* of the lens. In general making lenses bigger improves resolution at the focus but reduces the depth of focus. Thus a small lens can be used for weak focusing over a large depth range and a large lens for strong focusing over a narrow range (*Figure 21*).

Often lenses are fixed to the front face of probes or sometimes the front of the probe itself can be curved to achieve focusing. The thickness of this material (see *Figure 10*) can be set to equal one quarter of the wavelength of the ultrasound used. This theoretically ensures perfect transmission through the front epoxy plate and is known as *quarter wavelength matching*. In practice, because the pulse contains a range of frequencies, the matching is less than perfect, but still remains a significant improvement on the use of no matching.

3.4 *Absorption*

Even in the absence of losses due to reflection and scattering, the intensity of the ultrasound beam still falls off rapidly as it propagates through soft tissue. This is due to the process known as *absorption* by which some of the mechanical energy of the ultrasound beam is converted directly into heat. In soft tissue this is almost always the dominant source of energy loss in the beam normally accounting for over 90% of all energy reduction.

The absorption falls off exponentially with distance. The same fraction of the incoming energy is lost in each unit distance travelled (*Figure 22*).

The level of intensity I in a distance x can be expressed as

$$I = I_0 \exp\left(-\mu_A x\right) \tag{14}$$

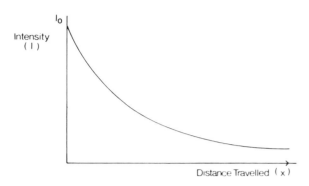

Figure 22. Graph of exponential absorption.

Table 4. Thickness of various materials required to reduce the intensity by half (half value thickness).

Material	Half value thickness (cm)	
	2 MHz	5 MHz
Air (NTP)	0.06	0.01
Bone	0.1	0.04
Liver	1.5	0.5
Blood	8.5	3.0
Water	340	54

Data from McDicken (1981).

where I_o is the initial intensity at $x = 0$ and μ_A, known as the Intensity Absorption Coefficient, is a constant for a given wave in a specified tissue but it increases steadily with frequency, roughly as

$$\mu_A = kf^{1.2}$$

where k is a constant which is tissue specific.

Because of the exponential nature of the absorption it is sometimes convenient to define a 'half-value thickness' for each tissue. This is the thickness of that tissue which would be required to reduce the intensity of an ultrasound beam at the specified frequency by a factor of two. Some examples are given in *Table 4*. It can be seen that absorption in blood is low relative to liver but bone and air are much more absorptive.

3.5 Attenuation

The overall loss of intensity of an ultrasound beam passing through a medium can therefore be ascribed to a number of possible mechanisms. The combined effect of these and any other loss inducing mechanisms is known as *attenuation*. Unlike reflection and re-

fraction which occur at interfaces, absorption and scattering are tissue specific and can be combined to form an attenuation coefficient for that medium μ. Thus

$$\mu = \mu_A + \mu_s \qquad (15)$$

where μ_A is the absorption coefficient and μ_s is the scattering coefficient. The attenuation coefficient μ, strictly speaking called the *intensity attenuation coefficient*, depends upon frequency.

The loss in intensity, characterized by μ, is often described using the *decibel scale*. The attenuation in decibels, abbreviated to dB, is defined as

$$\text{decibels (dB)} = 10 \log_{10} \left[\frac{I_1}{I_2} \right] \qquad (16)$$

where I_1 and I_2 represent the intensities measured at two points spaced 1 cm apart along the beam.

It can be shown that the attenuation coefficient μ can be expressed as D decibels where

$$D = -4.3 \, \mu \qquad (17)$$

As a rule of thumb, soft tissue attenuation is roughly $1 \text{ dB cm}^{-1} \text{ MHz}^{-1.}$

The decibel is useful for comparing two signal intensity or power levels. It is very important to realize that decibels are ratios and *not* absolute units. Thus we can describe the gain of an amplifier in decibels meaning the ratio of the output power level to the input power level. A negative number of decibels as in equation 17 above implies a loss in signal rather than a gain. Loss in signal is precisely what attenuation means.

The value of the decibel scale lies in its use for comparing very different values of some quantity. For example the signals coming back from within the body may produce voltages at the transducer varying from 10 μV (microvolts) to 10 V, a ratio of a million to one! Since electrical power is proportional to the voltage squared this means a power ratio of 10^{12} or a million millions to one. In decibels, this becomes

$$10 \log_{10}(10^{12}) = 120 \text{ dB} \qquad (18)$$

120 dB is a relatively easy number to deal with.

For comparison, consider the half value thickness concept described earlier in this chapter. By definition, this implies a drop in intensity of 0.5, therefore:

No. of decibels =
$$10 \log_{10}(0.5) = 10 \times (-0.3) = -3 \text{ dB} \qquad (19)$$

Thus a half-value thickness means the thickness required to reduce the intensity by 3 dB.

4. Acknowledgements

Sincere thanks are due to Benita Slater for patiently typing this manuscript and Julia Evans for careful attention to the illustrations in this chapter.

5. Bibliography

Readers will find further information on the material of this chapter in the following books:

Hussey,M. (1985) *Basic Physics and Technology of Medical Diagnostic Ultrasound*. Macmillan, London.

McDicken,W.M. (1981) *Diagnostic Ultrasonics. Principles and Use of Instrument*, 2nd Edition. John Wiley & Sons, New York.

Wells,P.N.T. (1977) *Biomedical Ultrasonics*. Academic Press, London.

Chapter 2

Pulse–echo ultrasound

J.A.Evans

1. Principles of echo ranging

Anyone who has ever stood in front of a wall or a cliff-face and heard echoes of his voice will be aware of the basic underlying principles of diagnostic pulse–echo ultrasound.

As the distance between the source of the sound and the echo-producing surface or target increases, the delay between the generation of the sound and the reception of the echo also increases. This is because sound takes a perceptible time to travel over the round trip distance. As we saw in the previous chapter the speed of sound is high but nothing like as high as the speed of light. In air sound travels at approximately 300 m s^{-1} which is about five times slower than in soft tissue and by comparison light travels at about 300 000 000 (3×10^8) m s^{-1}.

We can easily work out the time T s taken for the sound emitted by the 'stick-man' in *Figure 1* to travel to the target and back, since time equals distance divided by speed;

$$T = \frac{2D}{c} \qquad (1)$$

where D is the one-way distance to the target and c is the speed of sound in the surrounding medium.

Equation 1 is a simple mathematical expression of the principle of echo ranging. If the speed of sound in the medium is known, then the distance to any target can be worked out from the time taken for the sound to travel there and back. For instance if the 'stick-man' is 150 m from the wall and the speed of sound in air is 300 m s^{-1} then T is given by

$$T = \frac{2 \times 150 \text{ m}}{300 \text{ m s}^{-1}} = 1 \text{ s} \qquad (2)$$

and delays of a few seconds are indeed typical in such cases. Of course, if the 'stick-man' moves too far away no measurement can be made. This is not because the

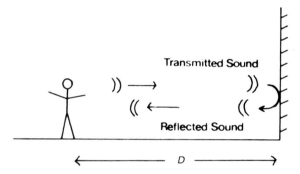

Transmitted Sound

Reflected Sound

D

Figure 1. Echo generation from a reflecting surface.

principle fails but rather because the echo signal is attenuated such that its amplitude falls below the background noise level and it cannot then be detected. If we repeat the above calculation for the case of pulse–echo ultrasound scanning, the result is quite different. For instance, if the echo is produced from a pulse of ultrasound travelling from the anterior abdominal wall to the posterior uterine wall in an obstetric scanning examination, the target distance might be say 7 cm. The speed of sound in soft tissue is normally assumed to be 1540 m s^{-1}, and so we have

$$T = \frac{2 \times 0.07 \text{ m}}{1540 \text{ m s}^{-1}} = 0.000909 \text{ s} \qquad (3)$$

This is obviously a very small number representing a very short time. For convenience the time units are therefore changed from seconds into microseconds (millionths of a second). Thus now

$$T = 0.000909 \text{ s} = 909 \text{ } \mu\text{s} \qquad (4)$$

So the basic principle underlying diagnostic pulse–echo ultrasound is the same as that for any sound echo measurement. The same is also true for the ranging technique of bats and the sonar system of submarines.

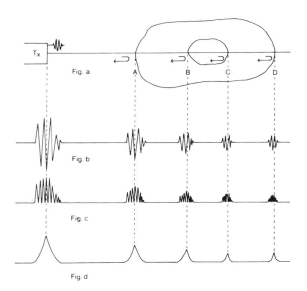

Figure 2. Production of an A-scan.

The complication in the medical ultrasound case is that the times involved are very short and typically measured in hundreds of microseconds. This adds to the technological complexity but in no way alters the basic simplicity and elegance of the underlying physics.

The process is thus conducted in four stages:

(i)   generation of a sound pulse;
(ii) reflection of pulse from target or targets;
(iii) detection of reflected echo(es);
(iv) calculation and display of range of target(s).

2. A-Scanning

2.1 *Detection, smoothing and filtering*

The simplest instrument of practical use in medical ultrasound which combines the above four functions is the A-scanner. In the detection stage the complex pulses are also *smoothed* and *rectified* (see later this Section) so that the net result is a relatively simple series of spikes which do not go below the baseline.

In *Figure 2* a hypothetical structure is being examined by ultrasound. The short pulse is emitted by the *transducer* Tx and encounters interfaces at A, B, C and D. These may be boundaries between different sorts of tissue but they are also abrupt changes in *acoustic impedance*. Thus at each of these points, some of the energy is reflected (see Chapter 1). The

remaining energy is transmitted and travels on to the next interface and thus four echoes are produced for every one transmitted pulse.

It is customary, although not essential, to use the same transducer to receive and transmit, and so, if the voltage on the transducer of *Figure 2a* is monitored in time, the result will be as shown in *Figure 2b*. This shows the initial transmitted pulse on the left and the four successive echoes of much smaller amplitude following but separated from each other in time. In practice *Figure 2b* is not drawn to scale since the transmitted voltage is typically tens or hundreds of volts whereas over the largest echoes it will only be a few hundred millivolts and many will be much smaller.

Having received these signals the A-scanner proceeds to process them further. The first stage is normally *rectification* which means not allowing the signals to cross the baseline. Any signals that would have done so are inverted as shown in *Figure 2c*.

After the rectification stage, it is common to smooth or *filter* the signal. This removes the fast 'spikes' in the signal replacing them by a single envelope as in *Figure 2d*, the smoothing being performed by an electronic filtering circuit. The result is an envelope which displays a time-average value of the signal instead. However, it is important to perform this step after rectification since if the time-averaged values of the signal of *Figure 2b* were obtained, they would be zero, since the signal spends as much time being positive as being negative. This early stage in the receiver in which the signal is varying rapidly in time and before smoothing is called the radiofrequency (RF) stage.

The two processes of rectification and smoothing are responsible for a loss of information. The resultant envelope is not unique and may have been obtained from several different incoming waveforms. The reasons for discarding information in this way are largely due to historical technical limitations. Once rectification and smoothing have been completed, the waveform has lost much of its higher frequency components and can therefore be processed by lower frequency electronics. In the past, the technical problems of handling such high frequencies were often insurmountable. Modern digital scanners are not limited in this way and provide many different signal processing options. It is now at least feasible to attempt tissue characterization with such machines (see Chapter 3, Section 6.8).

Thus the outcome of those initial stages is a series

of smoothed envelopes. The envelopes or spikes are separated on the horizontal axis by time and the time separation can then be used to calculate range as discussed above. The vertical height of the spikes is directly related to the height or *amplitude* of the original incoming signal. Larger amplitude signals produce bigger spikes. For this reason the display is known as the A (for Amplitude)-scan. This is the type of display in *Figure 2d*.

2.2 *Time gain compensation (TGC)*

One extra stage of processing needs to be applied to the A-scan signals before they are suitable for display, this is *time gain compensation* (TGC). This is also termed *swept gain* on some instruments. The purpose is to compensate for the attenuation of the tissue overlying echo-producing interfaces.

As the ultrasound pulse propagates through tissue (as discussed in Chapter 1) it loses energy and therefore amplitude mainly because of absorption. Thus two identical interfaces at different depths will not receive equal amplitude signals and therefore will not produce equal amplitude echoes. Further as the echoes themselves travel back to the transducer they will also be subject to absorption. Thus signals from deep targets will be much smaller than those received from identical, more superficial targets. TGC attempts to correct for this and is discussed fully in Chapter 3.

The complete A-scan layout is shown schematically in *Figure 3*. Note that the trigger signal controlling the transmitter is also sent to the TGC and the display. This is to ensure the whole system is maintained in synchronous timing.

2.3 *A-scan applications*

The electronic processes required to produce an A-scan are also found in the more common sophisticated scanners which will be discussed later. Thus the A-scan circuitry is an essential 'building-block' in the make-up of a modern scanner, although the actual visual display is often omitted from current machines.

The principle application of A-scanning is in the accurate measurement of structures lying along the ultrasound beam. Indeed, if accuracy of measurement is paramount, then the A-scan would normally be the technique of choice.

The chief disadvantage of A-scanning is that it gives a one-dimensional section through the target. Thus, unless the anatomy is known very precisely before the scan, is simple and is not prone to significant normal

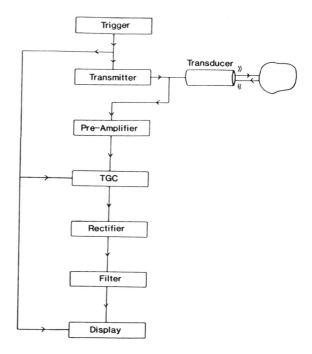

Figure 3. Schematic diagram of an A-scanner.

variations, the interpretation of the scan becomes very difficult. Searching for a fetal head midline echo without knowing the fetal presentation or orientation is, in many cases, an impossible task. Certainly, the A-scanner requires a highly skilled and experienced operator if it is to be used reliably, and for these reasons, some form of two-dimensional imaging is almost always required for useful clinical practice. Therefore the problems with the A-scanner used alone are predominantly those of display. An instrument is needed which takes the same raw information and uses it to display recognizable two-dimensional anatomical sections. This is the B-scanner.

3. B-Scanning

In the A-scanner, one of the axes, the horizontal one, is used to represent time. Since this is related directly to the distance travelled, it also represents one physical dimension of the target, that is depth. The other axis, the vertical, is used to display amplitude and it is necessary to use this for a second physical dimension in the target if a two-dimensional anatomical section

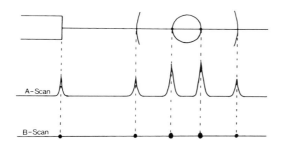

Figure 4. Formation of a B-scan line.

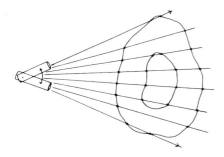

Figure 5. Formation of a 'sector' scan.

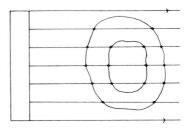

Figure 6. Formation of a 'linear' scan.

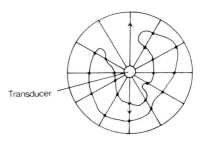

Figure 7. Formation of 'PPI' scan.

is to be displayed. The question of how to deal with the amplitude information is therefore vital. The solution in the case of the B-scanner is to use the brightness of the screen to display the amplitude (*Figure 4*).

Thus each spike on the A-scan is reduced to a bright dot on an otherwise invisible line. The greater the amplitude, the greater the brightness of the spot. Thus this mode is known as B (for brightness)-mode. We can see from *Figure 4*, that the A-scan information is now condensed into a single line, thus liberating the vertical axis for display of another physical dimension.

Some means now needs to be found of moving the B-scan beam in a regular fashion in order to scan systematically the target of interest. For example, if the transducer itself is rocked backwards and forwards the beam directions trace out a 'fan' or *sector* shape (*Figure 5*).

Every point at which a scan line intersects an inter-face results in a dot being produced. Provided some means exists to store or remember the dot positions, a 'join-the-dots' exercise can be conducted by the eye and an entire section built up. The scanner can then go back to its original scan line and start again. New

information can then be obtained to refresh or modify the existing stored image.

3.1 *Categories of B-scan*

The particular format shown in *Figure 5* is that of a sector scanner. In this case, all the scan lines can be traced back to a single point. This is by no means the only geometrical arrangement possible. Another very popular arrangement is to make all the scan lines parallel thus producing a rectangular field of view (*Figure 6*). Such scanners are often described (rather unhelpfully) as having a linear format. The choice of which format of B-scanner to adopt depends very much on the clinical application.

In obstetrics, for example, the rectangular or linear format works well because it provides a large anterior field which is compatible with the size of pregnant uterus until very late in pregnancy. On the other hand, cardiological investigations need to be conducted through a small scanning 'window' to the left of the sternum. In this case a probe with a small front face producing a field which diverges rapidly is ideal. Many intermediate formats have been used at various times. However, the only one which differs fundamentally from the sector/linear options is the so-called PPI (plan position indicator) format (*Figure 7*).

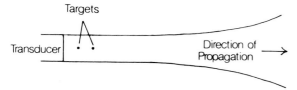

Figure 8. Definition of axial resolution.

In this case the transducer is at the centre of the field and the scan lines radiate outwards from it in all directions. This format is used for the more recent endoscopic applications of ultrasound for imaging from within oesophagus, rectum, vagina, etc.

3.2 *Resolution of B-scanners*

The quality of the image from any scanning system is due to many factors but resolution is obviously of great importance. There are many sophisticated definitions of resolution available but for the present it will be sufficient to define resolution as 'the ability to distinguish two targets placed close together'. In quantitative terms the resolution can then be expressed as the minimum separation of two targets which permits them to be distinguished in the image.

Not surprisingly, the above definitions have complications associated with them. One of these is that in an ultrasound system, the resolution depends critically on the orientation of the targets. Resolution is divided into *axial resolution* and *lateral resolution*.

Axial resolution refers to targets lying along the same scan line (*Figure 8*). The ultrasound beam strikes one target before the other, and the two targets are therefore not at the same distance from the transducer. Fundamentally, axial resolution depends upon pulse length. If the trailing portion of the pulse from the front target overlaps with the leading portion of the pulse from the second (*Figure 9a*), then the two targets will cease to be resolved (*Figure 9b*). On the other hand if the output power or gain of the system is reduced then the same two targets may then be resolved (*Figure 9c*).

The pulse length itself is often 2−3 cycles in modern scanners and this is normally independent of the probe frequency. Since $c = f\lambda$ (Equation 3 of Chapter 1) and the speed of sound is almost constant with frequency, higher frequencies mean shorter wavelengths and hence shorter pulses. Thus, all other things being equal, higher frequencies tend to lead to better axial resolution. However, the nature of the signal processing used is critical here and some modern scanners

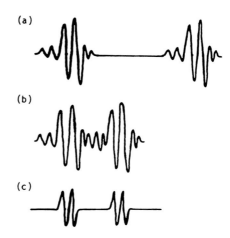

Figure 9. Effect of gain on axial resolution. (**a**) Targets clearly resolved. (**b**) Echoes from target overlapping—probably not resolved. (**c**) Same target separation as in (**b**) but lower gain targets clearly resolved.

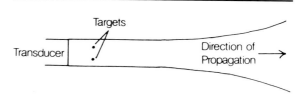

Figure 10. Definition of lateral resolution.

deliberately use long pulses and still maintain good axial resolution by using more sophisticated signal processing. Axial resolution will also be influenced by the nature of the targets chosen; more reflective targets are more easily resolved. Values for axial resolution are typically 0.5 mm or better in good modern scanners and this should be virtually independent of depth.

Lateral resolution in contrast refers to targets lying 'across' the scan lines (*Figure 10*). Ideally a scan line strikes only one of the targets with the second target being struck by a nearby scan line. These targets are equidistant from the transducer, and so it is immediately clear that lateral resolution depends upon the beam width. Narrower beams will allow targets closer together to be resolved.

The beam width itself may be very depth dependent. In Chapter 1, we saw how the ideal beam consists of near field and far field regions. Higher frequencies lead to longer near field lengths for a given transducer diameter and so smaller transducers can be used at

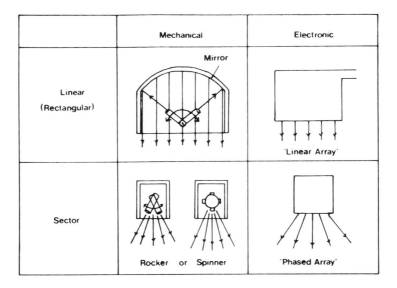

Figure 11. Possible real-time scanner types.

those frequencies. This, in turn, leads to narrower beamwidths and ultimately to better lateral resolution. For a full discussion of resolution and its measurement, the reader is referred to Chapter 5.

4. Real-time scanners

We have seen that B-scanners use a number of scan lines to build up a two-dimensional image of closely spaced dots. Each individual scan line must be stored for display before the scanner proceeds to acquire the next one. When all the scan lines have been completed, the scanner can then go back to the beginning and start again. One complete group of scan lines is called a *frame* and the rate at which individual frames are produced by the scanner is the *frame rate*.

In the original B-scanners the scanning action was manual and hence controlled by the operator. Such scanners are called *static B-scanners* because the time taken to build up one frame manually is several seconds and hence the image obtained is essentially static. If anything moves during the time taken to build up a frame, the result is a blur on the image. However with modern automatic scanners, the scanning action is faster than almost all movement experienced, and hence a moving structure can be displayed as a dynamic image. We can see the movement as it

happens and this is therefore called *real-time scanning*. The constraints on the scanning speed and frame rates will be discussed in Section 4.2.

If we take the 'linear' and 'sector' formats as the two main standards, then it is also possible to categorize scanners in terms of the means by which the beam is scanned. This would produce two families of scanners, viz: electronic and mechanical. The situation is summarized in *Figure 11*. Ultrasonic real-time scanners can thus be categorized using two descriptions, for example electronic linear scanner or mechanical sector scanner. Each of the four categories will be examined in turn.

4.1 Mechanical sector scanners

In many ways, the mechanical sector scanner is the simplest of all ultrasound scanners. As shown in *Figure 11*, it can exist in at least two different forms. In the 'rocker', a single transducer is made to rock backwards and forwards by an electric motor and the pivot point for the motion can be varied depending upon the desired field width anteriorly. In the other form, the 'spinner', a number of transducers (typically four) are fixed in a paddle-wheel arrangement. As each transducer in turn faces the active face of the probe it is turned on and used to scan across the sector.

The advantage of the paddle-wheel is that the whole assembly is kept rotating at a constant speed and

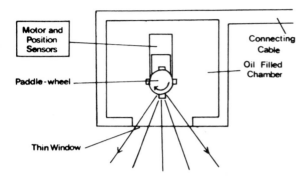

Figure 12. Schematic diagram of a mechanical sector scanner.

should, in principle, be easier to balance and make smooth running. The 'rocker' probe is required to stop and change direction frequently with all the consequent accelerations and decelerations. On the other hand, the 'paddle-wheel' design calls for good matching of all the transducers and this can be very difficult in practice.

The whole assembly in which the mechanical transducer is housed needs to be filled with some fluid (typically oil) to couple the ultrasound out of the probe (*Figure 12*) and small air bubbles in such probes can seriously degrade performance. Although the probe housing needs to be thick and fairly rugged, it is necessary to include a thin plastic window somewhere in order to let the ultrasound pulses through in both directions. It is also necessary to house the electric driving motor and position-sensing electronics in the probe head. It is important that the scanner 'knows' the orientation of the probe at any instant in order to display the correct line of information on the display monitor.

4.1.1 *Frame rate of real-time scanners*

The number of frames per second and the number of scan lines per frame are important parameters in ultrasonic scanner design. They are however subject to several constraints which are quite fundamental.

The concept of a real-time scanner involves sending out a pulse corresponding to one scan line and then waiting for echoes to come back from that pulse before firing another pulse. If the second pulse is fired too quickly after the first the machine will be unable to distinguish late arriving echoes due to deep structures encountered by the first pulse from early echoes arriving from the second pulse interacting with

superficial structures. Thus the greater the depth which the scanner interrogates, the longer the time required for each scan line.

The other constraint is the need to avoid 'flicker' in the display. If the eye is presented with a rapid succession of images of a moving object then the brain will register this as a continuous image. Such is the case with a normal domestic television set in which 50 frames per second are presented to the viewer. If the frame rate falls to too low a level, the image is perceived to flicker as in early 1920s silent movies. It is often stated that at least 20 frames per second are needed if flicker is to be avoided.

We can now perform a calculation based upon these constraints. Suppose that it is required to image up to a depth of 20 cm in tissue. The go and return time t, for pulses and their echoes from the maximum depth can be found if a speed of sound of 1540 m s^{-1} is assumed:

$$t = \frac{\text{distance}}{\text{speed}} = \frac{2 \times 20 \text{ cm}}{1540 \text{ m s}^{-1}} = \frac{0.4 \text{ m}}{1540 \text{ m s}^{-1}} \quad (5)$$

$$= 0.000259 \text{ s} = 259 \text{ } \mu\text{s}$$

This figure of 259 μs is the time for each scan line. Now the slowest practical frame rate is 20 frames per second, and in fact, it is often preferred to use 25 frames per second since this is half the mains frequency (50 Hz). Twenty five frames per second means 0.04 s or 40 ms per frame. This means that the maximum number of lines per frame is 40/0.259 or 154 lines. If the required depth is increased then the number of lines in the image must be decreased if flicker is to be avoided. It is possible to synthesize other 'fill-in' lines to make the image more appealing to the eye as discussed in Chapter 3, but the constraint on the number of real acoustic lines per frame is fundamental.

4.2 *Mechanical linear scanners*

We can now return to the discussion of real-time scanner types. The mechanical sector scanner described above is likely to have a sector angle of between 60 and 180° with about 100 lines of real information. Special 'small part' scanners have also been built with high frequency high resolution transducers for imaging thyroid, breast or testes. The depth of penetration of such scanners may be less than 10 cm, so more lines per image and/or higher frame rate can be achieved in such cases.

One problem with such sector scanners is that the

scan lines 'bunch' close to the probe and leave gaps at long distance. The linear format has therefore been adopted in some cases to avoid these problems and others mentioned earlier. One way of achieving this format is to fire the transducer 'backwards' towards an ultrasonic mirror (*Figure 11*). If either the transducer or mirror is rocked, then the geometry can be arranged such that all reflected beams emerge parallel.

One advantage of a system such as this is that a wide aperture lens and mirror system can be used and strong focusing at some depth can be achieved. Although the diagram shows a sealed probe system, it is also possible to incorporate the arrangement into a large open tank in which the patient is also positioned. This arrangement has been used for breast scanning since resolution is critical in such cases.

However, mechanical linear systems are not generally popular, their probes being large and cumbersome and often difficult to couple to the patient. Also strong focusing is useful only at one depth and so the systems are not suitable for general purpose use.

4.2.1 Mechanical scanners – advantages and disadvantages

The main advantage of the mechanical sector scanner is its relatively cheap simplicity. Although the precision engineering of the probe head is difficult, the rest of the system is straightforward. Resolution in these systems should be at least reasonable over a wide depth range. In terms of value for money, they are often excellent choices as general purpose abdominal scanners.

The disadvantages lie primarily with the probe itself. As a complex, precision moving mechanical component it is prone to wear and tear. Designs have improved significantly in recent years but intrinsically the moving scanhead will always be a 'weak-link' in reliability terms. Some, but not all, mechanical systems have also had problems relating to air bubbles or leaks in the probe. Another problem which is specific to 'paddle-wheel' scanners is the need to align the images from all the individual transducers. Failure to do so results in a serious 'jitter' in the image.

In principle, an all electronic system should have several intrinsic advantages.

4.3 Electronic linear scanners

4.3.1 Simple arrangement

The simplest electronic scanner arrangement consists

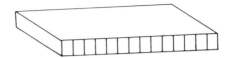

Figure 13. Simple linear array probe.

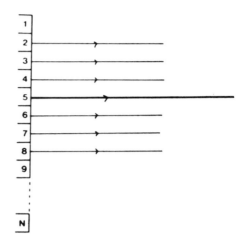

Figure 14. Firing elements in groups. In this case scan line 5 is the central axis.

of an array of small independent transducers arranged in a row forming a composite probe (*Figure 13*). Each small transducer (or element) in the array can then fire its own scan line in turn as in *Figure 14*.

In this way a rectangular or linear format can be generated. The first generation of linear array scanners was indeed designed in this way but they suffered from very poor lateral resolution and an examination of their dimensions shows why. If typical figures of 64 elements in an overall array length of say 12 cm are assumed, then the width of each element is 12/64 or 0.19 cm. If a working frequency of 3 MHz is assumed then the wavelength will be roughly 0.5 mm. We can now try to calculate the length of the near field. Unfortunately the calculation can only be approximate since each individual element is rectangular rather than circular. Nevertheless, if we assume that the element approximates to a circle of diameter 2 mm, then the near field length (see Chapter 1) is given by $D^2/4\lambda$ = $(2)^2/2.0 = 2$ mm.

Thus at all depths over 2 mm, the elements will be working in the far field with rapidly diverging beams. Thus lateral resolution would be expected to be very

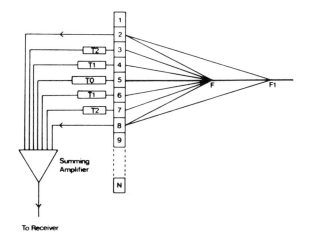

Figure 15. Focusing using delay lines.

poor in such scanners as indeed proved to be the case.

The designer is thus on the horns of a dilemma. If fewer elements of greater width are used, then the near field length increases but fewer scan lines are obtained and the image becomes obviously coarsened. If the element size and number are increased, the overall array length becomes so large that the probe itself becomes cumbersome and impractical. The solution to this apparently intractable problem was found in the ingenious technique of firing in groups.

4.3.2 Group firing

The idea of group firing is explained in *Figure 14*.

Suppose that the scan line from element 5 is required at some time in the imaging process. Instead of firing element 5 on its own, a number (say three) of elements on either side are all fired together. In this case, elements numbered 2, 3, 4, 5, 6, 7 and 8 would all be fired. The result is that a composite transducer is produced with a width of seven times the width of any single element. In consequence the near field length ($= D^2/4\lambda$) is increased by $(7)^2$, that is 49 times.

Having fired on elements $2-8$ inclusive to obtain a scan centred on element 5, the next step is to fire on elements $3-9$ centred on element 6. In this way the whole sequence progresses one element at a time along the array producing a series of overlapping lines. The result is to increase the line density without sacrificing resolution.

4.3.3 Focusing

The so-called third generation of linear array scanners made use of group firing to introduce a focusing action. In *Figure 15*, imagine that a target exists at point F along scan line 5. If echoes are produced at F then they will travel to the nearby elements and will reach all the elements numbered $2-8$ in the diagram. However the distance the echo from F travels to reach elements 2 and 8 is greater than that to element 5 and so it will not arrive at the same time. Therefore there will be a blurring of the echo information from that point. Echoes from F to intermediate elements will have intermediate travel times. However if the echo information arriving at element 5 is delayed by the right amount, then it can be added to the signals from 2 and 8 at exactly the right time to reinforce it. This can be done electronically by means of delay lines. All of the delayed signals are then added in a suitable amplifier and sent to the receiver and display. Thus in *Figure 15*, the longest delay T0 is applied to the central element and zero delay is applied to the outer groups. This process has the effect of focusing the echoes received from depth F.

The next stage is to step the firing sequence one element along as described earlier. Thus now elements $3-9$ will be fired with the scan centred at element 6. The delays now have to be altered so that T0 is applied to 6, T1 to 5 and 7, etc. In this way, the focusing action is maintained at the same depth as the array steps along.

It is also possible to focus the transmitted signal in the same way. In this case the excitation pulses to each transducer also pass through delays, so that the central elements are fired later than the outer ones.

4.3.4 Swept (dynamic) focusing

With the advent of focused linear array scanners, the lateral resolution of the machines began to become acceptable. Obviously the complex problem of rapidly controlling and switching between the various delay lines is only tractable using modern microelectronics and this carries a cost penalty. However still more sophistication is possible in the form of swept focusing techniques. Consider the point F1 in *Figure 15*. By altering the values of the delays T0, T1 and T2 it is possible to make the group of elements $2-8$ focus there instead of F. However the echoes from point F do not arrive at the transducer at the same time as those from F1 and therefore there is an opportunity to focus at F while echoes are turning from there and then

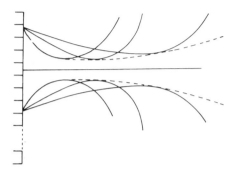

Figure 16. Dynamic focusing using three focal zones. The dotted line represents the cumulative effect.

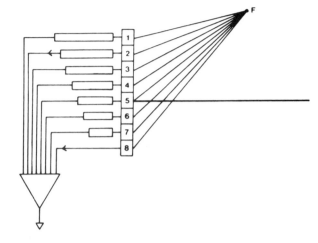

Figure 17. Focusing off-axis for an electronic sector scanner.

switch the focus to F1 a short time later when echoes from that depth are arriving. In fact, the system can be split into several focal zones, with rapid switching between different delay line combinations in time with the echoes returning from different depths (*Figure 16*). This technique is known as *swept or dynamic focusing* and is very common on modern scanners.

Swept focusing works on reception only since the echoes from different depths are separated in time. On transmission there is only one pulse per line and it can therefore only be focused at one point. The tendency therefore is to produce a weak focusing effect on transmission using relatively few elements to stimulate a small composite aperture and then change to multiple strongly focused zones on reception using more elements to simulate wider apertures.

There is one final option open to linear array designers and this is to use multiple zone focusing on transmission. One transmitted pulse can be sent out focused at a short range and on reception only echoes from that range accepted. The electronics can then send a second transmitted pulse along the same line, this time focused at a greater depth and then accept only echoes from that depth and so on. However there is a hidden penalty in that if two or more pulses are transmitted per line then the frame rate must fall. Swept transmission focus systems therefore have low frame rates typically 5 or 10 frames per second and special interpolation techniques are required to give the operator's eye – brain combination the impression that there is no flicker.

Linear arrays have become one of the commonest versions of ultrasound technology available. They are characterized by high reliability and good resolution and are ideal for most routine obstetric examinations.

Their main limitation lies in their size which makes them unsuitable for many upper abdominal investigations, in which the rectangular field of view is undesirable.

4.4 Electronic sector scanners (phased arrays)

The sophisticated use of delay lines to achieve focusing can also be extended to look preferentially at echoes off-axis.

Figure 17 shows a transducer array system in which it is desired to focus at point F which is off-axis. The element 1 is closest to F and will therefore receive echoes from there soonest. Element 8 will be the last element to receive echoes. Thus the delays required to ensure that all echoes from F reach the receiver at the same time must gradually increase from zero at element 8 to a maximum at element 1. The major difference between this case and the linear array case is that now the delays are not in pairs symmetrically placed around the central element.

We can now see that delays have been used effectively to swing the beam and it is possible to arrange for transmission and reception to focus almost anywhere in 180° in front of the probe. There is now no need to step along the array and all elements are used every pulse for both transmission and reception. With suitable switching and control circuitry, swept focusing is possible on these scanners too.

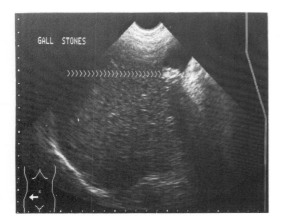

Figure 18. Shadowing due to gall stones.

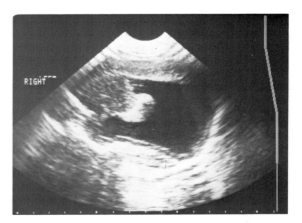

Figure 19. Flaring through a fluid-filled region. Note the saturation along the posterior uterine wall.

Such scanners, known as phased arrays, have small probes and produce sector-shaped scanning patterns. They are popular for echocardiography and some other applications. However, their complexity does make them relatively expensive and many have been prone to side-lobe artefacts. This means that diffraction effects at the transducer can sometimes produce 'hot-spots' in the beam where they were not intended and this leads to image of targets being misplaced by the scanner. Many of these problems do now appear to have been resolved.

5. Artefacts in B-scanning

An artefact can be defined as false or misleading information introduced by the imaging system. In the ultrasound case this can mean displayed echoes where there are no targets or vice versa or perhaps distorting the spatial relationships between structures. Some of the more important and common artefacts will be discussed here. The reader is referred to the bibliography for more detailed reading.

5.1 Shadowing

It is often the case that several targets of interest lie along the same scan line, that is behind one another. Provided too much energy is not reflected by the proximal target, then there will be sufficient energy left to interact with subsequent targets along the line and thereby produce echoes. Sometimes however, the proximal target is strongly reflective and insufficient energy is transmitted to produce detectable echoes from distal structures. These structures are said to lie in the 'shadow' of the strong reflector. The image displays a dark space behind the reflector where in fact, there should be echoes.

This effect is found distal to air and bone interfaces and effectively precludes scanning large regions of the gut and lung fields. There are situations however in which the effect can be used to advantage such as in the diagnosis of gall stones. Here the shadow posterior to the target identifies it as a strongly reflecting object which is much more reflective than the posterior gall bladder wall (*Figure 18*).

If shadowing is a major impediment to the scan then there are certain manoeuvres which can be tried. The first is to scan through a different plane and thereby avoid the reflector if possible. Secondly, it may be possible to replace reflective gas pockets with liquids, for example a liquid meal helps stomach imaging. Thirdly the relative anatomy may be altered by some procedure, for example inhaling helps to lift the diaphragm and helps liver imaging.

5.2 Flaring

The inverse of shadowing is flaring. If some part of the scanning section has a lower attenuation path to a target than other parts, then the brightness of the echoes from that target will vary. In the worst cases, echoes will be so bright that they saturate the display (*Figure 19*). Such saturated regions may contain lost information and this can be recovered by reducing the TGC. However this may mean that the TGC is wrongly set for the rest of the scanned section. Some modern scanners have automatic TGC. This means that the scanner selects the TGC curve and permits different curves to be used in different parts of the image. In such cases, flaring should almost never be a problem. Again, the artefact may be used diagnostically. If a clear structure is found in an image, this may be either a highly homogeneous solid or a liquid-filled structure, it is normally a low attenuation region and therefore flaring would be expected at its distal surfaces.

5.3 Reverberations

Another process which can occur at strongly reflecting interfaces is reverberation. In this case, two such interfaces are needed, although one of them may be the transducer itself. Echoes from the strong reflector travel back to the transducer in the usual way but are sufficiently energetic to be reflected from the transducer face and set off on another round trip. They then undergo a second reflection at the strong reflector and return to register a second time at the transducer. The scanner interprets this as an echo from another interface at twice the distance of the first and thus an entirely spurious interface is recorded on the display (*Figure 20*). In fact, three or four further round trips might take place resulting in a series of equally spaced identical surfaces on the display.

Reverberations are often seen in the urinary bladder (*Figure 20*) due to a strong reflection from the anterior wall of the bladder. They are readily recognizable because of the precise equality of their spacing which is quite unnatural (*Figure 21*).

5.4 Other artefacts

Many other ultrasound artefacts have been recorded, often due to changes in beam direction in layers of unusual sound velocity or due to the finite beam width. If an artefact is suspected, then it is always wise to change the scanning section and image the target in some other place. This is often sufficient to confirm suspicion or otherwise.

6. Distance measurement by ultrasound

It is often the case that a measurement will be required at some stage in an ultrasound examination. The

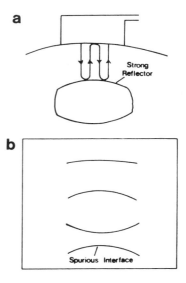

Figure 20. Reverberation artefact. Formation of a spurious interface by multiple reflection (**a**) schematic (**b**) display image.

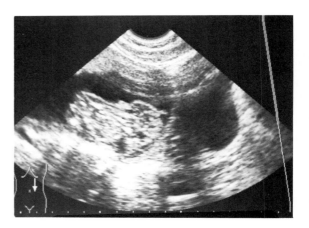

Figure 21. Fetal scan showing reverberation in the anterior portion of the scan. Note the spurious echo which is continuous through bladder and uterus.

commonest of these is the fetal biparietal diameter (BPD) measurement. Measurement facilities are almost always available on modern scanners and are termed ultrasonic calipers.

One of the most frequently encountered forms is that of a pair of white crosses which can be directed anywhere on the screen using some control which may be a joystick or 'trackerball'. The display will normally

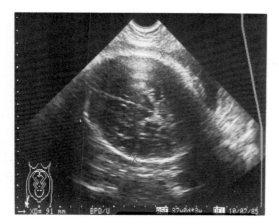

Figure 22. Fetal BPD measurement showing crosses to mark the measurement line.

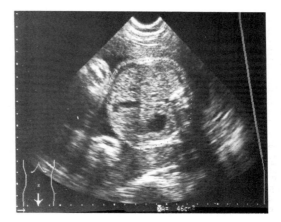

Figure 24. Fetal abdominal transverse section showing area measurement made inside the outline shown by dotted lines.

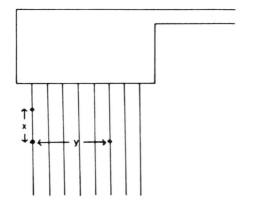

Figure 23. Caliper measurement *x* distance along a scan line *y* distance across several scan lines.

show the distance between the crosses, or cursors as they are sometimes called, as a distance in millimetres (*Figure 22*). The equipment, however, will only show the correct value if it is correctly calibrated and this is where the assumption of a 1540 m s^{-1} value for the speed of sound comes in.

For measurements made along a single scan line (*Figure 23*) the distance between two echoes is measured by assessing the difference in arrival time of the two corresponding echoes. Thus technically it is more correct to quote BPD measurements in microseconds than in millimetres. The conversion into millimetres is the result of the scanner performing a speed/time calculation on the time measured. In order

to perform measurements across scan lines, it is first necessary to establish the scale of the display from the speed of sound calculation made along the scan line. In other words 1 mm on the display is found to be equivalent to *x* mm in soft tissue. This scale must then be fed into the computer to correctly assess distance across the scan lines.

The accuracy of measurement along a scan line depends upon:

(i) the accuracy of placement of the cursors;
(ii) the correct calibration of the equipment;
(iii) the time speed of sound in the tissue being close to 1540 m s^{-1};
(iv) the axial resolution of the scanner.

It should be possible to reliably measure axial distances in soft tissue with modern scanners to ± 0.5 mm or better. In practice, the caliper display is often rounded to the nearest millimetre making this impossible.

Best reproducibility and reliability in an axial measurement of this kind is obtained by measuring from the beginning of one echo to the beginning of the next. This measurement, often called 'leading edge to leading edge', is least susceptible to variation in TGC settings.

If the measurement is made across a number of scan lines then the main source of error is likely to be the beam width of the transducer. The separation of scan lines is likely to be less than the beam width and so there is uncertainty as to which scan lines mark the beginning and end of the target.

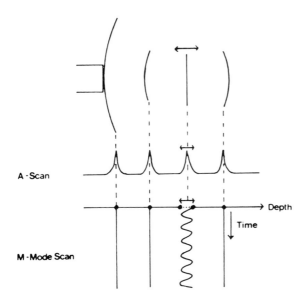

Figure 25. Formation of an M-mode scan.

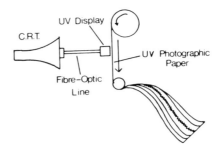

Figure 26. Schematic diagram of a fibre-optic chart recorder.

7. M-mode and recording

This chapter has so far discussed the use of A- and B-mode formats in ultrasonic imaging. A special requirement in echocardiography is the display and recording of the motion of some moving structures. In this case, one dimension is used for depth in tissue and the other represents time (*Figure 25*). This is called M (for motion)-mode scanning, or sometimes T-M (time – motion) mode.

Thus static structures are represented by straight lines and moving ones as curved tracings depending upon the movement concerned. The length of the record representing time is almost limitless. Thus tracings of the movement of heart valves for many minutes can be faithfully recorded (see Chaper 9). The problem is to find a suitable form of hard copy.

The conventional hard copy devices utilize photographic transparencies or prints recorded from a single television image which is normally a frozen frame from a real-time scanner. If many cardiac cycles are recorded in this way, the record becomes impracticably squashed and detail is lost. The solution is to use a fibre-optic chart recorder. In this machine a long strip of paper is used to represent the time axis and the depth is displayed as a horizontal distance across the paper. The recording is made by rolling the paper steadily across a small rectangular fibre-optic coupled display to generate a latent image on the special photographic paper (*Figure 26*).

8. Acknowledgements

The author would like to thank Benita Slater for typing this manuscript and Julia Evans for patient production of the drawings.

Machines are now often equipped with facilities for measuring distances along arbitrary curves or enclosed perimeters. The uncertainties in such measurements are more difficult to assess but it is unlikely that an accuracy of better than ± 1 mm will be achieved.

Area and even volume measurements are also facilitated in modern scanners (*Figure 24*). The operator must bear it in mind that any errors involved in taking a one-dimensional measurement will be greatly increased in two-dimensional and further increased in three-dimensional. Thus volumes of cysts or gestation sacs estimated in this way may well have absolute errors of ± 30%. The fact that a measurement is reproducible to within a small variation does not mean that it is accurate to that extent.

Scanners also come equipped with computerized 'look-up' tables for the commonest fetal measurements. Thus a BPD measurement of x cm can be instantly converted to A weeks and B days gestation. This is highly convenient but the operator must be aware of which 'look-up' table is written into the scanner and also must beware of quoting estimates to too high a degree of accuracy. For example, a BPD of 92 mm may correspond to a 'mean' gestational age of say 36 weeks and 5 days but the uncertainty (± 2 SD) is probably ± 4 weeks and so the 'days' figure displayed is quite meaningless.

9. References

1. Hussey,M. (1985) *Basic Physics and Technology of Medical Diagnostic Ultrasound.* Macmillan, London.
2. McDicken,W.N. (1981) *Diagnostic Ultrasonics: Principles and Use of Instruments.* 2nd edition. John Wiley, New York.
3. Wells,P.N.T. (1978) *Biomedical Ultrasonics.* Academic Press, London.

Chapter 3

Signal processing and display

R.A.Lerski

1. Introduction

The function of the signal processing in an ultrasonic imaging system is to take the fundamental radio-frequency (RF) echo signals detected by the transducer, amplify them, modify them to produce the video signal, store them for the B-scan image and pass them to the display monitor (*Figure 1*). In Chapter 1 the nature of the signals collected by the ultrasonic transducer was described, that is RF waveforms of widely varying amplitude. Although, in the early days of ultrasound, the first of these features caused some difficulty in handling the high frequencies, it is the second, the wide range of amplitudes, that still causes some problems but is also to a large degree responsible for the diagnostic usefulness of ultrasonic grey-scale imaging. This wide range is called the *dynamic range* and is often expressed in decibels. It arises because the largest ultrasonic echoes, coming from specular reflections close to the transducer, are many times larger than the small scattered echoes, particularly those produced deep in the body and thus strongly attenuated before reception. This difference between the smallest and largest can be by as much as a ratio of 1 million times which is a voltage decibel ratio of 120 dB. Such a large range cannot be maintained on passage through the electronics of the ultrasonic scanner nor meaningfully shown in the display (the human visual system could not appreciate it) and it must be *compressed* in the early part of the electronics to a more reasonable level. There are several methods employed to carry this out and these are critical to the success of ultrasound.

2. RF Amplification

The RF echoes picked up by the transducer are generally very small and must be subjected to amplification in the RF amplifier before they can be usefully handled in the later stages of the signal-processing chain. Sometimes this amplifier is tuned to the frequency of the transducer with a restricted bandwidth with the purpose of limiting the noise fed through to the later stages. This RF amplifier often has an adjustable overall gain control which has the simple effect of providing a uniform amplification of all the echoes, regardless of the depth or position from which they emanated. In clinical practice it is better to use this gain control rather than a variation of the

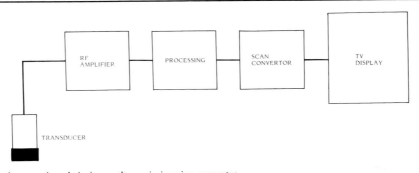

Figure 1. The signal processing chain in an ultrasonic imaging apparatus.

transmitter output to adjust the 'brightness' of the image. Scanning should always be performed with the minimum transmitter output necessary for diagnostic images.

3. Compression

3.1 *Time-gain-compensation*

Time-gain-compensation (TGC) or swept gain is the first of the techniques employed to *compress* the dynamic range of the ultrasonic RF echo signals. It occurs generally in the RF amplifier or pre-amplifier (see Figure 3 of Chapter 2) and is necessary because echoes from similar interfaces throughout the depth of scanning will not be of the same amplitude on reception due to the effects of attenuation as the ultrasonic beam travels through the body. Clearly this increases the dynamic range and is undesirable in that on the final image interpretation will be aided if similar interfaces produce echoes of equal amplitude no matter where they are located. Although knowledge of attenuation can be useful diagnostically (see Section 3.1 of Chapter 8) it is generally desirable to cancel out its effect on the displayed image.

TGC operates by increasing the gain of the RF amplifier as a function of time after the transmission pulse so that echoes received from superficial regions of the body are subject to a relatively low gain whereas echoes from a depth benefit from much greater gain. The net result is a tendency to equalize the echo amplitudes over the received depth. *Figure 2* illustrates the situation both with no TGC and with a TGC applied. An effective TGC may be judged to have equalized all echoes from similar interfaces or scattering centres so that attenuation is cancelled out. The general controls supplied to adjust TGC are as follows. An *initial attenuation* varies the amount by which the gain of the RF amplifier is reduced at the beginning of the reception period; a delay controls the time (or depth) which passes before the gain is made to rise again (the purpose of this is to suppress echoes from superficial body layers of no interest); a *slope* adjusts the rate at which the gain rises (quoted in dB cm^{-1}) and a *far gain* sets the level of final gain provided by the RF amplifier. Obviously different equipment may use different labelling or indeed may miss out some of the controls. The overall purpose is, however, the same.

A common variant is the use of sliders that control the gain at pre-set depths. These may be manually

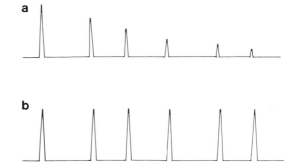

Figure 2. The effect of time-gain-compensation (TGC). (**a**) No TGC. (**b**) TGC applied showing equalization of echo amplitudes.

adjusted to optimize the B-scan image visually and provide a greater degree of flexibility than the linear variation described above. The TGC function finally selected by such a system may have a very complex shape. As with the previous system there is always a compromise in B-scanning in that the same TGC function will not be appropriate for all directions of view through the scanning field. When scanning complex arrangements of internal organs, for example in the abdomen, the TGC function can only be chosen to be an average.

3.2 *Automatic TGC*

A solution to the problem just stated is to design the electronics to automatically attempt to equalize echo amplitudes over the scan depth. Clearly, the design has to allow for the natural variation in specular and scattered echoes and is subject to the problem that information contained in such genuine variations may be suppressed by the automatic control reducing the diagnostic data. However, the use of this system does remove the possibility that diagnostic data may be lost through operator mis-setting of controls and has recently been gaining in popularity.

3.3 *Compression characteristics*

It is also common to design the RF amplifier so that, apart from its TGC function, it compresses the dynamic range of the signals input to it. Such a function may be described by the amplifiers characteristic (*Figure 3*) in which the input dynamic range is plotted against the output dynamic range. An amplifier which displays no compression has a linear characteristic, whereas a common compression charac-

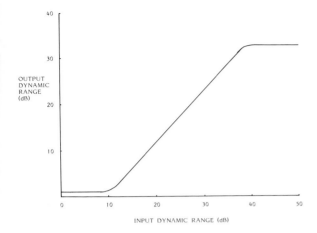

Figure 3. The compression characteristic of the receiver RF amplifier.

Table 1. Binary number system.

Normal (denary)	Powers of two 2^5 2^4 2^3 2^2 2^1 2^0						Binary
1						1	1
2					1	0	10
3					1	1	11
4				1	0	0	100
7				1	1	1	111
15			1	1	1	1	1111
31		1	1	1	1	1	11111
32	1	0	0	0	0	0	100000
63	1	1	1	1	1	1	111111

teristic is S-shaped. Here, the high and low echoes are compressed in the non-linear regions into a smaller dynamic range on the output. As will be discussed later, these characteristics are often selectable so as to accentuate or compress particular parts of the dynamic range.

4. Demodulation

After the RF amplification and TGC stages it is usual to demodulate the RF signal leaving a video signal suitable for input into the digital scan convertor. Figure 2 of Chapter 2 showed the various stages in this process. The bandwidth of the signal is considerably reduced from 2 or 3 MHz down to say 0.1 MHz in the course of this smoothing procedure. The stronger the demodulation the greater the tendency to degrade the resolution, and a compromise must be sought between good resolution and a less cluttered display.

5. Analogue-to-digital conversion

A common reason to employ compression in the early stages of the RF amplifier is that digital techniques are used in the storage and display of the B-scan image and hence the signal must be converted to a stream of digital numbers. The larger the dynamic range of the signals the more expensive and difficult is this *digitization* process.

5.1 *Binary arithmetic*

The basis of the computer storage of information is the binary number system. This is because digital memory may only store a pattern of 0s or 1s and everything—data, programs—must be converted into such a format. Instead of numbers being represented in the normal denary or powers of 10 system they are converted to a binary or powers of two system. A number is converted to a sum of powers of two. An example is shown in *Table 1*. It can be seen from the table that generally more binary digits are necessary to store a number than in the denary system which is only to be expected since only 0 and 1 are available to represent the number rather than $0-9$. For example 15 needs four binary digits (1111). These binary digits are called bits, and large numbers of bits are required to store large numbers. Computers group these bits into bytes and words. A byte is 8 bits and a word may be 8, 16 or 32 bits.

5.2 *Digital sampling of a signal*

A signal that has to be converted from its basic analogue form (continuous time varying values) (*Figure 4a*) to a digital stream of numbers must be subject to analogue-to-digital conversion. The number of bits that are chosen for this conversion is a fundamental matter in determining the faithfulness with which the signal will be represented in the computer memory. The frequency at which it is sampled is also important but is basically a separate choice.

If, for example, only 2 bits are chosen then only four levels will be retained in the signal (*Figure 4b*) and it will be subject to considerable loss of detail. The more bits that are used then the greater the fidelity of the conversion, as for 6 bits (*Figure 4c*). For

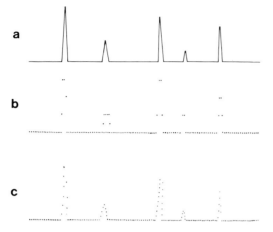

Figure 4. Analogue-to-digital conversion—the variation of the number of bits. (**a**) Continuous analogue signal; (**b**) 2-bit digitization; (**c**) 6-bit digitization. (In these figures the sampling frequency has been chosen to be an optimal value.)

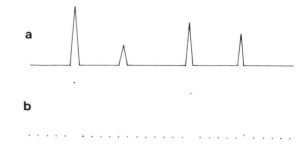

Figure 5. Analogue-to-digital conversion—the variation of the sampling interval. (**a**) Continuous analogue signal and (**b**) under-sampled digitized signal showing gross loss of information.

ultrasonic signals with a wide dynamic range this means that the number of levels and hence number of bits must be large. This is because the smallest echo must be represented by the lowest level and the largest by the full-scale maximum level. In *Figure 4c* the largest echo would represent 63 units and the smallest level 1 unit. This is a dynamic range of 63 or in decibels $20*\log_{10}(63) = 36$ dB. (A good rule to remember is that 1 bit gives a voltage dynamic range of 6 dB).

In *Figure 5* the effect of varying the sampling frequency is shown. If insufficient samples are taken then the signal will not be recorded properly—no matter how many bits are used. Clearly, the sampling frequency must be selected so that all the detail in the signal is recorded and, in fact, it must be at least twice the highest frequency present in the signal (this is called the Nyquist condition).

The selection of the number of bits in the analogue-to-digital convertor is something of a compromise, as also is the choice of the sampling frequency. A large number of bits and a high sampling frequency are both desirable but greatly increase the expense of the electronic system. It is true, however, that with the costs of electronics reducing all the time these restrictions and the need for large compression in the dynamic range are becoming less necessary. Recent equipment (e.g. Acuson) has taken advantage of this.

After analogue-to-digital conversion the binary values may be stored in random access memory (RAM) and read out at a later time, either at the same frequency or at a different frequency. This is the basis of the digital scan convertor.

6. Scan conversion

The ultrasonic scanner, whether static B-scan or real time B-scan, constructs the image in a wide range of different ways. The scan pattern (i.e. the direction that the beam takes to gather echo information from the imaging field) is different not only for every type of scanner (e.g. linear or sector) but also for every model of scanner. In all cases, for economy in display systems, this scan pattern must be converted to a different scheme for display and it is usually chosen to make the display system a standard closed circuit television (TV) format. This allows the use of a whole range of standard components from monitors to cameras to video tape recorders.

To *convert* the scanner scan pattern to TV format it is necessary to utilize an intermediate electronic storage system and a different reading and writing rate. In the early days of grey-scale ultrasound this was accomplished through an analogue storage tube using a special silicon target (the analogue scan convertor) but this is now obsolete. It has been replaced by the digital scan convertor where the storage is now provided by an array of digital memory (*Figure 6*). The individual elements of the memory are called *pixels*. Clearly, in order that they will not be visible on the display, leading to a grainy appearance, there should be a large number of pixels in the image. Numbers like 256×256 and 512×512 are typical in modern systems. The B-scan image is written into

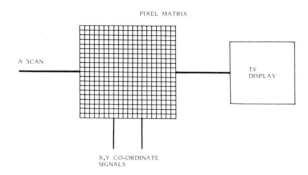

Figure 6. The digital scan convertor.

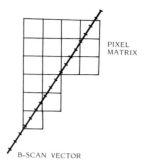

Figure 7. The B-scan vector directed across a pixel matrix. Sampling points shown by lines on the vector.

this memory at a rate determined by the ultrasound velocity, the scan pattern and the frame rate and read out at standard TV (video) rates. Since the data is stored digitally, the input may be stopped at any time and the image still read out non-destructively. This is called *frame-freeze*. Due to the digital nature of this procedure several basic features may be selected to suit certain applications and a number of image enhancement techniques may be used.

6.1 Pixel sampling

As described in a previous section, the demodulated ultrasonic signal is sampled at regular intervals by the analogue-to-digital convertor. These digital values must then be entered into the digital scan convertor memory in locations appropriate to form the B-scan image. In general (*Figure 7*) the B-scan vector will cross the array of pixels at an angle and the sampling is chosen so that several samples occur within each pixel. It is therefore necessary to select an appropriate value to enter into the pixel storage location in digital memory. There are many possibilities. Firstly, it is usual to weight the value entered depending on how near the B-scan vector passed to the pixel centre. If it passed exactly through the centre then the whole value would be entered whereas if it only just catches the edge of the pixel then a small fraction of the sampled value would be entered. Secondly, the number of samples falling within the pixel must be combined in some way. The maximum of the values may be used or the average chosen either of a single vectors' values or of all the values falling within the pixel during a total scan. This latter is sometimes called *compound* mode as distinct from *survey* mode where simply the latest value to be found is entered (i.e. continuous updating). This latter is useful for rapid searching for a region of interest.

6.2 Zooming

In digital scan conversion it is possible to *zoom* the image in two ways. Firstly, providing the sampling of the ultrasonic signal is fast enough that several samples normally occur within each pixel then the scale factors may be adjusted up to the point that only one sample is left per pixel and the image *write* zoomed. Clearly in this case the number of pixels in the final image is unchanged. The second possibility is simply the zooming of the output image (frame frozen or static B-scan) so that the TV display examines a smaller part of the pixel matrix (a *read* zoom). In this case the pixels will eventually become visible and the image will be grainy. The first method is therefore more desirable but has the disadvantage that it involves re-scanning the patient.

6.3 Line interpolation

Again, due to the digital nature of the scan convertor device it is simple to perform some processing of the output image. For example, an image with 256 pixel lines may be expanded to 512 lines simply by inserting an extra line in between each pair of original lines (*Figure 8*), the values used being the average of the lines on either side. This provides no new information but serves to improve the visual appearance of the image.

6.4 Pre-processing

As indicated previously (see *Figure 3*) it is common to compress the dynamic range of the ultrasonic signals before they are input to the scan convertor system. In some machines this compression is selectable in form with the intention of emphasizing or de-emphasizing particular parts of the dynamic range

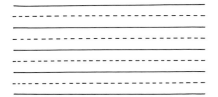

Figure 8. Line interpolation in digital scan conversion. Dotted lines represent the values calculated between the actual data lines.

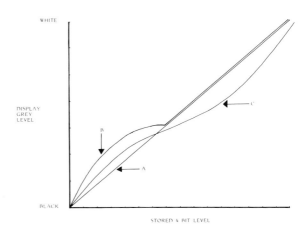

Figure 9. Post-processing characteristic curves for a 4-bit digital scan convertor. **Curve A** is a linear transfer between the stored level and the display grey scale; **curve B** accentuates the lower levels; **curve C** compresses the middle range and emphasizes the lower and upper ranges.

as it is recorded in the digital memory. Clearly if the pre-processing is changed then the scan must be retaken to observe the effect of this change. Pre-processing is less necessary if more bits are available in the analogue-to-digital convertor.

6.5 Post-processing

In an analogous manner to the provision of pre-processing, it is common to provide the facility to transfer the stored digital data in a variable manner to the display monitor. This is called post-processing and it also may be described in a characteristic transfer curve (*Figure 9*). If the scan convertor possesses a reasonable number of bits (e.g. eight) then post-processing may be an extremely useful facility since it is possible to change the contrast and brightness levels and emphasis within the *stored* dynamic range and thus study more closely a particular part of the echo amplitudes. This may allow the identification of a suspicious area within, for example, liver after the scan has been taken.

With the advent of analogue-to-digital convertors with even more bits and wider dynamic range, then the importance of post-processing may be expected to grow. It is also useful as a means of matching the grey-scale of the displayed image to the photographic characteristics of the recording film in the event that the same monitor is used for visual display and photography (see Section 7.5).

6.6 Frame-freeze

Particularly in a real-time system it is usual to provide the facility to freeze the recorded image so that the image stored in the scan convertor memory at a particular time is read out repeatedly to the display monitor for study or for measurements to be made from it. In a digital scan convertor this is a very simple operation to carry out technically.

It is notable that the quality of a real-time image frozen in this way is substantially poorer than that perceived on the monitor when the image is moving. The reasons for this are 2-fold. Firstly, the eye tends to integrate a number of real-time frames in the moving image so that it appreciates a greater quantity of information than is present in a single frame. Secondly, there is tendency for the echoes from fast moving structures (e.g. the heart) to be smeared in the frozen frame.

6.7 Measurements from images

The facility is often provided on modern scanners to measure selected features from the frozen image. The simplest and most frequent is a linear dimension from two calipers, this being used for biparietal diameter or fetal crown−rump measurement. Also possible, because of the digital nature of the stored image, is an outlining and measurement of areas which can also be useful for fetal measurement. With either of these a pre-programmed table can provide a direct readout of gestational age from these measured dimensions. Care must be taken to ensure that the table used is appropriate for the patient population under study.

6.8 Tissue characterization

In addition to the ease of measuring dimensions directly from the digitally stored image, it is also possible to quantitate the echo pattern from different tissue regions with the aim of identifying either tissue types or disease states of a single tissue. Research in this area has been continuing for several years with focus being made on the echo amplitude (1), scattering behaviour (2) and texture (3). Usually the liver has been used since it provides a large area of diffuse scattered ultrasound to concentrate on. Although in each of the above areas reasonable success has been reported, as yet no technique has come into widespread clinical use.

7. Image recording

The video images output from the digital scan convertor may be recorded in a whole series of different ways. The use of video tape recorders or simple Polaroid cameras is possible but perhaps the highest quality images may be obtained from multi-format cameras.

7.1 Photographic recording

The fundamental principle of recording grey-scale ultrasonograms faithfully is that the photographic exposure characteristics of film are such that it is necessary to use different brightness and contrast settings on the monitor to take photographs from those used for viewing.

A typical exposure characteristic of film is shown in *Figure 10*. [For a discussion of exposure characteristics see (4)]. To utilize the available grey-scale optimally it is necessary to ensure that the light from the monitor covers the linear part of the characteristic only.

7.2 Video tape recorders

The video tape recorder (VTR) has become a ubiquitous household product. It operates by recording the video signals magnetically on tape coated with oxides of magnetic material. A signal is passed through a coil (or head) over which the tape passes at high speed (usually involving a rotation of the coil as well as tape movement to provide a helical path over the tape) and thus the oxide is magnetized to store the video information. On playback the opposite takes place with the magnetic tape inducing a signal in a coil over which it passes. It is possible to read a single image or frame repeatedly by stopping the tape but still allowing the head to rotate over the region in which the frame is

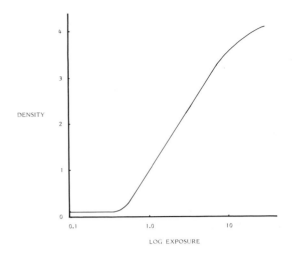

Figure 10. Exposure characteristic of typical photographic film.

stored. An audio track may be added as commentary to the tape, in particular for teaching purposes.

Domestic quality VTRs (e.g. VHS) may be used for the storage of ultrasonograms, although professional units (e.g. U-MATIC) will give better results. The disadvantage of all VTR systems, particularly for the storage of static or frame-freeze B-scans is the difficulty of locating a particular image within a long tape. Archival properties are only good if the tape is run through at intervals, since there is a tendency to 'print-through' from adjacent layers of tape. Video-tape recordings display a good grey-scale.

7.3 Polaroid photography

Instant (diffusion-transfer) Polaroid photography has been very popular since it provides a very rapid hard-copy record of the ultrasonic image. However, it does suffer from the significant disadvantage that its grey-scale rendition is limited and cost is high compared with the use of transparent rapid process film. However, the hardware to produce Polaroid pictures, that is the photographic monitor and Polaroid camera, is fairly inexpensive.

7.4 Multiformat cameras

A multiformat camera is a system in which a specialized photographic monitor is provided with a lens system to enable the recording of images on sheets of transparent film held in special cassettes. Often the cassettes may be loaded in a daylight system and developed in a daylight processor. It is common to provide the facility to select the format of images on the film (e.g. 4 images, 9 images, 16 images etc.). The films are viewed like X-rays so that they fit in well to the routine of the X-ray department.

The advantage of this type of system is that transparent single-sided (emulsion) film has an exposure characteristic far superior to Polaroid film in terms of grey-scale rendition. Fundamentally it is possible to view more grey shades in a transparent film than may be seen in a picture viewed by reflection (a positive print). Against these features is the necessity to process the film in a processor and to view it on a viewing box which may not be convenient in say, an obstetric department.

7.5 Setting-up of photographic systems

The exposure characteristics of film are such that when the TV monitor is correctly adjusted for optimal grey-scale rendition then it appears lacking in contrast to the eye. Thus it is an advantage to utilize separate monitors for viewing and photography, and this is partly why the use of multiformat cameras has increased over the last few years. Additionally, the spectral response of film and the human visual system is different, so that a blue phosphor gives the best results in the case of film, but is difficult to appreciate by eye.

The aim of the setting-up procedure for a photographic monitor is to ensure that the brightness and contrast controls are set so that the exposure falls on the straight line part of the films characteristic (*Figure 10*). The effect of a brightness increase is to move the exposure up the characteristic, whereas a contrast increase widens the range of exposure. Clearly, both controls have to be adjusted for optimal results, and since they are generally interactive, the process must be done by repetitive trials. The process is made far easier if a grey-scale test pattern can be provided on

the monitor, although final tuning must be done on a good quality clinical image (with the same post-processing settings as normally used).

8. Conclusion

In common with all other types of electronics, ultrasound is becoming more and more based on digital techniques. For some time the scan convertors used to translate between the ultrasonic scan pattern and the TV raster pattern have been based on digital methods. Now the analogue-to-digital conversion of the ultrasonic signals is tending to take place earlier in the signal processing chain with all the advantages of stability and flexibility that this entails. It is also now possible to build the electronics in a microcomputer controlled fashion so that changes to the mode of operation can be made simply by changing the controlling programme (or software). Updates and improvements can thus be made very simply. All these advances can lead to significant gains in image quality as witnessed by the very latest equipment and will eventually lead to less expensive scanners.

9. References

1. Mountfield,R.A. and Wells,P.N.T. (1972) Ultrasonic liver scanning: the A-scan in the normal and cirrhosis. *Phys. Med. Biol.*, **17**, 261–269.
2. Nicholas,D. (1979) Ultrasonic diffraction analysis in the investigation of liver disease. *Br. J. Radiol.*, **52**, 949–961.
3. Lerski,R.A., Smith,M.J., Morley,P., Barnett,E., Mills,P.R., Watkinson,G. and MacSween,R.N.M. (1981) Discriminant analysis of ultrasonic texture data in diffuse alcoholic liver disease. 1. Fatty liver and cirrhosis. *Ultrasonic Imaging*, **3**, 164–172.
4. Meredith,W.J. and Massey,J.B. (1977) *Fundamental Physics of Radiology*. John Wright publishers, Bristol.

Chapter 4

Doppler ultrasound

David H.Evans

1. Introduction

An increasingly important use of ultrasound in medicine is that of monitoring, measuring or imaging moving structures, and particularly blood flow within the body, using the Doppler technique. The Doppler principle was first described in the nineteenth century and now has many applications in astronomy and physics. Essentially it states that when an observer moves towards a source of waves, the frequency he measures is higher than that transmitted by the source, and when he moves away the frequency he measures is lower. A common everyday example of the Doppler effect is the decrease in pitch of a siren that is heard as an ambulance passes an observer.

The Doppler effect is modified slightly in its use in medical ultrasound where the source and the observer of the ultrasound (the Doppler transducer or transducers) are at rest with respect to each other, but the waves of ultrasound are reflected from targets that are moving with respect to both (*Figure 1*). In such a case it can be shown that the difference between the transmitted frequency f_t, and the received frequency f_r is given by:

$$f_d = f_t - f_r = 2\,V f_t \cos \theta / c \qquad (1)$$

where V is the velocity of the target, θ the angle between the ultrasound beam and the direction of the target's motion, and c the velocity of ultrasound in the tissue. The most important point to note about this equation is that f_d is proportional to the velocity of the target and thus provides a means of studying motion. It is very convenient that, for normal velocities encountered in the body, f_d tends to lie in the audio-frequency range (0−20 kHz) and this allows an operator to monitor the Doppler signal simply by listening to it, although some further processing of the signal is usually desirable.

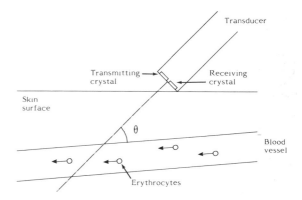

Figure 1. Schematic diagram of the use of a continuous wave Doppler probe to measure flow velocity in an artery. One crystal transmits a beam of ultrasound into the tissue. This is scattered by the moving erythrocytes and received by a second crystal. Large echoes also return from the stationary tissues but these are rejected by the Doppler electronics.

2. Continuous wave Doppler devices

The simplest type of Doppler device is the continuous wave (CW) Doppler that continuously both transmits and receives ultrasound. Because the transmission is continuous rather than pulsed such instruments have no depth resolution except in the sense that signals from targets close to the transducer experience less attenuation than those from a distance. In practice the lack of depth resolution is usually not a problem since in most cases there are a limited number of moving targets along any given line from the transducer, and continuous wave methods have significant advantages over pulse wave methods in terms of simplicity, ability to detect high velocities, and a reduction of ultrasonic dose.

Figure 2 is a block diagram of a simple non-directional CW Doppler unit. The master oscillator normally has a frequency of between 2 and 10 MHz although higher frequencies may be used in specialized applications. The signal from the oscillator is amplified and used to drive the transmitting side of the transducer. CW Doppler transducers differ from pulse–echo transducers used for imaging in two respects; they employ separate transmitting and receiving crystals, and the crystals have little or no damping applied to them so that they tend to 'ring'. The Doppler transducer is coupled to the skin in the normal way, and the ultrasound beam directed at a region of interest within the body. The returning ultrasound signal, which contains both shifted and non-shifted signals, is received by a second crystal and amplified before being mixed with a signal from the master oscillator. The process of mixing produces signals with frequencies of $(f_t + f_r)$, and the required Doppler shift frequency $(f_t - f_r)$. The mixed signal is low-pass filtered to remove the unwanted high frequency components, amplified, and either used to drive a loudspeaker or sent for further processing.

Most modern Doppler instruments have additional circuitry that determines the direction of the flow and separates the forward and reverse components, but the details of how this is achieved is beyond the scope of this chapter and the interested reader is referred to an article by Coghlan and Taylor (1) for further information.

3. Pulsed wave Doppler devices

In many instances it is useful to be able to control the depth within the tissue from which the Doppler signal is gathered. This is achieved by pulsing the ultrasound and only accepting echoes whose times of flight correspond to the range of depths of interest.

The basic design of a pulsed wave (PW) Doppler instrument is shown in *Figure 3*. The signal from the master oscillator is gated under the control of the pulse repetition frequency (PRF) generator. Fairly high PRFs are used in Doppler work because of the Nyquist limitation which requires that a signal (in this case the

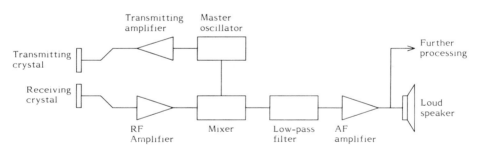

Figure 2. Block diagram of a simple non-directional continuous wave Doppler unit.

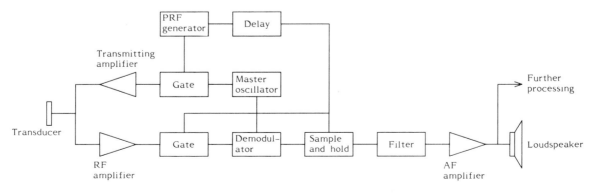

Figure 3. Block diagram of a simple non-directional pulsed wave Doppler unit.

Doppler signal) be sampled at at least twice its maximum frequency component for its correct interpretation. The length of time the gate remains open depends on the range of depths from which signals are required and is usually sufficient for a number of complete cycles to be transmitted. The amplified pulse is used to drive a transducer, which is similar in construction to those used for pulse−echo imaging work, and which unlike CW transducers contains only one crystal. Signals returning from the tissue are amplified and then fed to a receiving gate which is opened after an operator-determined delay to admit signals to the mixer. It is this delay between transmission and the opening of the gate which determines the depth from which the signals are gathered, whilst the time for which the gate is left open, taken together with the length of the transmitted pulse, determines the sample volume length. The output from the mixer is sampled during the time the receive gate is open and then filtered, amplified and sent for further processing.

PW units are used in two distinct fashions. Either the sample volume is made sufficiently large to encompass the entire blood vessel (or other region containing movement), or it is made much smaller than the vessel size to enable flow in just one part of the lumen to be probed. In the former case the range discrimination is used to reject signals from other nearby structures, whilst in the latter the high spatial resolution is used to extract information about flow in a specific part of the vessel. It is also possible to use a number of short gates across a vessel lumen to build up images of the instantaneous velocity profiles it contains (2).

4. Duplex scanners

Duplex scanners are devices that combine a pulse−echo B scanner with Doppler equipment. When used in the Duplex mode the B-scan image is used primarily to guide the Doppler beam and to place the Doppler sample volume at a known anatomical site. A cursor is superimposed on the B-scan image (*Figure 4*) which shows the direction that the Doppler beam is pointing or will point and, if the Doppler is pulsed wave, the position of the sampling gate. Once the operator is satisfied with the position of the sample volume, the image is frozen and the Doppler turned on. Depending on the type of scanner in use the operator may be able to make adjustments to the position of the

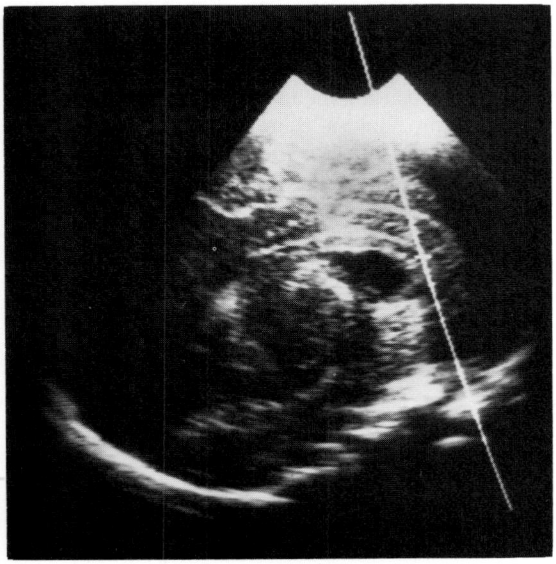

Figure 4. Image from a duplex scanner of para-sagittal section through a neonatal brain. The direction in which the Doppler beam will point, and the depth from which the Doppler signals will be extracted are shown by cursors. .

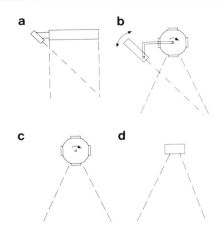

Figure 5. A variety of configurations of duplex probes. Probe (**a**) consists of a linear array imaging transducer with a fixed off-set CW Doppler transducer; probe (**b**) is a mechanical sector scanner with an off-set Doppler probe; probe (**c**) has both imaging and Doppler probes built into the same mechanical scanning head; probe (**d**) is a phased array which is used both for imaging and Doppler studies.

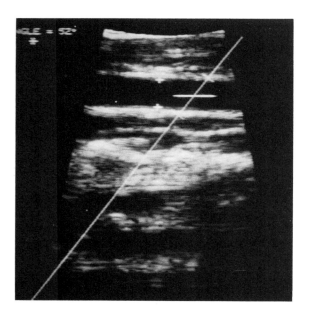

Figure 6. Image of an *in situ* femoro-popliteal vein graft produced using the type of probe illustrated in *Figure 5b*. Note that the Doppler cursor does not have the same origin as the image (cf. *Figure 4*).

sample volume either using the frozen image or a regularly updated image. At present it is difficult to achieve both real-time images and good Doppler signals simultaneously.

Duplex scanners have revolutionized the use of Doppler ultrasound, both because they allow the choice of the exact site from which a signal is recorded, and because they allow the measurement of the angle θ which the ultrasound beam makes with the flow axis. In the case of larger vessels the imaging facility may also be used to measure vessel diameter. The ability to choose the sampling site under visual control allows the operator to gather signals from sites which previously could not be reproducibly selected. The ability to measure the angle θ allows the Doppler frequencies to be converted to velocities (using Equation 1), and if the vessel diameter can be measured and its cross-sectional area calculated, the measurement of volumetric flow is also possible.

Duplex probes are manufactured in a variety of configurations. The imaging transducer may be a mechanical sector scanner, a phased array or a linear array system. The Doppler transducer may be separate from the imaging transducer or may use the same crystal. Some typical configurations are shown in

Figure 5. The simplest type (*Figure 5a*) consists of a linear array transducer with a CW Doppler transducer clamped at one end at a fixed angle. The entire transducer assembly must be moved in order to position the Doppler beam in the required position. These transducers have been used mainly in obstetric applications. Another type of transducer with an off-set Doppler probe is illustrated in *Figure 5b*. In this case the imaging transducer is of the mechanical variety, and the Doppler transducer pulsed wave. The angle the Doppler probe makes with the imaging probe is easily adjustable and the cursor on the Duplex screen is automatically adjusted to correspond to the correct line through the tissue. An image from a scanner using such a transducer arrangement is shown in *Figure 6*. *Figure 5c* shows the design of a mechanical spinning sector scanner in which three of the four elements are imaging crystals of different frequencies, whilst the fourth element is a PW Doppler crystal. *Figure 4* was produced with a scanner of this variety; note how the imaging and Doppler beams originate from the same direction, in contrast to different directions shown in *Figure 6*. The final type of Doppler head illustrated (*Figure 5d*) is a phased array device that uses the same elements to produce both the imaging and Doppler beams (indeed some such devices actually use the same pulses).

Different transducer configurations have different advantages. In general the optimal angle for imaging blood vessels (i.e. 90° to the vessel) is not the same as that for measuring the Doppler shift (i.e. 30–60°), and therefore unless there are other compelling reasons, off-set Doppler probes, as shown in *Figure 5a* and *b*, have advantages over other designs. Where an organ is insonated through a small acoustic window (e.g. the neonatal brain through the anterior fontanelle, or the heart through a rib space) then the type of probe illustrated in *Figure 5c* has obvious advantages. Probes which use the same elements for imaging and Doppler measurements must compromise performance in one or other area and often have excessively high output powers in the Doppler mode due to the heavy damping necessary to produce short pulses.

5. Doppler colour flow mapping

A recent, and exciting development in Doppler methodology is that of colour flow mapping. This technique combines pulse−echo B-scanning with Doppler mode B-scanning. The pulse−echo infor-

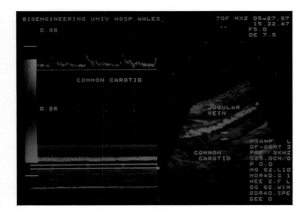

Figure 7. Image produced by a Doppler colour flow mapping system of the common carotid artery (red) and jugular vein (blue) (Courtesy of Professor J.P.Woodcock).

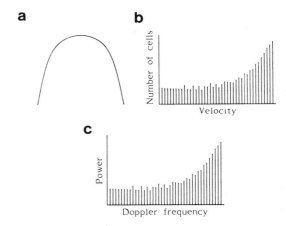

Figure 8. (a) Velocity profile in a normal common femoral artery at one point during the cardiac cycle, and (b) the corresponding velocity histogram. (c) The idealized Doppler spectrum that would be recorded from (a).

mation is used to produce an ordinary grey-scale image, whilst the Doppler shift information from all or part of the scan area is superimposed as a colour image (*Figure 7*). Different shades of blue and red represent different velocities away from and towards the probe respectively. The scanning may be mechanical or electronic, and the same transducer is used to collect both the imaging and Doppler information. The method has, so far, had most impact in cardiac studies where it has speeded up conventional Doppler examinations, and has been particularly valuable in defining the severity of regurgitant lesions, and shunt sizes in patients with intracardiac communications. Cardiological applications of the technique have recently been reviewed by Switzer and Nanda (3). Colour flow mapping may also have applications elsewhere in the body, particularly in the assessment of extracranial carotid artery disease.

It is important to realize that colour flow mapping is subject to the same physical limitations as conventional Doppler studies. The detected Doppler shift is of course proportional to the cosine of the angle between the ultrasound beam and the direction of motion (Equation 1) and therefore the same colour in different parts of a Doppler image does not necessarily correspond to the same velocity. Also aliasing, resulting from insufficiently high pulse repetition frequencies, will result in complete misrepresentation of both the direction and speed of flow.

Colour flow mapping seems destined to be a technique of importance, but it is as yet too early to evaluate the impact it will have on the use of other Doppler methods.

6. The Doppler signal

The simple equation relating the Doppler shift frequency to velocity (Equation 1) was presented in the introduction to this chapter. In practical Doppler applications there are always a multitude of reflectors or scatterers in the Doppler sample volume and therefore the Doppler shift signal consists not of a single frequency as suggested by Equation 1, but a spectrum of frequencies. In the case of blood flow, the frequency spectrum would ideally correspond in shape to a velocity histogram of the blood cells within the sample volume (*Figure 8*), but there are a number of mechanisms which may reshape this 'ideal' spectrum. Amongst these are non-uniform vessel insonation, attenuation, intrinsic spectral broadening, filtering, and aliasing.

6.1 Non-uniform insonation

The degree to which a vessel is uniformly insonated will significantly influence the relationship between the Doppler power spectrum and the velocity distribution within the vessel. If the vessel is uniformly insonated, then the two should be similar in shape; if the interrogating ultrasound beam is narrower than the vessel then the slow moving blood cells at the periphery of the vessel will tend to be under-represented in the sample and the higher Doppler frequencies accentuated; if a short pulse length is used with a PW unit then the power spectrum will represent the velocity distribution in a small 'region of interest'. It is important that the user is aware of these sampling effects because they can lead to incorrect measurement of blood flow velocity (4) and to the misinterpretation of spectral shapes (5).

6.2 Attenuation

Attenuation may distort the Doppler spectrum in a number of ways. The most important of these are related to the effects of attenuation on vessel insonation. If a structure with a particularly high or low attenuation overlies a part of a vessel, then signals from that part are either under- or over-represented in the Doppler spectrum (this is the Doppler analogue of shadowing in pulse−echo work). Further, even under ideal circumstances, an ultrasound wave must traverse more tissue (and less blood) to reach the lateral edges of a blood vessel. Since tissue has a higher attenuation coefficient than blood, signals from the edge of the vessel are more highly attenuated than those from the centre. Large insonation angles and high frequencies augment the degree of distortion that occurs.

6.3 Intrinsic spectral broadening

Intrinsic spectral broadening (ISB) is an increase in the spectral width of a Doppler signal unconnected with an increase in velocity distribution such as that which occurs due to turbulence. A number of mechanisms may give rise to ISB, but the most important is the broadening that occurs when very short ultrasound pulses are used to interrogate a small sample volume. Short pulses cannot be said to have a single frequency, rather they contain a spectrum of frequencies. Because of this even a single target moving with a constant velocity generates a spectrum of Doppler shift components. ISB only becomes important for very short pulse lengths (the fractional broadening is approximately given by the reciprocal of the number of cycles present in the pulse), but it imposes a fundamental limit on the degree of accuracy with which range and velocity may be measured simultaneously.

6.4 Filters

In addition to wanted components, the Doppler spectrum inevitably contains unwanted components, particularly low frequency high amplitude signals which arise from bulk tissue movement, and in particular from vessel wall movement. These signals may be much larger than those arising from blood flow and therefore must be rejected by a high-pass filter. Such filters, often called wall thump filters, usually roll off at about 50−200 Hz. There is no way to distinguish between wanted and unwanted low frequency signals and therefore signals from slowly moving blood are also rejected. This type of distortion of the signal is particularly troublesome in areas where normal velocities are low, such as in the fetus and neonate.

6.5 Aliasing

Aliasing has already been mentioned in an earlier section on PW Doppler. It is the misinterpretation of the frequency of a Doppler shift signal due to the use of an insufficiently high pulse repetition frequency (PRF). A familiar example of aliasing is the apparent backwards motion of the wheels of a cart seen on the television when the frame rate is inadequate to correctly reproduce the speed and direction of motion.

Aliasing can only be overcome by increasing the PRF, but this may lead to range ambiguity if there is more than one pulse with a significant size propagating in the tissue at the same time. Aliasing is particularly troublesome in cardiology where there is a conflict of requiring a high PRF to measure the Doppler shift resulting from high velocity jets, and requiring a low PRF because of the relatively large distance from which measurements must be made. To prevent aliasing some Doppler units deliberately introduce multiple range gates, whilst others have a CW option that can be used when it occurs.

7. Signal processing techniques

The Doppler signal consists of a continuously changing spectrum of frequencies from which information must be extracted. The easiest method of doing so is simply to listen to the signal which conveniently falls within the audio range. Whilst this is attractive in some qualitative applications it is highly subjective, and it is usually better to apply further processing to the signal, either so that it can be visually displayed, or numerical parameters derived from it.

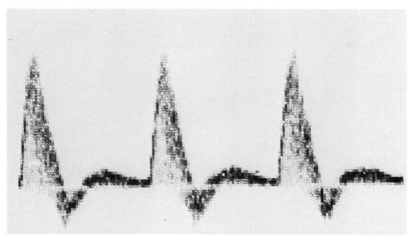

Figure 9. Sonogram of a signal recorded from a normal common femoral artery. Time is represented along the horizontal axis, Doppler shift frequency along the vertical axis and the power of the Doppler shift signal at a particular time and frequency by the blacknesss of the paper.

Maximum velocity

Time ⟶

Figure 10. Maximum frequency outline of signals recorded from the anterior cerebral artery of a neonate.

7.1 *Spectral analysis and the sonogram*

The best method of analysing a Doppler signal is to perform a full spectral analysis on it. As explained in the previous section, the Doppler spectrum is related to the velocity histogram of the blood cells within the ultrasound beam. By calculating the spectrum of the Doppler signal at regular intervals (usually every 5−20 ms) and suitably displaying the results it is possible to produce a pseudo three-dimensional display of the signal (*Figure 9*) usually called a sonogram. In this type of display time is plotted along the horizontal axis, Doppler shift frequency (or target velocity) along the vertical axis; the intensity (or colour) of each element of the display (i.e. each pixel) is related to the number of targets moving with the corresponding velocity at the corresponding time. The vast majority of Doppler signals are now analysed in this way, and whilst there are a number of ways of deriving the spectrum, they are increasingly calculated using the fast Fourier transform method. The details of this are beyond the scope of this chapter (the interested reader is referred to ref. 6), but it is important to understand

some of the properties of all Fourier-based methods. Perhaps the most important of these is the trade-off that occurs between time resolution and frequency resolution. Time resolution is dictated by how often an analyser generates a new (independent) power spectrum. If the signal is changing very rapidly (as for example with some signals from the heart) then it may be necessary to calculate a new spectrum every 5 ms, whilst if slowly varying venous signals are being analysed a new spectrum every 40 ms may be adequate. Frequency resolution is dictated by the length of the Doppler signal used in the analysis and is given by its reciprocal. Therefore if only 5 ms of data are used in a transform, the best spectral resolution that can be achieved is 200 Hz (i.e. the frequency 'bins' will be 0−200 Hz, 201−400 Hz, 401−600 Hz etc.). If 40 ms of data are used then the spectral resolution will be 25 Hz. It follows that if an *independent* estimate of the spectral content of a signal is required every 5 ms, frequency resolution cannot be better than 200 Hz. At present most analysers offer little or no choice as to the interval between estimates.

A further important property of an analyser is that

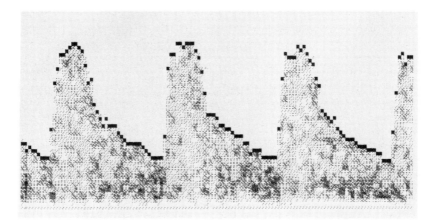

Figure 11. Maximum frequency envelope superimposed on a sonogram recorded from a normal common carotid artery.

the maximum frequency that can be analysed is dictated by the rate at which the signal is sampled—another example of the Nyquist limitation mentioned earlier. For the case of a real signal the sampling frequency must exceed twice the maximum Doppler shift frequency, otherwise aliasing will occur. The sampling rate also increases the number of points that must be Fourier transformed for a given data length and this may limit the rate at which an analyser can produce new spectra.

7.2 Frequency followers

Spectrum analysers are still relatively expensive pieces of equipment and therefore many simple and cheap Doppler devices contain circuits which produce an envelope signal which follows the instantaneous mean or maximum Doppler shift frequency (*Figure 10*). More complex devices are also able to produce such envelope signals, though in this case they are usually calculated from the spectral information. Envelope signals are of particular value for waveform analysis and for calculating mean velocity or volumetric flow.

The simplest type of envelope detector is the zero-crossing detector. These were at one time to be found in virtually all small Doppler units, but are now generally regarded as being unsatisfactory for all but the most qualitative applications. Under ideal circumstances zero-crossing detectors produce an output which is proportional to the root mean square Doppler frequency. Some of the drawbacks of these devices have been discussed by Lunt (7) and Johnston (8).

True (intensity weighted) mean frequency pro-cessors are now much preferred to zero-crossing detectors, and have the important property that their outputs are proportional to mean velocity, provided conditions of uniform insonation are achieved. The mean frequency of the Doppler signal may be derived in a number of ways, either by digital calculation from the spectrum or using analogue methods such as the well known Arts and Roevros (9) circuit. The output from a mean frequency follower is influenced by any mechanism which influences the Doppler spectrum itself, the most important of these being non-uniform vessel insonation, high-pass wall thump filters, and finite signal-to-noise ratios. The utility of this type of detector is further discussed in the section on blood flow measurement.

Maximum frequency followers are also popular, and may be implemented in a number of ways using both analogue and digital techniques. In practice most devices derive a so-called composite maximum envelope which is found by summing the maximum and minimum frequency envelopes, otherwise during periods of reverse flow the maximum is found to be zero. The major advantage of maximum frequency followers is that they are particularly immune to noise, and are not generally influenced by wall thump filters or the shape of the ultrasound beam. Full advantage of the noise-resistant properties of this type of follower can only be obtained if the maximum frequency envelope is superimposed on a sonogram (*Figure 11*). This allows the operator to ensure that the processor has not been misled by signals from an unwanted source and may allow very noisy signals to be satisfactorily analysed. Maximum frequency followers are particularly suited to waveform analysis applications, but may also be the follower of choice for some velocity and flow measurements.

For further discussion of the relative merits of various frequency followers the reader is referred to articles by Gill (10) and Evans (11).

8. Applications of Doppler ultrasound

Doppler ultrasound is now used in the investigation of blood flow and blood flow disorders in virtually every part of the human body. It has proved to be useful in the investigation of the fetus, the neonate, children and adults alike. In principle most Doppler applications can be classed as: (i) flow detection; (ii) velocity or flow measurement; (iii) waveform analysis and other pattern recognition methods; (iv) imaging.

8.1 *Flow detection*

Flow detection is the simplest application of the Doppler method. A simple Doppler unit, usually with just an audible output, is used to establish whether the flow in a particular artery or vein is present or absent. The most common use of this method is to determine blood pressure, particularly systemic blood pressure in the neonate, and local blood pressure in the lower limb of patients with suspected peripheral vascular disease. In these cases a pneumatic cuff is placed around a limb proximal to the Doppler insonation site and inflated until the Doppler signal disappears. The cuff is then slowly deflated until flow resumes, and the pressure in the cuff noted and taken to be equal to systolic pressure.

The presence of flow in a particular vein may usually be established by placing a Doppler probe over the vein and either asking the patient to perform a valsalva manoeuvre, or gently squeezing the tissue the vein drains.

8.2 *Blood velocity and flow measurements*

One of the most exciting applications of Doppler ultrasound is the non-invasive measurement of blood flow velocity and volumetric blood flow. There are a number of approaches to these measurements [see Evans *et al.* (12) for a complete survey], but at present the vast majority of measurements are made using Duplex scanners and so it is these methods that will be considered here.

Velocity and flow may be calculated either from the mean frequency or maximum frequency envelope of the sonogram, but whichever is used the basic scanning method is the same. The vessel is first imaged using the pulse−echo facility of the Duplex scanner, and the Doppler sample volume placed in an appropriate part of the vessel (*Figure 6*). If the mean frequency envelope is to be used to derive the velocity information, it is most important that the sample volume should encompass the whole vessel. The Doppler signals from a number of cardiac cycles are acquired and stored in a digital memory. The operator then measures the angle θ, between the axis of the ultrasound beam and the axis of the blood vessel, and the diameter of the blood vessel at the point from which the Doppler measurements were taken.

The most usual method of calculating the flow from these data is the 'uniform insonation method'. This relies on the fact that if the vessel is uniformly insonated then Equation 1 may be re-written in terms of mean Doppler shift frequency $\bar{f}_d$, and mean velocity $\bar{V}$, i.e.

$$\bar{f}_d = 2\,\bar{V}\,f_t \cos\theta/c \tag{2}$$

Rearranging this equation gives:

$$\bar{V} = \bar{f}_d\,c/2\,f_t \cos\theta. \tag{3}$$

The mean Doppler shift frequency may be found by averaging the intensity weighted mean frequency envelope over an integral number of cardiac cycles, the velocity of sound in tissue and the transmitted ultrasound frequency are known, and the angle θ has been measured. Mean blood flow velocity may be converted to volumetric flow by multiplying it by the cross-sectional area of the vessel. Most usually this is calculated from the measured vessel diameter by assuming it to be circular in cross-section.

The alternative method of using the maximum frequency envelope to calculate flow is not always suitable as the relationship between time-averaged maximum velocity and time-averaged mean velocity is in general complex, but there are two situations in which it may be used (4). The first is when the velocity profile in the vessel is known to be flat, as for example, in the aortic arch. In this case the time-averaged maximum velocity is approximately the same as the time-averaged mean velocity. The second is when the time-averaged velocity profile is parabolic, as for example in neonatal cerebral vessels. In this case the mean velocity may be calculated by halving the time-averaged maximum velocity.

Of the two methods of calculation the uniform

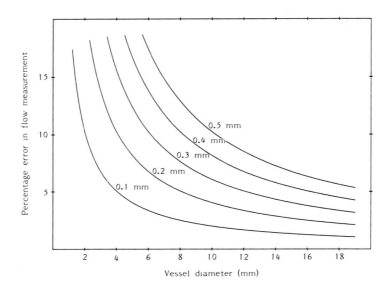

Figure 12. Percentage error in flow measurement due to the uncertainty in the diameter measurement (± 0.1 mm to ± 0.5 mm) plotted as a function of diameter.

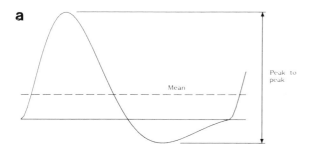

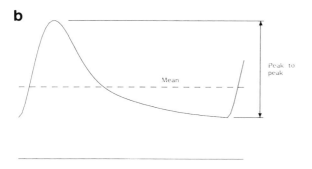

Figure 13. Pulsatility index is defined as the peak to peak excursion of the waveform divided by its mean height.

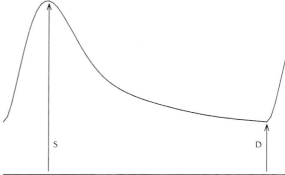

Figure 14. Resistance index is defined as (S−D)/S.

insonation method is the method of choice, but where uniform insonation is impossible, or the Doppler spectrum is greatly influenced by the effects of noise or high-pass filtering, then the maximum velocity method is often better.

All flow measurement methods are subject to inaccuracies, and it is important that the operator of a Duplex machine is aware of the possible sources and magnitudes of errors so that they may be kept to a minimum, and so that undue reliance is not placed on values which are intrinsically inaccurate. Errors may arise from inaccuracies in measuring three quantities, viz the mean Doppler shift frequency, the cross-sectional area of the vessel and the angle of insonation θ. The errors from each of these have been discussed in detail elsewhere (13,14) and only the most important will be noted here.

The major sources of error in the measurement of mean frequency arise from non-uniform vessel insonation, poor signal-to-noise ratios and the effects of high-pass wall thump filters. If the ultrasound beam is narrower than the blood vessel, then the central, faster moving, laminae of blood tend to be overrepresented in the Doppler signal when compared with the slower moving blood at the edge of the vessel. This means that in general the flow in a vessel is overestimated when uniform insonation is not achieved. In extreme cases this may cause an error in flow measurement of as much as 33%. The effect of wall thump filters is also to produce an overestimate of mean velocity, because, in addition to reducing unwanted high-amplitude low frequency tissue movement artefacts, such filters also reduce the signals from slow moving blood cells, and yet do not affect the signals from faster moving targets.

The errors due to vessel size are often the largest of all, especially when the vessel is relatively small. Even assuming that the change in vessel diameter throughout the cardiac cycle is not too great, and that the vessel is virtually circular in cross-section and its area can therefore be calculated from a single diameter, large errors may occur due to difficulties in measuring the diameter accurately. *Figure 12* illustrates the percentage errors that occur due to small degrees of uncertainty in the measurement of vessel diameter for different sized vessels. It can be seen that even small errors in diameter measurement, particularly on small vessels, can lead to considerable errors in flow measurement.

The errors which result from the measurement of the angle θ can usually be kept relatively small if θ itself is less than about 60°. This is because for small values of θ, $\cos\theta$ changes very slowly with θ, but as θ approaches 90°, $\cos\theta$ varies rapidly; large angles of insonation should be avoided wherever possible.

Blood flow measurements with a good degree of accuracy can be made using the Doppler technique, but it is important that the operator is fully aware of error sources if reliability is to be achieved.

8.3 *Waveform analysis*

Although the measurement of volumetric flow is often regarded as the 'Holy Grail' of Doppler studies, there is often as much or more to be gleaned from changes in the shape of the Doppler sonogram. This is so for a number of reasons. Firstly, as was stressed in the last section, accurate flow measurements can be quite difficult to make. Secondly disease processes often disturb the waveform shape before they have any effect on volumetric flow (possibly because the body compensates to keep flow constant). Thirdly a reduction in flow alone does not give any information as to whether the circulatory changes causing the reduction are proximal to or distal from the site of the Doppler measurement, whereas the waveform shape may give clues to their location.

The sonogram from even a single cardiac cycle contains a vast amount of information, and therefore attempts to recognize and quantify abnormalities are usually restricted either to envelope signals (usually the mean or maximum frequency envelope), or to power spectra chosen from specific parts of the cardiac cycle, for example at peak systole. In each case a multitude of approaches have been attempted and only a few simple methods will be explained here by way of illustration. Once again the interested reader is referred to Evans *et al.* (12) for a comprehensive survey of available methods.

8.3.1 *Waveform analysis of envelope signals*

The most widespread methods of quantifying changes in Doppler waveform shapes are the 'pulsatility index' and the so-called 'resistance index'. Both are very easy to calculate and have proved to be useful in many applications. The pulsatility index was introduced to quantify the damping in the outline of a sonogram that occurs distal to arterial stenoses in the peripheral circulation. It was at first defined in terms of the Fourier components of the envelope waveform, but was later replaced by a simpler but similar pulsatility index (PI) defined as the peak to peak excursion of the waveform divided by its mean height (*Figure 13*), (15). In normal subjects the PI of Doppler waveforms progressively increase down the leg; in patients with peripheral vascular disease this is not so, and a significant arterial narrowing will cause a decrease in the PI of all waveforms recorded distal to it.

An even more simple index is the resistance index, originally used on waveforms from the common

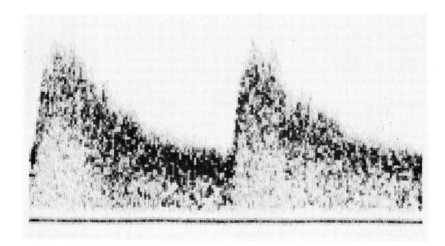

Figure 15. Sonogram from a normal internal carotid artery showing a clear window.

carotid as a method of detecting internal carotid stenosis (16). This index is defined as $(S-D)/S$, where S and D are the maximum and minimum values of the envelope waveform recorded during the cardiac cycle (*Figure 14*). In patients with severely stenosed internal carotid arteries the resistance index is significantly higher than normal.

8.3.2 Analysis of the power spectrum

Sonograms from some arterial sites normally exhibit a clear 'window' (*Figure 15*) which results from plug-like flow profiles during all or part of the cardiac cycle. It is a common observation that in sonograms recorded from the internal carotid arteries (for example) the window tends to become filled in if the vessel is diseased. This is thought to be the result of disturbed non-axial flow. Numerous indices for quantifying these changes have been described, the majority of which are defined as the ratio between combinations of the maximum, minimum and mean frequencies of spectra recorded at peak systole. One example of this class of index is the 'fractional width' defined by Johnston *et al.* (17) to be the 'ratio of the maximum minus the minimum frequency divided by the maximum frequency'.

8.4 Imaging

Doppler methods can be used to image vessels in a number of ways. The most simple is to perform a raster scan over the skin surface with a CW Doppler

probe attached to a position-sensing gantry (18). Using the Doppler shift information and the coordinates of the Doppler probe it is then possible to build up a type of C-scan of underlying blood vessels either as a bistable image or one which is colour coded in accordance with the size and direction of the Doppler shift. Scans in other planes may also be produced by the use of PW Doppler units (19), but in general these simple Doppler imagers have been superseded by other techniques. Duplex scanners combine high quality pulse−echo images with Doppler information, whilst Doppler colour flow mapping systems produce real-time Doppler images superimposed on conventional B-scan images. Both these devices were described earlier in this chapter.

9. References

1. Coghlan,B.A. and Taylor,M.G. (1976) Directional Doppler techniques for detection of blood velocities. *Ultrasound Med. Biol.*, **2**, 181−188.
2. Reneman,R.S., Van Merode,T., Hick,P. and Hoeks,P.G. (1986) Cardiovascular applications of multi-gate pulsed Doppler systems. *Ultrasound Med. Biol.*, **12**, 357−370.
3. Switzer,D.F. and Nanda,N.C. (1985) Doppler color flow mapping. *Ultrasound Med. Biol.*, **11**, 403−416.
4. Evans,D.H. (1985) On the measurement of the mean velocity of blood flow over the cardiac cycle using Doppler ultrasound. *Ultrasound Med. Biol.*, **11**, 735−741.
5. Van Merode,T., Hick,P., Hoeks,P.G. and Reneman,R.S. (1983) Limitations of spectral broadening in the early detection

of carotid artery disease due to the size of the sample volume. *Ultrasound Med. Biol.*, **9**, 581−586.

6. Cochran,W.T., Cooley,J.W., Favin,D.L., Helms,H.D., Kaenel,R.A., Lang,W.W., Maling,G.C., Nelson,D.E., Rader,C.M. and Welch,P.D. (1967) What is the Fast Fourier Transform? *IEEE Trans. Audio Electroacoust.*, **AU-15**, 45−55.

7. Lunt,M.J. (1975) Accuracy and limitations of ultrasonic Doppler blood velocimeter and zero crossing detector. *Ultrasound Med. Biol.*, **2**, 1−10.

8. Johnston,K.W., Maruzzo,B.C. and Cobbold,R.S.C. (1977) Errors and artifacts of Doppler flowmeters and their solution. *Arch. Surg.*, **112**, 1335−1341.

9. Arts,M.G.J. and Roevros,J.M.J.G. (1972) On the instantaneous measurement of blood flow by ultrasonic means. *Med. Biol. Eng.*, **10**, 23−34.

10. Gill,R.W. (1979) Performance of the mean frequency Doppler modulator. *Ultrasound Med. Biol.*, **5**, 237−247.

11. Evans,D.H. (1985) Doppler signal processing. In Altobelli,S.A., Voyles,W.F. and Greene,E.R. (eds), *Cardiovascular Ultrasonic Flowmetry*. Elsevier, New York, pp. 239−261.

12. Evans,D.H., McDicken,W.N., Skidmore,R. and Woodcock,J.P. (1988) *Doppler Ultrasound—Physics, Instrumentation, and Clinical Applications*. John Wiley, Chichester.

13. Gill,R.W. (1985) Measurement of blood flow by ultrasound: accuracy and sources of error. *Ultrasound Med. Biol.*, **11**, 625−641.

14. Evans,D.H. (1986) Can ultrasonic duplex scanners really measure volumetric flow? In Evans,J.A. (ed.), *Physics in Medical Ultrasound*. IPSM, London, pp. 145−154.

15. Gosling,R.G. and King,D.H. (1974) Continuous wave ultrasound as an alternative and complement to X-rays in vascular examinations. In Reneman,R.S. (ed.), *Cardiovascular Applications of Ultrasound*. North-Holland, Amsterdam, pp. 266−282.

16. Pourcelot,L. (1976) Diagnostic ultrasound for cerebral vascular diseases. In Donald,I. and Levi,S. (eds), *Present and Future of Diagnostic Ultrasound*. Kooyker, Rotterdam, pp. 141−147.

17. Johnston,K.W., deMorais,D., Kassam,M. and Brown,P.M. (1981) Cerebrovascular assessment using a Doppler carotid scanner and real-time frequency analysis. *J. Clin. Ultrasound*, **9**, 443−449.

18. Reid,J.M. and Spencer,M.P. (1972) Ultrasonic technique for imaging blood vessels. *Science*, **176**, 1235−1236.

19. Fish,P.J. (1972) Visualising blood vessels by ultrasound. In Roberts,V.C. (ed.), *Blood Flow Measurement*. Sector, London, pp. 29−32.

Chapter 5

Performance checks

Peter D.Clark

1. Introduction

Two classes of performance checks for diagnostic ultrasound scanners are described in this chapter. They are first comparative performance checks and secondly quality assurance performance checks. The former are quantitative checks of complete ultrasound scanners for comparing them and for giving improved understanding which may influence the purchase and use of the scanners. The latter are qualitative checks of one complete scanner at regular intervals for maintaining a high standard of operation throughout the lifetime of the scanner. Other kinds of performance check which are not covered in this chapter include acceptance checks, safety checks and checks of individual scanner components.

The comparative performance checks take at least a day to perform, the exact time depending on the number of transducers and the number of permutations of control settings that are tested. They are likely to be performed once only by technical personnel. The time taken for the quality assurance checks is about the same as for scanning a patient. It is hoped that these tests will be performed weekly or even daily by the users of the scanner. The bulk of the chapter is devoted to comparative performance checks but if your interest is only in a quality assurance routine then you need only read Sections 3, 4.2 and 8.

2. Sources of information

The International Electrotechnical Commission (1,2), the American Institute of Ultrasound in Medicine (3,4) and the Hospital Physicists Association (5) have produced guidance documents for measuring the performance of ultrasound scanners. All the documents refer to static, single element B scanners whereas modern equipment is usually real-time and can have multi-element transducers such as linear, curved linear, phased or annular arrays. Many of the performance tests are for a part of the scanner and access to the scanner's cabling or circuits is then required. The results of such tests are more easily understood by people of a technical background.

With these points in mind the Cardiff Test System (CTS) was developed at the University Hospital of Wales by McCarty and Stewart (6,7). It was devised to be applicable to the range of modern real-time scanners and to test the complete ultrasound system from the transducer to the monitor (or even to the hardcopy image) without requiring access to the internals of the scanner. In addition, the CTS is more easily understood by clinicians because its results are produced by scanning and measuring techniques like those used on patients.

The problem many have found with the CTS has been understanding the instruction manuals. The concepts of using the system can be lost in the special terminology used. However the special terminology is appropriate because the reference points and conventions of the CTS are different from others. The performance checks in this chapter utilize the CTS and the policy adopted has been to retain terminology used in the rest of the book up to Section 6, which acts as a link to the Cardiff terminology. The CTS is currently under consideration as both a British and European standard for ultrasound performance testing.

In 1985 the Scottish Home and Health Department (SHHD) set up the Ultrasonic Equipment Evaluation Project (UEEP). There is national cooperation between the SHHD who fund and oversee the project and the DHSS who publish the reports. The evaluation reports are both technical and clinical. This chapter is based on the author's experiences as evaluation officer for the UEEP. The reports (8 – 14) are available from the DHSS and the SHHD.

A recent paper by Lunt *et al.* (15) is a useful guide to a quality assurance scheme that combines simplicity with utility. Section 8 is an interpretation of the scheme for the CTS.

3. Test objects and phantoms

The CTS is a set of test objects and instruction manuals. In this section the Cardiff Test Objects are related to the wide range of test objects and phantoms currently available. Many of the manufacturers are based in the USA and they adopt an American Institute of Ultrasound in Medicine (AIUM) convention which distinguishes a test object from a phantom. The AIUM defines a test object as 'a device specifically fabricated to test one or more parameters of the equipment. In no way is it meant to be equivalent to human tissue or to produce an image which resembles a patient scan. Test objects are reproducible and standardizable'. A phantom is also reproducible and standardizable but is defined as 'a device which is specifically fabricated to be equivalent to human tissue in terms of the final medical image'. Test objects are usually filled with water or another liquid acoustic medium whereas phantoms are filled with tissue mimicking material (TMM). A good TMM has four properties similar in value to tissue:

(i) density;
(ii) velocity of ultrasound;
(iii) attenuation of ultrasound, including frequency dependence of attenuation;
(iv) scattering of ultrasound, in particular backscatter coefficient.

Some test objects including the widely used AIUM 100 millimetre test object (16) can be purchased filled with TMM and therefore the distinction between test object and phantom is not always according to the AIUM definition. The British manufacturers of the Cardiff Test Objects, which are filled with TMMs, have not adopted the American convention. The acoustic properties of the Cardiff TMMs are given in Section 4, except for scattering information which is not yet available. The advantage of using TMMs is that the performance check can be made with the scanner at settings similar to clinical use.

Madsen and his team at the University of Wisconsin (17) have been developing phantoms for many years.

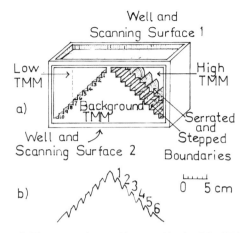

Figure 1. The grey scale test object. (**a**) Sketch of the GTO with steps numbered for cross-reference with *Figures 2* and *4*. (**b**) Profile of serrated boundary with peaks and troughs numbered for cross-reference with *Figures 3* and *5*.

Table 1. Acoustic properties of test object materials.

	Density $kg\ m^{-3}$	CW attenuation $dB\ cm^{-1}\ MHz^{-1} - dB\ cm^{-1}$ [a]	Velocity $m\ s^{-1}$
Background TMM	1087	0.86 – 0.1	1539.0
Low TMM	1044	0.16 – 0.25	1539.8
High TMM	1107	1.16 – 0.05	1542.7
LDPE Window	849	5.21	2203

[a]Slope and intercept of a linear fit of a plot of attenuation in dB cm^{-1} against frequency in MHz over the frequency range 2 – 8 MHz. CW = continuous wave.

An anthropomorphic breast phantom has been produced which mimics individual tissues including skin, subcutaneous fat, retromammary fat, pectoral muscle and ribs. A range of spherical TMM fat lesions of various sizes and contrasts have been incorporated into the phantom to test image quality.

4. Apparatus

The two Cardiff Test Objects that will be used and other apparatus are described below. The test objects are available from Diagnostic Sonar Ltd (see ref. 7).

4.1 Grey scale test object

A sketch of the grey scale test object (GTO) is shown in *Figure 1*. The test object container is a 1 cm thick

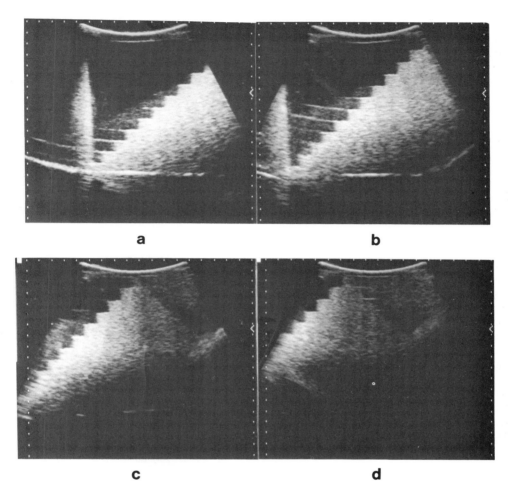

a　　　　　　　　　　　　　b

c　　　　　　　　　　　　　d

Figure 2. Grey scale test object stepped boundary imaged from surface 1 using a 3.5 MHz curved linear array transducer. The time gain compensation is set to compensate for the background TMM attenuation. (**a**) Image centred above step 8, low side. (**b**) Image centred above step 5, low side. (**c**) Image centred above step 0 (middle of test object). (**d**) Image centred above step 2, high side. Contrast loss can be seen as the ghost image to the left of the main image.

Perspex box. The top and bottom surfaces of the box have a 1 mm thick low-density polyethylene (LDPE) scanning window at the bottom of a shallow well. Within the box are three types of TMM namely low attenuation TMM, high attenuation TMM and background TMM. The acoustic properties of the materials are shown in *Table 1*. The low attenuation TMM is a gel substance which is both acoustically and optically clear. The background and high TMMs have in turn increasing concentrations of graphite particles. The size of the particles is of the order of tens of microns. The graphite makes the jelly black and opaque and

acoustically scattering. The boundary between the TMMs is stepped for half the width of the test object and serrated for the other half. The steps are 1 cm in width and depth. *Figures 2−5* show real-time B scans of the test object using a 3.5 MHz curved linear transducer. *Figures 6−8* are real-time B scans produced by a 10 MHz small parts transducer.

4.2 Resolution test object

A sketch of the resolution test object (RTO) is shown in *Figure 9*. It is contained in a Perspex box and has three scanning windows, one of which has a well. The

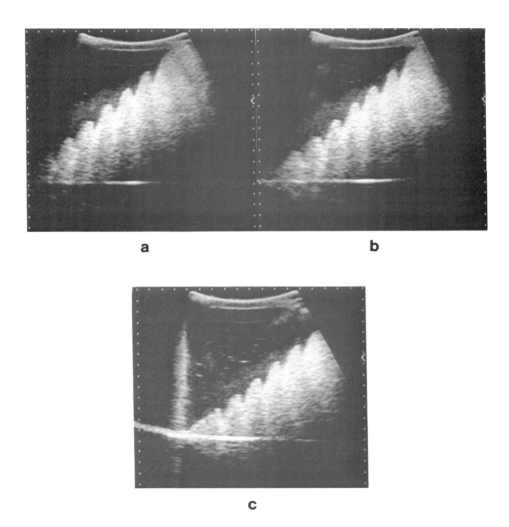

Figure 3. Grey scale test object serrated boundary imaged from surface 1 using a 3.5 MHz curved linear array transducer. (**a**) Image centred above 2nd trough, low side. (**b**) Image centred above 2nd peak. (**c**) Image centred above 5th trough. Contrast loss can be seen as the ghost image to the left of the main image.

box is filled with 'background TMM' and has six rod structures labelled respectively the A, B, C, D, E and F rods. The rods are made from type 302 stainless steel and are 0.15 mm in diameter except for the F rods which are 0.5 mm in diameter. The rods run horizontally in the direction of the width of the test object except for the F rods which run horizontally in the direction of the length of the test object. The A rods are spaced 10 mm horizontally and 15 mm vertically. The B rods are in a vertical line starting at 10 mm depth. The first five are spaced 5 mm apart

and the deeper ones at 15 mm. The C rods start at 1 mm depth and are spaced 1 mm vertically and 10 mm horizontally. The D rods are four groups of six rods. The vertical spacing within each group is 5, 4, 3, 2 and 1 mm and the rods are horizontally displaced to avoid acoustic shadowing. The bottom three groups of D rods have one rod in common with the A rods. The E rods are groups of six rods and each group is separated by 15 mm vertically. Each group is three pairs spaced 1.25, 2.5 and 5 mm horizontally and 4 mm vertically.

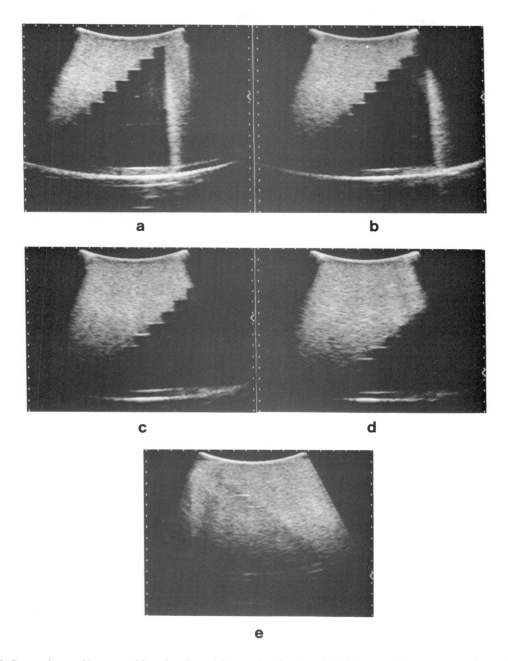

Figure 4. Grey scale test object stepped boundary imaged from surface 2 using a 3.5 MHz curved linear array transducer. (**a**) Image centred above 9th step, low side. (**b**) Image centred above 7th step, low side. (**c**) Image centred above 5th step, low side. (**d**) Image centred above 3rd step, low side. (**e**) Image centred above 7th step, high side.

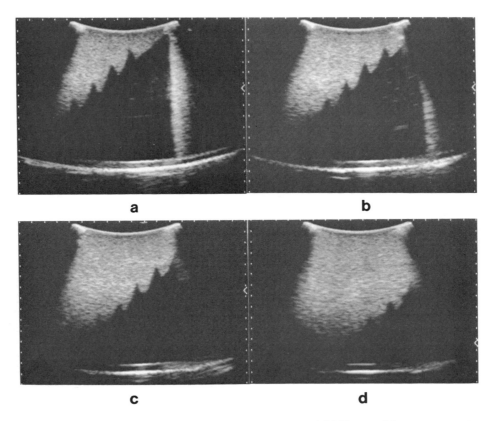

Figure 5. Grey scale test object serrated boundary imaged from surface 2 using a 3.5 MHz curved linear array transducer. (**a**) Image centred above 6th trough. (**b**) Image centred above 5th trough. (**c**) Image centred above 3rd peak. (**d**) Image centred above 1st peak.

Figures 10–12 are real-time B scans of the RTO using 3.5 and 10 MHz transducers. The gain has been turned down to enhance the rod images.

4.3 Other apparatus

4.3.1 Photometer

A photometer is used to quantify grey scale measurements for comparative performance checks. Commercially available photometers are not suitable for measuring the brightness of small parts of video screens. A suitable video monitor photometer can be constructed as follows.

(i)   Mount a photodiode sensor inside a metal pen body having first removed the ink cartridge.

(ii) Choose a photodiode that has the same spectral response as the human eye such as the BPW 21 available from RS Components.

(iii) Mount the photodiode about 1 cm from the hole at the writing end of the pen which acts as a window of a few millimetres diameter.

(iv) Withdraw the insulated connecting wires from the top end of the pen and connect them to the circuit shown in *Figure 13*. The metal pen body can be used as a shield to reduce noise pickup.

(v)   Make the pen body light tight except for the window and ensure that the photodiode is securely fixed in place. A black biro pen top makes a useful window cover when setting the dark current at zero. The dark current varies with temperature and is either set to zero before each reading is taken or it is subtracted from each reading.

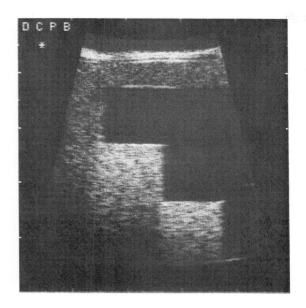

Figure 6. Grey scale test object imaged from surface 1 using a 10 MHz small parts transducer. The image is centred above step 1, low attenuation side.

4.3.2 Oscilloscope and receiving transducer

For measuring frame rate in comparative performance checks an oscilloscope and receiving transducer or hydrophone are required. Use a receiving transducer or hydrophone of about 1 cm diameter with centre frequency close to the centre frequency of the transducer being assessed that can be connected to a BNC input. If on the other hand you only have one transducer available it will probably do the job although it may not be ideal. Correct termination of the BNC cable from the receiving transducer to the oscilloscope gives the required bandwidths (usually 75 or 50 Ω termination).

5. Comparative performance

The purposes of this section are to review the meaning and significance of ultrasound performance parameters in general terms, and to act as background for the introduction of the Cardiff terminology in Section 6. Performance parameters for comparing diagnostic

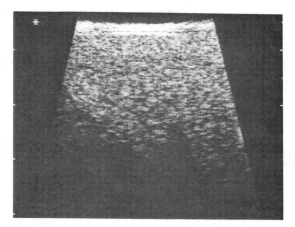

a

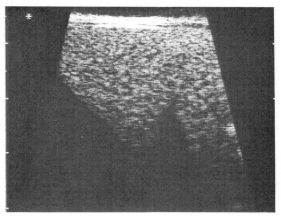

b

Figure 7. Grey scale test object imaged from surface 2 using a 10 MHz small parts transducer. (a) Image centred above peak 6, high side. (b) Image centred above peak 6, low side.

ultrasound scanners are listed in *Table 2*. Ultrasound scanning is a medical imaging modality, and the image quality is of first importance to the user. Thus two general categories of parameters can be distinguished namely image quality parameters and other parameters.

5.1 *Image quality parameters*

X-ray image quality has been specified in terms of three parameters: contrast, resolution and noise (18).

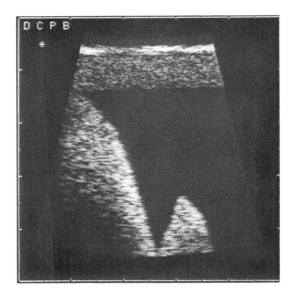

Figure 8. Grey scale test object imaged from surface 1 using a 10 MHz small parts transducer. The image is centred above trough 1 on the low attenuation side.

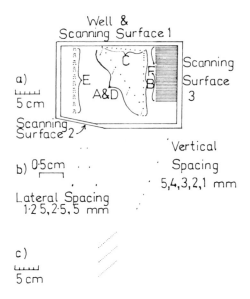

Figure 9. Resolution test object. (**a**) Profile of the test object showing the A, B, C, D, E and F rod structures. (**b**) Detail of the E and D rod structures. (**c**) End view of F rods through scanning surface 3.

Ultrasound image quality requires at least five parameters: contrast, resolution, sensitivity, artefacts and registration or magnification uniformity. Tissue characterization ability is a sixth parameter that may be important but is not easily quantified at the present time.

5.1.1 Contrast

Contrast is the difference in brightness on the monitor of echoes of different size. It is essential for distinguishing organ boundaries or tissue boundaries which produce larger or smaller echoes than their surroundings. Scanners that are designed to have a greater dynamic range of echoes at each depth in the image compress the greater echo range into the same monitor brightness range and thereby have less average contrast per dB of echo range. The variation of contrast over the echo range is determined by the pre- and post-processing curves of the scanner. The monitor contrast and brightness controls, when at optimum setting, allow the full dynamic range of echoes to be seen but the controls can be used to soften or harden the contrast and to compensate for the level of room illumination.

5.1.2 Resolution

Resolution has three orthogonal components: axial resolution, lateral resolution and slice thickness. All three vary with depth in the image. Axial and lateral resolution are measures of the minimum distance in the patient between echoes which can be separately identified in the image. Axial resolution is in the direction along the ultrasound beam and lateral resolution is at right angles in the image plane. The ultrasound image loses clarity as a result of the slice thickness of the ultrasound beam perpendicular to the image plane because each point in the image plane is produced by the interference of echoes from within the slice. For a symmetrical transducer the slice thickness is the same as the lateral resolution, but not for linear, curvilinear or phased arrays. Lateral and axial resolution distances are dependent on the relative sizes of the echoes to be distinguished. This is because the beam is wider at lower levels. The beam width at which similar sized echoes can just be distinguished is normally taken at the half amplitude (-6 dB) level, but dissimilar echoes must be further separated to be distinguished. Typical plots of acoustic pressure

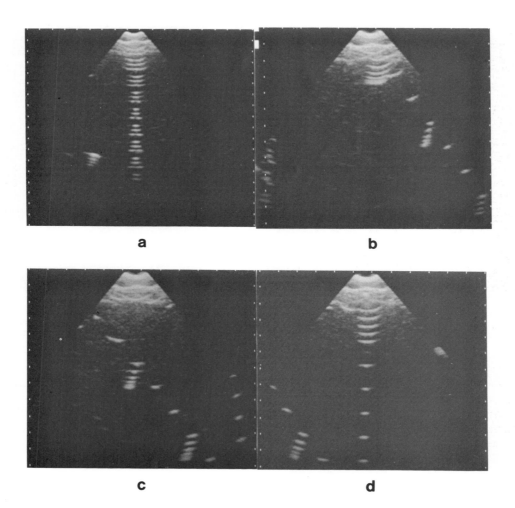

Figure 10. Resolution test object imaged from surface 1 using a 3.5 MHz oscillating sector transducer. (a) Group E rods. (b) Group A and D rods with some of group E. (c) Group A and D with some of group B. (d) Group B rods with some of group A and D.

against distance in the lateral and axial directions are shown in *Figure 14* for a 3.5 MHz transducer. The plots can be used to determine lateral and axial resolution at a range of levels.

The width of the beam at the −50 dB level has been called the contrast resolution (19). Contrast resolution is a measure of the distance over which contrast is lost when there is a bright reflector bounding on a region of reflectivity that is below the dynamic range of the scanner. Such a low reflectivity region should appear completely dark but brightness spills over into it

because of the width of the bright echo at the low level. Contrast loss can be seen in *Figures 2* and *3*.

5.1.3 *Sensitivity*

The maximum sensitivity is set by the smallest echo that can be detected by the scanner. It is usually expressed as a fraction of the largest echo which is obtained from a perfect reflector at optimum distance from the transducer. Sensitivity is usually limited by the noise introduced to the received echo from the scanner's amplifier or from the hospital mains supply.

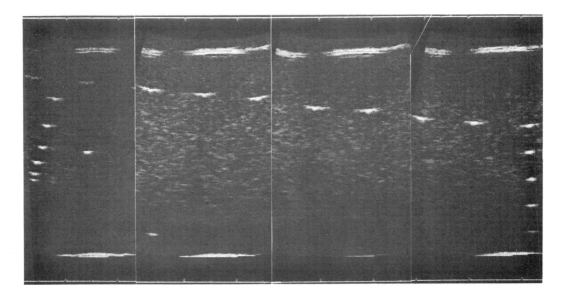

Figure 11. Composite picture of four images of the resolution test object group C rods taken with a 10 MHz small parts transducer.

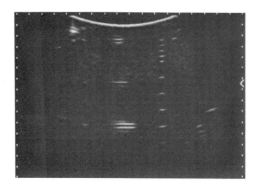

Figure 12. Resolution test object group F rods imaged from surface 1 using a 3.5 MHz curved linear array transducer. The scan is in the same plane as those of *Figures 10* and *11*, although right and left are reversed.

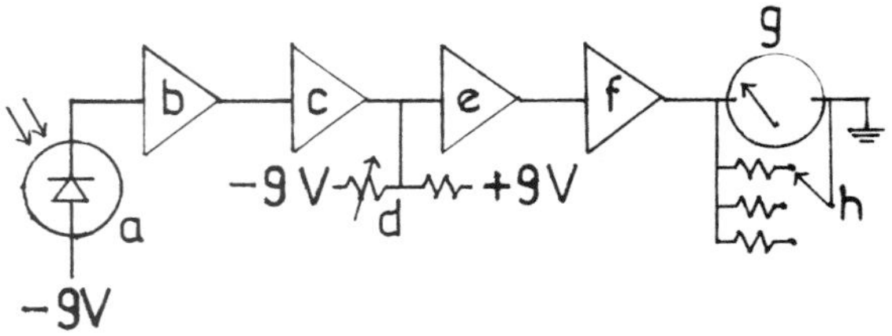

Sensitivity increases with transmitter power and receiver gain and is dependent on the transducer used and the distance of the echo source from the transducer.

5.1.4 *Artefacts*

Artefacts are common to all ultrasound images. They include multiple reflections, refraction, acoustic shadow-

Figure 13. Schematic of video monitor photometer. (**a**) Photodiode connected photoconductively. Use for example a BPW 21 available from RS Components, PO Box 99, Corby, Northants, UK. (**b**) Current to voltage convertor. (**c**) Low pass amplifier (DC to 30 Hz). (**d**) Dark current zero adjust. (**e**) Summing amplifier. (**f**) Meter driver. (**g**) Microammeter. (**h**) Range selector.

Table 2. Performance parameters for diagnostic ultrasound scanners.

A. Contrast
B. Resolution
C. Sensitivity
D. Artefacts
E. Registration/magnification uniformity
F. Frame rate
G. Measurement accuracy

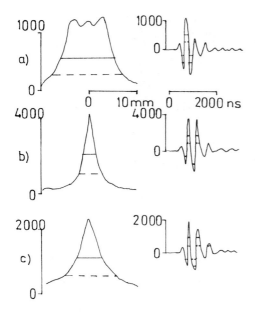

Figure 14. The left hand plots are peak positive acoustic pressure in KPa against lateral displacement from the beam axis in mm, and the right hand plots are acoustic pressure on the beam axis in KPa against time in nanoseconds. The waveforms were obtained in water for a 3.5 MHz circular transducer using a Nuclear Enterprise NPL Ultrasound Beam Calibrator. The depths of measurement are (**a**) 2 cm (**b**) 7 cm (**c**) 12 cm. Note that times measured from the right hand plots can be converted to axial resolution by dividing by the velocity of ultrasound in water (1480 m s^{-1} at 20°C). The solid and dashed lines represent resolutions measured at −6 and −12 dB respectively.

ing and acoustic enhancement. These are due to the physical nature of ultrasound and its interaction with tissue. Other artefacts are caused by design factors and will vary from scanner to scanner and from transducer to transducer. Although a number of such artefacts are possible, dead zone is perhaps the easiest to quantify for purposes of comparison. Dead zone is either due to multiple reflections within the transducer assembly or to amplifier overloading. The dead zone of the image is near the transducer and is masked by bright artefactual echoes.

5.1.5 *Registration/magnification uniformity*

The registration of a compound static B scanner is the ability to locate correctly in the image the position of each echo so that echoes from the same point in the object remain coincident in the image when the transducer is moved around.

Magnification uniformity is the analogous quality for real-time scanners which do not have the possibility of superimposing images when the transducer is moved. For uniform magnification the whole object is related to the whole image by a magnification factor that is independent of position and direction within the image. Magnification is used here in the general sense that the magnification factor may be greater or less than unity.

The effect of registration error and magnification uniformity error is that the image is a distorted representation of the object.

5.1.6 *Tissue characterization ability*

Different tissues, such as normal and tumour tissue, can sometimes be distinguished in the image because of a difference in tissue texture even when there is no contrast difference. Tissue texture is the pattern of speckle produced by the sub-wavelength scatterers in the tissue. The pattern may vary in fineness or grouping of the speckle for different tissues. However it is only partly due to the nature of the tissue and has a strong dependence on the local variations of the ultrasound field with position and time. Transducer and scanner design factors will influence these variations and so there is the possibility that one scanner may produce better distinguishing patterns for tissues than another. Quantifying this tissue characterization ability for the range of tissues encountered in patients is a very difficult thing to do at present. Clinical comparison of image quality does include tissue characterization but does not separate it from the other image quality factors.

5.2 *Other parameters*

5.2.1 *Frame rate*

For a real-time image the time taken to produce one image frame is of interest and importance. Its inverse, the frame rate, is the usual form in which it is quoted. For abdominal examinations high frame rates are not required, but they are useful in cardiac and vascular studies.

5.2.2 *Measurement accuracy*

For all examinations where measurements are made the accuracy of the measurement calipers is important.

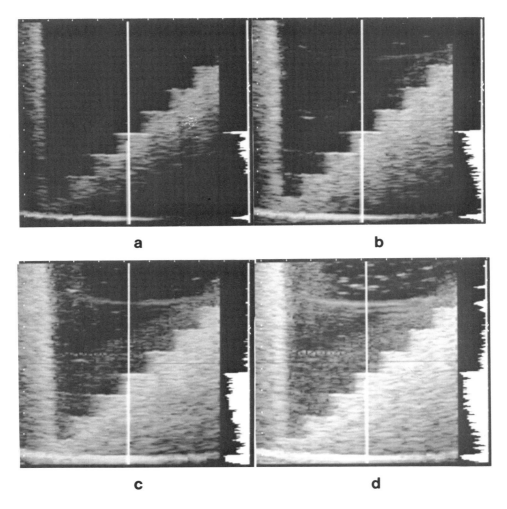

Figure 15. The 30% white criterion. Gain is increased in sequence (**a**), (**b**), (**c**), (**d**) to set the 30% white criterion at the square on the boundary at step 8. An A mode scan along a vector through step 8 is shown at the right of the image.

Distance, circumference and area measurements are made in obstetrics and volume measurements are made in abdominal and cardiac examinations but there is a wide variation in the use of measurements.

6. Cardiff test system terms and conventions

The CTS terms used in Section 7 that differ from those used in Section 5.1 are introduced below.

6.1 *IAR and GTC (contrast)*

IAR stands for imaged acoustic range and GTC for grey scale transfer curve (see Section 5.1.1).

The IAR is the range of echo sizes, excluding swept gain compensation, that can be distinguished in the image. The effects of swept gain can be removed by specifying a fixed depth for the echoes. Thus the IAR is the dynamic range at fixed depth of the whole imaging process from the transducer to the monitor. Criteria are needed for defining the limits of both ends of the range. They are the 30% white criterion and the 70% dark criterion. The criteria refer to the images of the GTO as shown for example in *Figure 2*. At fixed depth in the image the attenuation, and therefore the speckle brightness, varies with the transducer's lateral position. An imaginary 1 cm square grid can be constructed by projecting the edges of the stepped boundary within the background TMM. The attenuation within each 1 cm square is approximately constant. A square that has attenuation corresponding to the top end of the echo range has 30% of its area at peak white and a square that has attenuation corresponding to the bottom end of the range has 70% of its area at peak dark. The choice of the figures 30% and 70% is to some extent arbitrary but note that they add up to 100% so that in a bistable image the same square will satisfy the 30% white and 70% dark criteria. It is not easy to set exactly the criteria without the aid of an image processing computer. By eye you can probably judge somewhere between 10% and 50% white and somewhere between 50% and 70% dark. This is reasonable accuracy. An example of setting the 30% white criterion is shown in *Figure 15*.

The GTC is a plot of relative image brightness against echo size in dB. Like IAR, the GTC includes all the processing effects on the echoes between the transducer and the monitor and therefore includes all pre- and post-processing effects. A photometer is used to measure the brightness and the 30% white and 70% dark criteria are not required. Usually the IAR measurement represents a greater dynamic range than the GTC measurement because the eye is more sensitive than the photometer at the dark end of the range. The relative difference in brightness between any two points of the GTC is the contrast in the image of the corresponding echo sizes. In most cases the GTC is non-linear and the region of greatest slope indicates the echo levels that have best contrast.

6.2 LIBP, AIPP and ST (resolution)

LIBP stands for lateral imaged beam profile, AIPP for axial imaged pulse profile and ST for slice thickness (see Section 5.1.2). AIPP corresponds to axial resolution and LIBP to lateral resolution. The word 'imaged' in the terms denotes that measurement is made on the image and therefore includes any effects that processing the echoes may have on the resolution. The word 'profile' denotes that the resolutions are measured at a number of depths giving a profile of resolution against depth. The same two points apply to the ST measurement.

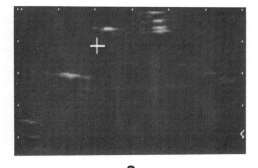

a

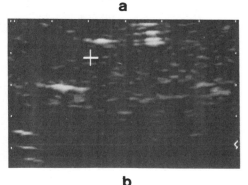

b

Figure 16. To illustrate setting the 70% dark criterion for a resolution measurement at the rod beside the cross. Image (**a**) is too dark but image (**b**) is about right.

The reference points for the Cardiff resolution measurements are quite different from those described in Section 5.1 where the beam width is measured on the transmitted beam at a certain level, often −6, −12 or −20 dB. In the Cardiff measurements the reference level on the image for the width of the rod echoes is the level of echoes from the background TMM. The 70% dark criterion is used to set the width of the rod echoes at the level of the background TMM as shown in *Figure 16*. The relative size of rod and TMM echoes is not fixed and can vary with depth between −10 and −30 dB. As an example consider that the relative size of the echoes near the transducer is −10 dB and at the focus of the beam is −30 dB. The effect of this is that the imaged beam width near the transducer is narrower than a transmitted beam width at the −20 dB level whereas at the focus the imaged beam width is wider than the transmitted beam width at −20 dB. Since this effect occurs for identical rods and TMM at different depths then it also occurs for identical tissue

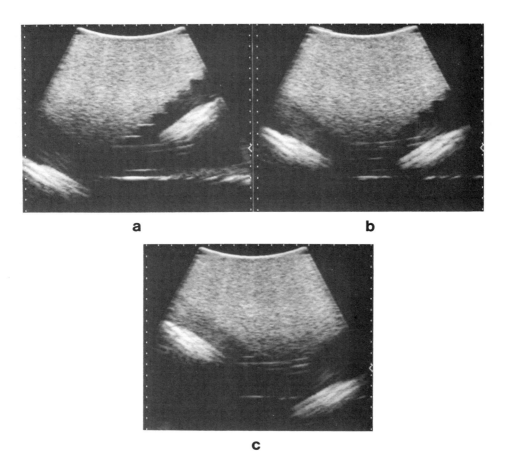

Figure 17. The artefacts produced by reflections from the long axis ends of the well when it is filled with water. (**a**) Transducer nearer right hand end of the well. (**b**) Transducer centrally positioned in the well. (**c**) Transducer nearer the left hand end of the well.

boundaries at different depths. In this sense the imaged beam widths have more clinical relevance than the transmitted beam widths.

6.3 LCP and HCP (sensitivity)

LCP stands for low contrast penetration and HCP for high contrast penetration (see Section 5.1.3). Increased penetration results from increased sensitivity. In the method of measuring sensitivity already mentioned in Section 5.1 the reference echo comes from a perfect or near perfect reflector located in water at the most sensitive depth (near the focus of a focused transducer or near the last axial maximum of an unfocused transducer). The water medium is almost non-attenuating and the minimum detectable echo is found by electrical

attenuation in the signal pathway.

The reference echo for the Cardiff LCP measurement is the backscattering of the background TMM in the resolution test object. In the case of HCP the reference echoes are the rod echoes. The method of attenuation is acoustic at a rate of 1.72 dB MHz^{-1} per cm of penetration.

HCP in the test object models the penetration limit set by strong reflectors such as tissue boundaries in a patient. LCP models the penetration limit for distributed scattering such as that from within tissues. In a patient the average attenuation will usually be less than that of the background TMM, so that penetration will be greater. Nonetheless the penetration in the test object is a means of comparing scanners in a relative way.

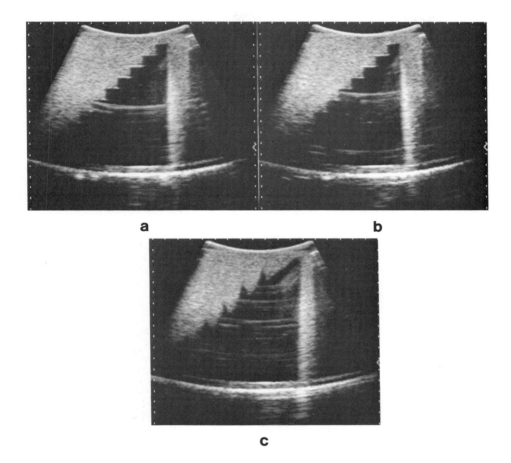

a　　　　　　　　　　　　　　**b**

c

Figure 18. The artefacts produced by reflections from the short axis ends of the well when it is filled with water. (**a**) Reflection from one long wall. (**b**) Reflection from one long wall plus reverberation. (**c**) Reflections from two long walls plus reverberations.

7. Comparative performance measurements

7.1 *General points*

(i) Before commencing any of the measurements fully acquaint yourself with the function and operation of the scanner controls by reading the operator's manual and scanning the GTO.

(ii) Scan from surface 2 of the GTO and observe the effect of the controls on the image. Often the manufacturer will recommend initial settings for the controls in the operator's manual or the scanner will switch on with some controls pre-set.

(iii) When you are familiar with the controls set them to produce an even mid-grey image over as much

as possible of the depth of the image as in *Figures 4* and *5*.

(iv) Note that a well full of water is a more convenient acoustic couplant than acoustic gel because the gel spreads, thins and develops unremovable air bubbles as the transducer is moved around. In addition water at room temperature introduces less refraction errors in sector scanning images than acoustic gel (20). However, the water in the well can produce another kind of artefact as shown in *Figures 17* and *18*. Surface waves reflected from the long axis ends of the well can generate the artefact shown in *Figure 17*. The corresponding short axis artefact is shown in *Figure 18*.

(v) The artefacts can be removed by placing four absorbent baffles immediately around the trans-

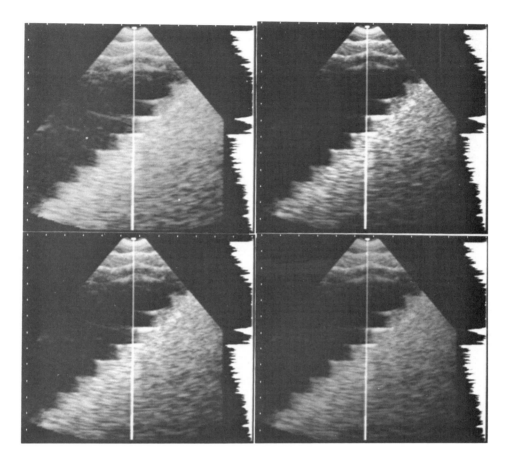

Figure 19. Examples of the 30% white criterion for four post-processing curves on the same scanner. An A scan is taken through the square at 30% white to show that the echo sizes are the same in each case. The 30% white square is adjacent to the background TMM boundary and is crossed by the A scan vector. In all cases gain is at maximum.

ducer in the well. Rolled up paper towels, stuck with sellotape and cut to size, make suitable baffles.

(vi) Check the transducer face and scanning surfaces for air bubbles and wipe clear if necessary.

7.2 Imaged acoustic range

7.2.1 Influence of scanner controls

The pre- and post-processing controls and the transducer selected may affect the IAR value. In particular controls labelled compression, log compression or dynamic range have an effect.

7.2.2 Measurement procedure

(i) Scan the GTO from surface 2 and set the swept gain controls to produce an even mid-grey image. With the same swept gain settings, turn the test object over and scan from surface 1 viewing the stepped boundary at the low TMM side.

(ii) Optimize the monitor contrast and brightness as follows. Set both controls to minimum. Increase brightness to the point where the scan lines are just visible. Increase the contrast to make the peak white parts of the image as bright as possible without losing the distinction from the second brightest level and without defocusing. Because

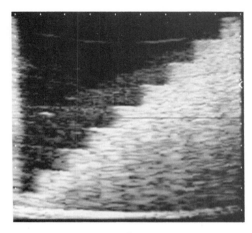

a

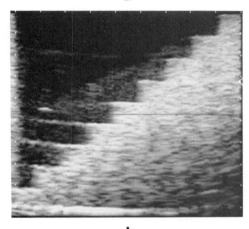

b

Figure 20. To illustrate optimum orientation of the transducer. (**a**) Non-optimum. (**b**) Optimum. The bright extension of the tops of the steps is greatest at optimum orientation.

the contrast and brightness controls sometimes interact, it may be necessary to repeat adjustment of brightness and contrast to obtain full use of the range of brightness available on the monitor.

(iii) Adjust the overall gain controls (transmit or receive) to obtain the 30% white criterion in one of the imaginary squares of the background TMM (see Section 6.1). It is easier to set 30% white on a magnified image. The square need not be adjacent to the stepped boundary. At 3.5 MHz the

step above the square will probably be step 7, 8 or 9 (see *Figure 1* for numbering). At greater frequencies nearer steps will be chosen. If transmit focus is available set the focus as close as possible to the depth of the 30% white square and if necessary tune the gain to re-set 30% white. Note that the appearance of 30% white can be different depending on the scanner processing. *Figure 19* shows the 30% white criterion for four post-processing curves of the same scanner.

(iv) Move the transducer to bring the 30% white square to the horizontal centre of the image, tuning the gain if necessary to re-set 30% white.

(v) Orient the transducer to give maximum strength echoes from the tops of the steps. At optimum orientation the extension of the tops of the steps will be greatest and the echoes brightest (*Figure 20*). Again tune the gain for 30% white.

(vi) Maintaining optimum orientation of the transducer, and without any further adjustment of gain, slide it towards the high TMM and watch for a square at 70% dark that is at the horizontal centre of the image and also is at the same depth as the 30% white square. *Figure 21* shows an example sequence of photographs moving from 30% white to 70% dark. The caliper cursor is used as a depth marker and is positioned just to the left of the horizontal centre of the image.

(vii) The difference in attenuation of the ultrasound beam between the 30% white square and the 70% dark square gives the IAR. The attenuation difference is calculated from the number of centimetres travelled by the ultrasound beam through the low, background and high TMMs to each square. There are two cases.

Case 1: 30% white and 70% dark both on low TMM side.
$$IAR = 1.4 \times (W - D) \times f.$$

Case 2: 30% white on low side, 70% dark on high side.
$$IAR = (1.4 \times W \times f) + (0.6 \times D \times f)$$

Where: W is the step directly above the 30% white square (i.e. the number of centimetres travelled by the ultrasound beam through the low TMM); D is the step directly above the 70% dark square on the low or high TMM side; and f is the centre frequency of the transducer in MHz.

Figure 21 is an example of Case 1 with W = 8, D = 1 and f = 3.5, giving an IAR of 34 dB.

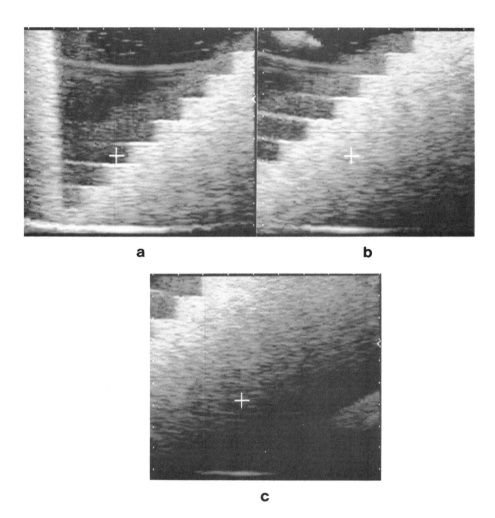

Figure 21. To illustrate Case 1 imaged acoustic range measurement. (a) 30% white square to the right of the cross and below step 8. (b) Moving across towards 70% dark square. (c) 70% dark square to the right of the cross. Note that the left edge of step 4 can be used along with the centimetre scale markers at the top of the image to determine the horizontal movement from (a) to (c). The horizontal movement is 7 cm which implies that the 70% dark square is below step 1 on the low attenuation side.

Figure 22 is an example of Case 2 with W = 8, D = 4 and f = 3.5, giving an IAR of 48 dB.

7.2.3 *Typical results*

The minimum, mean and maximum values of IAR found for all kinds of scanners are respectively 34, 43 and 58 dB. Some individual scanners can be adjusted over this full range.

7.3 *Grey scale transfer curve*

7.3.1 *Influence of scanner controls*

This is the same as in Section 7.2.1.

7.3.2 *Measurement procedure*

The measurement procedure is the same as in Section 7.2.2 down to and including step (vi), which should

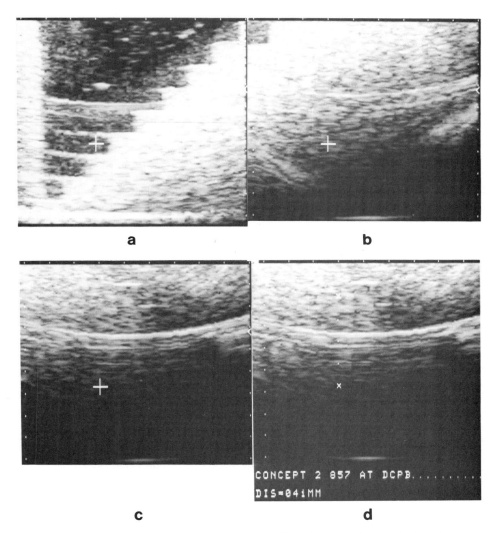

Figure 22. To illustrate Case 2 imaged acoustic range measurement. The 30% white square is immediately to the right of the cross in (**a**) at the level of step 8. The caliper cursor is used to indicate the measurement depth across through (**b**) to the 70% dark square to the right of the cross in (**c**). Image (**d**) illustrates one way of determining which step is above the 70% dark square. The distance up to the step is about 41 mm and therefore the step on the high attenuation side is 4 cm above the level of step 8, namely step 4. In the general case it may be necessary to reduce the magnification of the image to determine the step above 70% dark.

then be continued as given below.

(vii) In moving from the 30% white square to the 70% dark square take a brightness reading for each square at the measurement depth after it has been horizontally centred. Use the freeze frame facility of the scanner to facilitate the measurement

process after ensuring that the transducer is optimally oriented.

(viii) Switch off the room lights, draw the curtains and thus reduce the background illumination to a minimum. (The photometer picks up light emitted and reflected from the screen but you only want to measure the emitted part.)

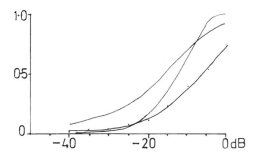

Figure 23. Three examples of measured grey scale transfer curves. Relative image brightness on a linear scale is plotted against relative echo size in dB. Typical raw data for one curve is shown by the dots (3.5 MHz transducer).

(ix) Place the photometer pen normal to the monitor screen at the square of interest and note the reading.

(x) Place the photometer at a part of the monitor screen that is at peak black and subtract the reading from the first reading to compensate for thermal changes in the photometer current, and time variation of the brightness of the monitor. Note that the monitor brightness varies for the same size echo signal at different parts of the screen. The variation can be as much as 25%. For this reason it is advisable when measuring the brightness of the next square to move the transducer so that the square is at the horizontal centre of the image. An additional reason for working always at the horizontal centre of the image when using a sector scanner is that the ultrasound path lengths will be different for other squares.

(xi) After all the squares between 30% white square and 70% dark square have been measured, repeat the whole process once or twice to reduce transducer and photometer pen positioning errors.

(xii) Once the brightness readings for each square have been obtained the mean relative brightness is plotted against the relative echo size of each square using the fact that the echo ratio between adjacent squares on the low TMM side is $1.4 \times f$ dB and on the high TMM side is $0.6 \times f$ dB.

7.3.3 Typical results

Three measured GTCs are shown in *Figure 23*. The S shape and half U shape are common shapes but many other shapes are possible.

7.4 Dead zone
7.4.1 Influence of scanner controls

Dead zone is dependent on the transducer selected. Gain settings also affect dead zone and the 70% dark criterion is used to standardize the gain.

7.4.2 Measurement procedure

(i) Scan the RTO from surface 1 and adjust the swept gain to produce an even mid-grey over as much of the depth of the image as possible. Reduce the overall gain to produce the 70% dark condition over the region close to the transducer.

(ii) Observe the B and C rods (see *Figure 9*). Start with the image centred on the B rod at 10 mm from the surface. Successively bring each C rod to the centre of the image subtracting 1 mm from the 10 mm for each C rod that can be clearly seen. The dead zone is the result of the last subtraction before a C rod is reached that is in any way obscured.

(iii) Note that in most RTOs the scanning surface is convex. This can mean that the top B rod is not in fact at 10 mm depth. In the author's RTO it is 14 mm deep (measured with a 10 MHz small parts transducer). *Figure 24* shows examples of dead zone measurement.

7.4.3 Typical results

Dead zones less than 4 mm cannot be measured with the author's RTO. For linear, convex and phased arrays the mean value measured is 5.6 mm with a standard deviation of 1 mm. The mean value for mechanical sector transducers is 12 mm with a standard deviation of 6 mm.

7.5 High and low contrast penetration
7.5.1 Influence of scanner controls

HCP and LCP are primarily affected by the gain controls. Greater transmit gain increases the penetration without increasing the noise level. Greater receive gain (including swept gain) increases the penetration up to the point where the noise is amplified to be within the IAR at the penetration limit. Thereafter increasing the receive gain will also increase the noise level and the penetration will not be increased. Bear in mind that greater transmit gain also increases the acoustic pressures and intensities delivered to the patient. Penetration varies with the transducer selected.

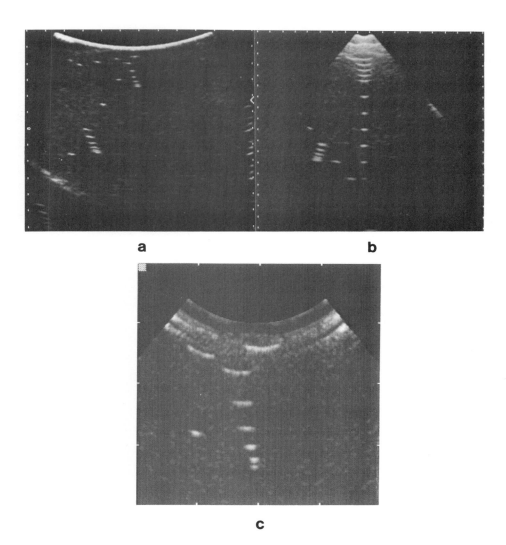

a   b

c

Figure 24. Examples of dead zone measurement. (a) 3.5 MHz curved linear array transducer: measured <1 mm; corrected <5 mm, (b) 3.5 MHz mechanical sector transducer: measured 10 mm; corrected 14 mm. (c) 5.0 MHz mechanical sector transducer: measured 2 mm; corrected 6 mm.

7.5.2 *Measurement procedure*

(i)   Image the RTO B rods from surface 1. Set the transmit gain to maximum. Set the receive gain and swept gain to produce an even grey image with maximum penetration. If the acoustic output is known to be high or if the manufacturer recommends a lower transmit gain setting, you may later wish to repeat the measurement for a lower transmit gain.

(ii) Using the B rods as a measuring scale down the centre of the image, determine the depth of the 70% dark level of the TMM echoes. Nominally assume the top B rod to be at 10 mm depth, the next four to be spaced at 5 mm intervals and those below at 15 mm intervals. Interpolate between the rods by eye to increase the accuracy of the measurement. The measured depth is the value of the LCP.

(iii) Determine the depth of the deepest rod which is

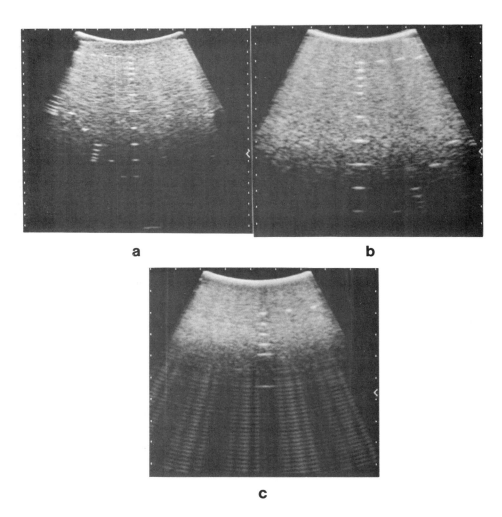

Figure 25. Examples of low contrast penetration and high contrast penetration measurement. (**a**) 3.5 MHz curved linear array transducer: LCP = 9 cm, HCP = 12 cm. (**b**) 5.0 MHz curved linear transducer: LCP = 8 cm, HCP = 10.5 cm. (**c**) 7.5 MHz curved linear array transducer: LCP = 4 cm, HCP = 4.5 cm.

visible. This depth is the HCP value.

(iv) Examples of LCP and HCP measurement are shown in *Figure 25*.

7.5.3 *Typical results*

Mean LCP values for 3.5, 5 and 7.5 MHz are respectively 9.4, 6.4 and 4 cm. Mean HCP values for 3.5, 5 and 7.5 MHz are respectively 13.3, 9.4 and 4.5 cm.

7.6 *Measurement accuracy*

7.6.1 *Influence of scanner controls*

Depth of view, magnification and the transducer

selected may affect the measurement accuracy of the calipers.

7.6.2 *Measurement procedure*

To check the accuracy of the calipers for small-scale measurements, large-scale measurements and area measurements various groups of rods of the RTO are used.

(i) Scan from surface 1 and adjust the swept gain, transmit and receive gains to obtain the 70% dark criterion over as much as possible of the depth of the image.

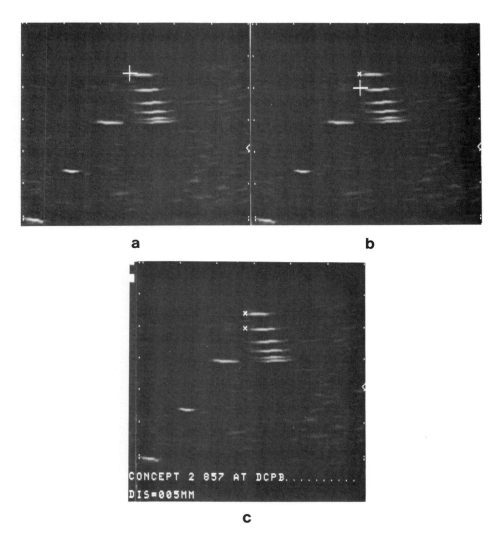

Figure 26. To illustrate the positioning of the caliper cursors for a check of small-scale vertical caliper accuracy. (**a**) Position the horizontal bar of the caliper cursor at the leading edge of the rod. (**b**) Position the other end of the caliper. On the scanner used the large plus changes to a small 'x' when the caliper is fixed. (**c**) The distance readout is 5 mm.

(ii) Choosing minimum depth of view and maximum magnification, observe the nearest set of D rods that can be clearly seen. Use the scanner's distance calipers to measure the vertical distance between each pair of D rods. Record each result against the nominal distances of 5, 4, 3, 2 and 1 mm. The 5 mm measurement is illustrated in *Figure 26*. Using the same scanner settings observe the nearest set of E rods that can be clearly seen. Record the measured distance against

the nominal distances of 5, 2.5 and 1.25 mm. The 5 mm measurement is illustrated in *Figure 27*.

(iii) Choosing maximum depth of view and minimum magnification observe the A and C rods. Make four vertical measurements at distances ranging from 15 mm up to the maximum within the LCP. *Figure 28* illustrates the two vertical caliper measurements. Record the measured values against the true values from a knowledge of the geometry of the test object (see *Figure 9*). Simi-

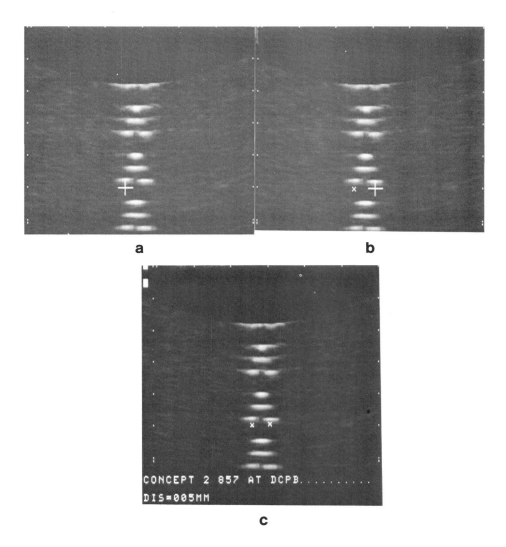

Figure 27. To illustrate the positioning of the caliper cursors for a check of small-scale horizontal caliper accuracy. (**a**) Position the tip of the caliper cursor at the centre of the brightest part of the rod echo. (**b**) Position the other end of the caliper. (**c**) The distance readout is 5 mm.

larly make four horizontal measurements over a wide range. *Figure 29* illustrates three horizontal caliper measurements. Diagonal caliper accuracy can be checked with the A rods.

(iv) To check the accuracy of area caliper measurements, draw straight edged figures using the A and B rods as vertices. *Figure 30* illustrates examples. It may improve drawing accuracy if the edges are drawn on the screen with a felt pen.

7.6.3 *Typical results*

The readout of results of some scanners is to the nearest millimetre and on others to the nearest 0.1 mm. In the latter case the least unit of measurement is limited to the size of one pixel in the image. This can be as small as 0.2 mm but is typically 0.6 mm. In 94% of measurements the measurement error has been found to be less than or equal to 1.0 mm.

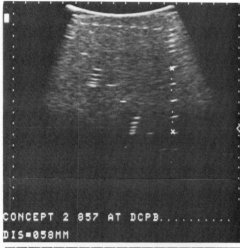

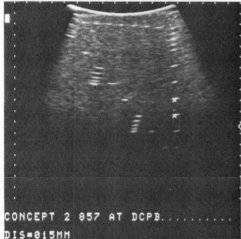

Figure 28. To illustrate large-scale vertical caliper checks. The measured values are 58 and 15 mm compared with true distances of 60 and 15 mm respectively.

7.7 LIBP and AIPP

7.7.1 Influence of scanner controls

LIBP and AIPP are dependent on the transducer selected and on the setting of any transmit focus controls. Gain is standardized at the 70% dark level.

7.7.2 Measurement procedure

(i)   Image the A rods of the RTO from surface 1. Adjust swept gain, transmit and receive gains to

obtain an even mid-grey image over as much as possible of the depth of the image.

(ii)   Bring the 1.5 cm deep A rod to the horizontal centre of the image and adjust the transmit or receive gain to obtain the 70% dark criterion in the region of the rod.

(iii)   Using maximum magnification (and minimum depth of view) position the caliper tip at one end of the rod image (the caliper tip is more convenient than the centre of the caliper because it obscures less of the features of interest). For this purpose the width of the rod is defined by all parts of the rod image that are brighter than the grey level of the background speckle. Fix the position of the caliper.

(iv)   Repeat (iii) for the other end of the rod image. See *Figure 31* for illustration.

(v)   Read out the width of the rod in millimetres from the scanner screen and note it down in an LIBP table of rod depth against rod width.

(vi)   Measure the axial length of the rod and note it down in an AIPP table of rod depth against axial length. *Figure 32* illustrates caliper positioning for the measurement.

(vii)   Repeat for all A rods down to the depth of LCP.

(viii) For transducer frequencies above about 7 MHz the LCP is reduced and few A rods are viewable. Use the C and B rods to measure LIBP and AIPP at 5 mm depth intervals.

7.7.3. Typical results

Table 3 shows minimum, mean and maximum results for 3.5 MHz transducers. In general terms, smaller results are expected for higher frequency transducers.

7.8 Slice thickness

7.8.1 Influence of scanner controls

The ST profile is dependent on the transducer selected and, for an annular array, on the selection of transmit focus setting. Note that for circular transducers, including annular arrays, the ST profile is the same as the LIBP and therefore need not be measured. Gain is standardized with the 70% dark criterion.

7.8.2 Measurement procedure

(i)   Image the RTO B rods from surface 1 and adjust swept gain and transmit and receive gains to give maximum penetration and an even mid-grey

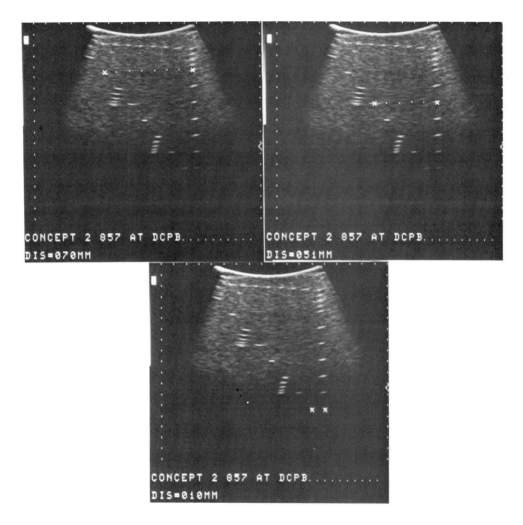

Figure 29. To illustrate large-scale horizontal caliper checks. The measured values are 70, 51 and 10 mm compared with true distance of 70, 50 and 10 mm respectively.

image. Slide the transducer across parallel to the long axis of the test object to centre the image over the F rods but keep the B rods at the side of the image as a depth gauge.

(ii) Slide the transducer parallel to the short axis of the test object to view the F rods at 1.5 cm depth and adjust the transmit or receive gain to produce the 70% dark criterion.

(iii) While keeping the scanning plane vertical and parallel to the long axis of the test object, slide the transducer small distances (~ 5 mm) parallel to the short axis of the test object and observe

the number of F rods visible at 1.5 cm depth. In particular note the minimum number of rods (p) visible. The slice thickness is $3 \times (p + 0.5)$ mm to the nearest 1.5 mm. Note in the ST profile table.

(iv) Repeat the measurement at depths corresponding to the other A rod depths down to the LCP depth. The measurement at 6 cm depth is illustrated in *Figure 33*. At some of the measurement depths the B rods and F rods cannot be simultaneously imaged. In such cases the scanner's calipers can be used to measure depth.

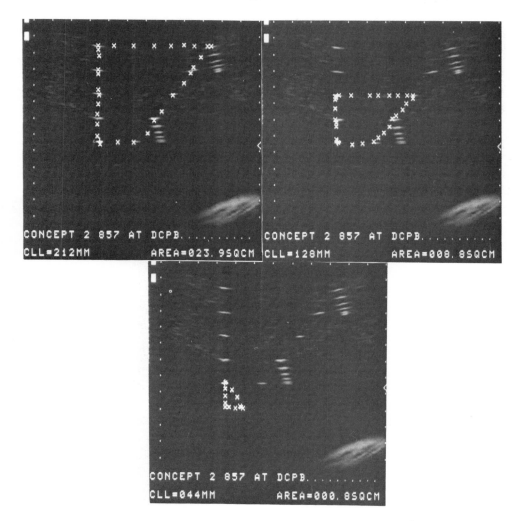

Figure 30. To illustrate area caliper checks. The measured areas are respectively 23.9, 8.8 and 0.8 cm² and the corresponding true areas are 24, 9 and 0.75 cm².

7.8.3 *Typical results*

Table 4 shows minimum, mean and maximum results for 3.5 MHz transducers.

7.9 *Registration/magnification uniformity*

7.9.1 *Influence of scanner controls*

The depth of view, the magnification and the transducer selected may affect the results.

7.9.2 *Measurement procedure*

Measure registration error only for static compound B scanners and measure magnification uniformity error only for real-time B scanners.

(i)   To measure registration error superimpose scans of the A, B and C rods taken from surfaces 1 and 3. The rod images should superimpose as crosses with the centre of each rod echo coincident. If the centres are not coincident then measure the displacement using the test scanner's calipers. Scan with the transducer in different orientations to obtain rods at the corners and centre of the image and repeat the measurement noting the results.

Performance checks

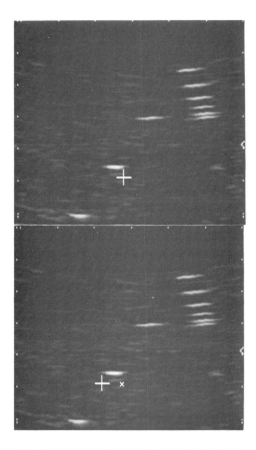

Figure 31. The two images illustrate caliper positioning for an LIBP measurement.

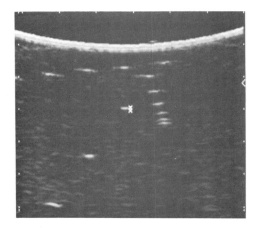

Figure 32. Caliper positioning for an AIPP measurement.

Table 3. LIBP and AIPP for 3.5 MHz focused transducers.

Depth (cm)	LIBP (mm)			AIPP (mm)		
	min	mean	max	min	mean	max
1.5	1	7.4	14	0.7	1.1	2
3.0	2.2	6.3	12	0.8	1.2	2
4.5	2.2	5.6	9	0.8	1.3	2
6.0	2.8	5.2	8	0.8	1.2	2
7.5	2.2	5.7	8	1	1.5	2
9.0	2.4	6.4	13	1	1.5	2
10.5	2.6	5.0	7	0.7	1.4	2

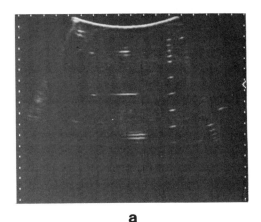

a

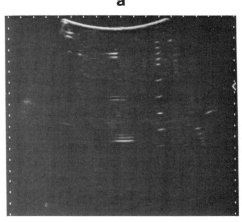

b

Figure 33. Slice thickness measurement at 6 cm depth using the B rods as a depth scale. (**a**) The minimum number of rods visible is one rod → ST = 3 × (1 + 0.5) = 4.5 mm. (**b**) It is useful to cross-check that the maximum number of rods visible is one more than the minimum number to ensure that the scan plane has been correctly aligned. In this case the maximum number of rods visible is two rods as expected.

Table 4. Slice thickness for 3.5 MHz focused transducers.

Depth (cm)	Slice thickness (mm)		
	min	mean	max
1.5	4.5	9.1	14
3.0	4.5	8.0	12
4.5	1.5	5.8	9
6.0	1.5	5.0	10.5
7.5	4	5.8	13.5
9.0	4.5	6.9	13.5

(ii) To measure magnification uniformity error image the A, B and C rods and freeze the image. Using a ruler on the monitor screen measure the horizontal distances between pairs of rods at the top, middle and bottom of the image and vertical distances at the left, middle and right of the image. From the geometry of the rods work out the true distance between each rod pair and divide measured distance by true distance to get the magnification at each part of the image. Reverse the left and right of the transducer and repeat.

7.9.3 Typical results

Registration error is typically less than or equal to 3 mm in all parts of the image. Magnification uniformity error is typically well within 5% over the image.

7.10 Frame rate

7.10.1 Influence of scanner controls

Depth of view, magnification and the transducer selected may affect the frame rate.

7.10.2 Measurement procedure

For a description of the apparatus see Section 4.3.2.

(i) Couple the receiving transducer/hydrophone to the scan face of the transducer with acoustic gel.
(ii) Ensure that the scanner image is not frozen.
(iii) Adjust the oscilloscope time base to around 10 ms cm^{-1}.
(iv) Adjust the trigger control to obtain a stable repeating pattern on the oscilloscope screen. Each blip on the screen corresponds to a scan line detected by the receiving transducer. The envelope of the scan line blips repeats once for every frame.
(v) Measure the period of the envelope. The inverse of this period is the frame rate.

For oscillating sector transducers the true frame rate corresponds to double the period of the envelope unless the receiving transducer has been placed symmetrically in the scan frame (consecutive frame envelopes of an oscillating transducer are mirror images).

7.10.3 Typical results

Measured frame rates vary widely from a few hertz to 50 Hz. The slower rates are suitable for abdominal and some small parts examinations and the faster rates for vascular and cardiac examinations.

8. Quality assurance routine

Many common faults have a very obvious effect on the image. Other faults can cause a gradual deterioration of the image or can affect the scanner's performance in a way that is not obvious on the image. The routine described here will not show up all possible faults but does detect those that affect scanner sensitivity (penetration), those that affect the caliper measurement accuracy and those that cause distortion of the image. This covers a broad spectrum of likely faults.

It is advisable to perform the routine at regular intervals and whenever the scanner has been serviced or altered in any way. Regular performance of the routine has the extra benefit of increasing the speed with which it can be performed. The apparatus used is described in Section 4.2 for which Section 3 provides background information. The quality assurance routine is as follows.

(i) Set the scanner controls at their normal settings for the transducer being assessed. The normal settings for each transducer are recorded in the quality assurance record book. Here are two suggestions for recording the position of uncalibrated controls: use the 'o'clock' notation for circular controls or make a sketch of the position of the controls.
(ii) Fill the well of the RTO with water at room temperature. De-ionized water is preferred to tap water. It is also important that the test object is at room temperature.
(iii) Image the B rods (see *Figure 10*) and adjust swept gain, transmit and receive gain to produce an even mid-grey image with maximum penetration. The settings of all the gain controls should be the same for each assessment and are recorded in the record

book. Typical images are shown in *Figure 25.* (Please ignore the figure legend unless you have also read the sections on comparative performance checks.)

(iv) Ensure that the B rods are in the centre of the image and that the transducer is oriented to give the brightest echoes from the rods. Take a hard copy of the image and compare it with previous images for the same transducer. Using the B rods as a measuring stick compare the depth at which the speckle fades out with previous images. The depth at which the speckle fades is the penetration and it should agree with previous images to an accuracy of about 5 dB. [The vertical distance between B rod echoes is 5 mm or 15 mm and 5 dB corresponds to a distance of $50/(1.72 \times f)$ mm, where f is the transducer frequency in MHz.]

(v) Position the transducer as in *Figures 28* and *29* and turn down the gain so that the speckle almost disappears. Freeze the image. Using the system calipers measure three vertical distances between an A and a C rod at the left, middle and right of the image. Repeat for three horizontal distances between an A and a B rod at the top, middle and bottom of the image. From the known geometry of the rods (*Figure 9*, Section 4.2) compare the measured values with the true values or with previously measured values. They should agree to within about 1 or 2 mm. Note that vertical distances are measured from leading edge to leading edge of the rod echoes and that horizontal distances are measured from bright centre to bright centre of the rod echoes.

(vi) For the six rod pairs measured in (v) above use a ruler on the monitor screen (with metal calipers if available) to measure the distances between the images of the rod pairs. Compute the magnification in each part of the image by dividing the imaged distance by the true distance. The magnification should be constant over the image to within 5%.

(vii) Repeat the above from (iii) for each transducer.

Under the date of each assessment keep a photograph of the penetration and a table of caliper accuracy (true distance against measured distance) and a table of magnification against screen position (top, middle, bottom, left, middle, right). If you find discrepancies greater than those indicated above then check that all the scanner controls are at the normal settings recorded in the quality assurance record book. Monitor small discrepancies daily but for large discrepancies contact service personnel.

9. Acknowledgements

I would like to thank Dr Malcolm Davison for his supervision of the Ultrasonic Equipment Evaluation Project, Mr Tom Duggan and Mr Rowland Eadie for sharing their experience of medical ultrasonics, and the Scottish Health Services Common Services Agency for their permission to write this chapter.

10. References

1. Brendel,K., Filipczynski,S., Gerstner,R., Hill,C.R., Kossof, G., Quentin,G., Reid,J.M., Saneyoshi,J., Somer,J.C., Tchevnenko,A.A. and Wells,P.N.T. (1976) *Ultrasound Med. Biol.*, 2, 343–350.

2. IEC Publication 854 (1986) *Methods of Measuring the Performance of Ultrasonic Pulse Echo Diagnostic Equipment.* Bureau Central de la CEI, 3 rue de Varembe, Geneve, Switzerland.

3. Goldstein,A. (1980) *Quality Assurance in Diagnostic Ultrasound.* AIUM Publication, 4405 East West Highway, Suite 504, Bethesda, USA.

4. AIUM (1981) Standard Methods for Testing Single Element Pulse Echo Ultrasonic Transducers. *Ultrasound Med.*, 1, Suppl. 7.

5. HPA (1978) *Methods of Monitoring Ultrasonic Scanning Equipment.* Topic Group Report 23, IPSM, 2 Low Ousegate, York, UK.

6. McCarty,K. and Stewart,W. (1982) *Ultrasound Med. Biol.*, 8, 4.

7. McCarty,K. and Stewart,W. (1985) *The Cardiff Test System Instruction Manual Parts A & C.* Diagnostic Sonar Ltd, Baird Road, Kirkton Campus, Livingstone, UK.

8. DHSS publication STB/86/6 (1986) *American Institute of Ultrasound in Medicine Meeting, Dallas.* (Enquiries from NHS to: DHSS, Diagnostic Imaging Group (STD5), Room 312, 14 Russell Square, London, UK: Enquiries outwith NHS to DHSS (Leaflets), PO Box 21, Stanmore, Middlesex, UK.

9. DHSS publication STD/87/8 (1987) *Assessment of Ultrasound Scanners: Dynamic Imaging Concept, Irex Meridian, Siemens Sonoline SX, Diasonics DRF 200, Hitachi EUB 240.* (See 8 for addresses.)

10. DHSS publication STD/87/18 (1987) *Evaluation of Acuson 128R Scanner.* (See 8 for addresses.)

11. DHSS publication STD/87/19 (1987) *Evaluation of GE RT3600 Scanner.* (See 8 for addresses.)

12. DHSS publication (1988) *Evaluation of ATL Ultramark 4 Scanner,* in press. (See 8 for addresses.)

13. DHSS publication (1988) *Evaluation of GL590 Scanner,* in press. (See 8 for addresses.)

14. DHSS publication (1988) *32nd Annual Convention of the American Institute of Ultrasound in Medicine, New Orleans,* in press. (See 8 for addresses.)

15. Lunt,M.J., Pelmore,J.M., Pryce,W.I.J. and Richardson,R. (1985) *Quality Control of Ultrasound Scanners; A Scheme for Routine Assessment.* In IPSM Report No. 47 Physics in Medical Ultrasound. IPSM, 2 Low Ousegate, York, UK.

16. AIUM publication (1974) *Standard 100 Millimeter Test Object Including Standard Procedures for its Use.* AIUM, 4405 East West Highway, Suite 504, Bethesda, MD, USA.

17. Madsen,E. (1987) Materials and Phantoms for Use in Performance Evaluation of Ultrasound Imagers. In *Proceedings of the 32nd Annual Convention of the American Institute of Ultrasound in Medicine.* AIUM, 4405 East West Highway, Suite 504, Bethesda, MD, USA.

18. Halmshaw,R. (1981) In *Physical Aspects of Medical Imaging.* Moore,B.M. (ed.), John Wiley, p. 32.

19. Maslak,S.H. (1985) In *Ultrasound Annual 1985.* Sanders,R.C. (ed.), Raven Press, New York, pp. 1−16.

20. Price,R. (1985) *Geometric Distortion in Quality Control Images from Sector Scanners.* IPSM Report No 47, IPSM, 2 Low Ousegate, York, UK.

Chapter 6

Obstetric ultrasound

J.P.Neilson and M.B.McNay

1. Introduction

It was the collaboration of Ian Donald and Tom Brown in constructing the first contact scanner which made diagnostic ultrasound a practical proposition during the 1950s. Donald was Regius Professor of Midwifery at the University of Glasgow and it was natural that early studies focused on gynaecological practice, to differentiate pelvic masses, and shortly thereafter, on a wide range of obstetric applications. Further fuelled by the findings of Alice Stewart of an increased incidence of childhood malignancy after pre-natal exposure to X-rays, obstetric ultrasonography evolved from identification of hydatidiform mole, multiple pregnancies and placenta praevia through fetal measurement to assess gestational age or growth, to become a screening technique applied to all pregnancies. As the quality of imaging improved with technical advances, so also did the ability of ultrasonographers to inspect fetal anatomy and identify malformations.

We will discuss the current status of the various applications of diagnostic ultrasound in obstetrics, from a practical standpoint and placed against the relevant clinical background. We will also discuss the controversies surrounding routine ultrasonography and the relatively recent application of Doppler ultrasound to obstetrics. The question of safety is, of course, of particular relevance to obstetric ultrasound and is discussed in Chapter 15. Our own view is that there is no evidence of harm having resulted from pre-natal exposure to ultrasound (1) and this chapter is written from this standpoint.

Ultrasonography is a unique investigative technique in pregnancy because the time taken, the detail inspected and perhaps the status and experience of the operator will vary with the reason for the ultrasound examination. To identify fetal presentation, for example, takes seconds whilst investigation of the cause of polyhydramnios may require considerable time and expertise. The two broad groups of ultrasound examination—specifically indicated and screening, should also be clearly distinguished. Screening examinations must be accomplished quickly for practical reasons and, inevitably, fetal malformation is then less likely to be recognized than when there are specific reasons to suspect its presence. The specific examination should be tailored to answer a precise question formulated by the referring physician. Some questions are readily answered—is this a multiple pregnancy? Is the fetus alive? Others are less readily answered—does this woman have an ectopic pregnancy? Interaction between obstetrician and ultrasonographer (these roles may overlap) is best when the ultrasonographer actively seeks feedback on outcome of pregnancies scanned and is aware of the practical influence of reports on clinical management. The obstetrician should be made aware of the breadth and limitation of available services and of the need for precise requests.

It is also desirable, in our opinion, that the ultrasonographer is able to communicate with the patient so that prospective parents do not miss out on an exciting, and possibly health-promoting (2), experience.

2. Fetal life

Ultrasound examination is the definitive technique of determining viability of the fetus. Life is established by observation of heart pulsation and death by its absence. There should not be any doubt about the diagnosis which may be made immediately, unlike radiological examination which required a delay of 48 h or more before development of Spalding's sign and other recognizable features. Ultrasound facilities

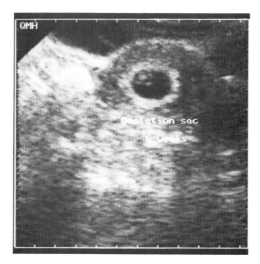

Figure 1. Intrauterine pregnancy at 6 weeks menstrual age.

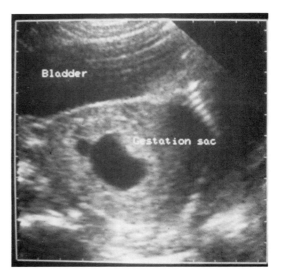

Figure 2. Blighted ovum: large, irregular sac without evidence of an embryo.

should be available at any hour of day or night in cases of suspected fetal death—it is not acceptable, for humanitarian reasons, to wait until the following day for a definitive answer.

3. Early pregnancy

With ultrasound, the early pregnancy may be identified as a ring-like structure at 6 weeks gestational age (4 weeks after conception) and the embryo within the ring by 7 weeks (*Figure 1*). As soon as the embryo can be identified it may be measured and its viability confirmed by detection of heart movement. Around 25% of pregnancies miscarry during the first 3 months and 60% of these are chromosomally abnormal. Usually arrest of embryonic development occurs weeks before miscarriage takes place, although the patient will often experience vaginal bleeding during this time. Management of threatened abortion has been rationalized by the development of ultrasound, because the experienced ultrasonographer can predict outcome with such certainty that the uterus may be evacuated if the pregnancy is demonstrably non-viable. Biochemical assessment is not sufficiently specific to give the same guidance.

Blighted ova (anembryonic pregnancies) constitute the largest group of early failures and are diagnosed by the absence of a fetal pole on careful examination (*Figure 2*). If the sac is small and the menstrual history is uncertain repeat examination will be necessary after 1 week to ensure that this is not a normal 6-week pregnancy. Missed abortions are more easily diagnosed—by the lack of heart pulsation in a fetal pole. Hydatidiform mole has a characteristic 'snowstorm' appearance which may also be seen in long-standing missed abortions and in degenerating fibroids. Although ultrasonography may identify features in the pelvis compatible with ectopic pregnancy, it is associated with both false positive and false negative rates. Often, the main contribution of ultrasound, in the difficult case, is the detection or exclusion of intrauterine pregnancy. The small group of 'live abortions' which constitute about 14% of early pregnancy loss (3) and usually occur in the second trimester, are not readily predicted although reduced amniotic fluid, diminished fetal activity and the presence of large intrauterine haematomas suggest a poor prognosis. Demonstration of the yolk sac as a small spherical structure within the gestation sac may not, despite earlier claims, provide useful prognostic information.

Most obstetricians faced with an unambiguous report of non-viable pregnancy will perform surgical evacuation of the uterus. If the ultrasonographer has any doubts about the diagnosis, these should be expressed and repeat examination arranged.

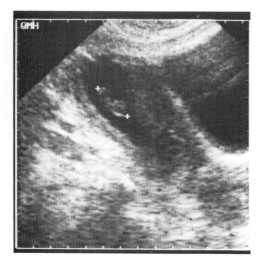

Figure 3. Crown−rump length measurement at 9 weeks.

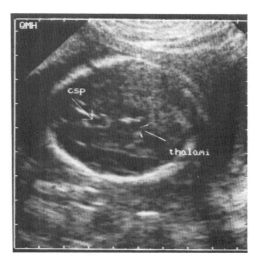

Figure 4. Transverse section of head suitable for BPD measurement (csp = cavum septum pellucidum).

4. Gestational age

The use of fetal measurement to assess gestational age hinges on the thesis that in early pregnancy fetal growth is rapid from week to week, there is little biological variation in fetal size and pathological growth retardation is uncommon. As pregnancy advances, there occurs a progressively larger scatter of size at any given week making gestational age assessment progressively less accurate. As a general rule, ultrasonography is useful before 24 weeks, less accurate between 24 and 29 weeks and of no value at all in assessing gestational age after 30 weeks.

During the first trimester, the crown−rump length (*Figure 3*) is measured and this is accurate to within 5 days in 95% of cases (4). It is essential that the patient has a full bladder and that the maximum length, and not an oblique section, is measured. Sometimes the yolk sac may be adjacent to one extremity of the fetus and this should not be included in the measurement. After 13 weeks varying flexion of the fetal trunk makes crown−rump length measurement less reliable and, instead, the biparietal diameter (BPD) or femur length are measured. BPD measurement, when performed before 24 weeks, will predict the date of delivery to within 10 days in 80% of pregnancies (5) and femur length measurement (6) is of similar accuracy and is a particularly useful option when BPD measurement is difficult, as when the head is in a direct occipito-anterior or occipito-posterior position.

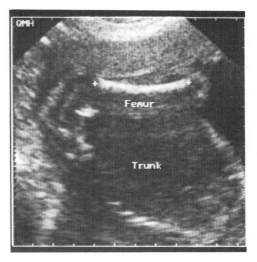

Figure 5. Femoral length measurement.

The stages involved in BPD measurement comprise identifying the longitudinal axis of the fetus, estimating the angle of asynclatism from the horizontal plane (with real time systems, the angle cannot be measured precisely), rotating the transducer through 90°, correcting for the angle of asynclatism and sliding the transducer up and down to identify the correct section of the head with the maximum transverse diameter (the BPD). Whilst the BPD can be quite correctly measured

Table 1. Patients in whom ultrasound assessment of gestational age is especially desirable.

Uncertain dates
Previous Caesarean section
Previous stillbirth
Previous small-for-dates baby
Hypertension
Diabetes mellitus
Multiple pregnancy
? All primigravidas

Table 2. Checklist for use when inspecting fetal anatomy.

Date
Head
Ventricles   VHR
Spine
4 chamber heart
Diaphragm
Abdominal wall
Stomach
Kidneys
Bladder
Limbs—upper R and L
          lower R and L
General movement
'Breathing' movement
Oedema
Ascites
Sex
Amniocentesis

Comments

in various planes, it is good practice to measure it in the plane of the occipito-frontal diameter in which the thalami and the cavum septum pellucidum should be identified as landmarks (*Figure 4*). This is discussed further when we consider cerebral anatomy.

Femur length should be measured on a thigh section in which soft tissue can be identified proximal and distal to the bone (*Figure 5*). If the femur is not, approximately, at right angles to the direction of the ultrasound beam, inaccuracies in measurement may occur.

Because gestational age cannot be reliably calculated from the menstrual history in a quarter of patients and because it is not possible to identify all patients in whom an accurate knowledge of gestational age will be vital later, there is a case for routine ultrasonography which will be discussed later. If a selective policy is employed, it is sensible to scan not only those with uncertain dates but also patients with high risk pregnancies and those in whom planned delivery (by induction of labour or elective Caesarean section) is likely (*Table 1*).

5. Fetal malformation

A sound knowledge of normal fetal anatomy and the changes which take place during fetal growth and development is essential for the correct interpretation of the ultrasound images obtained at any examination. The order in which a fetus is assessed and measured will vary according to its position and it is recommended that a written check list is recorded during each examination (*Table 2*). This should prevent wasting time in double checking, for example was the stomach seen or did the bladder fill? With experience one carries out a mental check list during the scan and records the results on concluding scanning.

A description of normal fetal anatomical structures at different stages throughout gestation and examples of commonly encountered abnormalities are illustrated. Fetal anatomy must be recognized no matter what the indication for the scan and the images obtained will be similar using both sector and linear transducers.

When to scan? For practical purposes, scanning specifically for fetal anomalies is performed from 16 weeks gestation onwards. Earlier in pregnancy only major anomalies such as anencephaly will be recognized.

Indications for anomaly scanning:

(i)   past history of a child with a structural defect;
(ii) family history of a structural defect;
(iii) known carrier of a disease with structural manifestations;
(iv) raised maternal serum α-fetoprotein levels;
(v)   clinical suspicion of oligohydramnios or polyhydramnios;
(vi) multiple pregnancy;
(vii) maternal disease such as diabetes, congenital heart disease, drug ingestion or alcoholism.

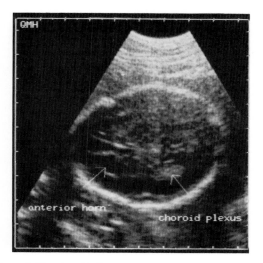

Figure 6. The choroid plexus.

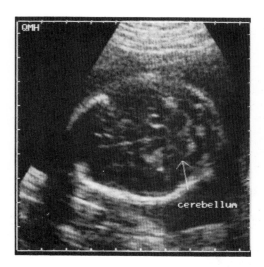

Figure 7. The cerebellum.

Table 3. Abnormalities of the head.

Anomaly	Features	Clues on scan	Commonly associated features
Anencephaly	Absence of the cranial vault	BPD impossible	Polyhydramnios
Hydrocephaly (*Figure 8*)	Dilatation of the lateral ventricles	Abnormal VHR	Neural tube defect (NTD) or X-linked or in isolation
Encephalocele (*Figure 9*)	Defect in skull	Swelling behind head	May be in association with Meckel's syndrome
Holoprosencephaly	Deficient midline facial and forebrain development	Abnormal shape of head and face. Extreme form is cyclops	May be in isolation or trisomy 13
Microcephaly	Abnormal head/trunk ratio	May be none because head looks normal	May be in isolation or part of a syndrome
'Lemon' sign	Collapse of frontal bones	The head has lemon shape	NTD but mild forms may be of no significance

5.1 Fetal anatomy

5.1.1 Head

The section required for measurement of the BPD has been described earlier. The thalami and the cavum septum pellicidum are identified (*Figure 5*). The anterior and posterior horns of the lateral ventricles are seen on the same section or with minimal adjustment and the bodies of the lateral ventricles appear as the transducer is slid away from the base of the skull. The ventricular/hemisphere ratio (VHR) is calculated from measurement of the anterior or posterior horns. The choroid plexus (*Figure 6*) occupies much of the lateral ventricle in its early development and in turn the lateral ventricles occupy much of the immature brain. The VHR progresses from around 50% at 16 weeks to less than 33% by late pregnancy. Sliding of the transducer below the BPD plane brings the cerebellum (*Figure 7*) and the structures of the base of the brain into view.

The abnormalities of the head, which can be examined by ultrasonography are listed in *Table 3*.

The facial features are often very easily identified from an early stage since the maxilla and mandible

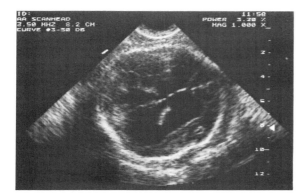

Figure 8. Hydrocephalic fetus: note the dilated lateral ventricles.

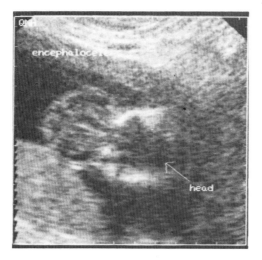

Figure 9. Fetus with encephalocele.

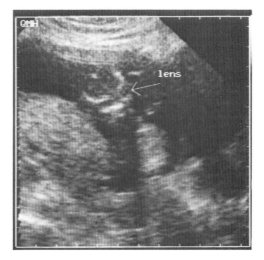

Figure 10. The lens.

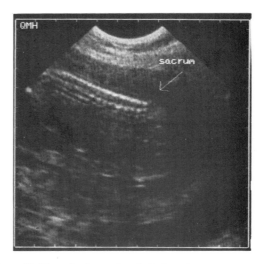

Figure 11. Normal spine on longitudinal section.

are among the first bones to ossify. The orbits should be seen and movement of the eye may be recognized by observing the lens (*Figure 10*). With the highest resolution scanners the eyelid and blinking movements are visible. The recognition of these eye movements has been of interest to researchers in the study of neurological development.

The nose and lips should be seen both in profile and face-on and in this way clefting may be recognized. Opening of the mouth, protrusion of the tongue and swallowing are all commonly noted. The latter is of clinical importance where oesophageal atresia is suspected. As the image quality of equipment con-

tinues to improve the detail which can be identified increases.

5.1.2 *Spine*

Three ossification centres are present in each vertebra. These will be visible on a transverse section of the spine but only two can be seen on a longitudinal view and appear as a pair of parallel lines diverging in the cervical area, widening slightly in the lumbar area

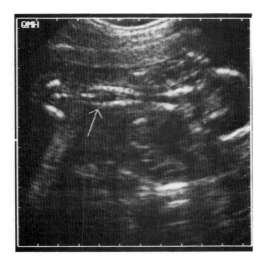

Figure 12. Longitudinal section of a fetus with thoraco-lumbar spina bifida (arrowed). The head is seen to the right, the iliac crests to the left.

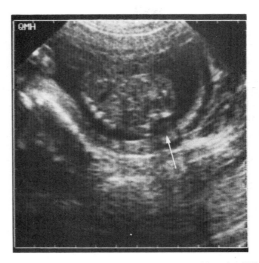

Figure 13. Oblique section of trunk of fetus with spina bifida (arrowed) showing 'open cup'.

before narrowing into the sacrum (*Figure 11*). Since neural tube defects constitute a high proportion of fetal abnormalities in the UK, recognition of the normal features of the spine is of great importance. Spina bifida may involve the whole spine or be restricted to only one or two segments. The former is easily suspected when the whole spine is in disarray but the latter may be a subtle deviation from the normal appearance. A defect is described as open when there is no covering to protect the spinal cord and closed when a covering of either meninges or skin is present. Open lesions are associated wth raised maternal serum α-fetoprotein levels. Examples of spina bifida are shown in *Figures 12* and *13*.

5.1.3 Neck

Cervical spina bifida is less common than a lumbo-sacral defect but soft tissue tumours are more often seen. These are usually predominantly cystic in nature (*Figure 14*) and should not be confused with an encephalocele (*Figure 9*). These cystic hygromata are often associated with chromosomal abnormalities particularly Turner's syndrome (45XO).

5.1.4 Thorax

The heart should occupy approximately one third of the thorax and the most useful view is a transverse section of the chest which enables the four chambers,

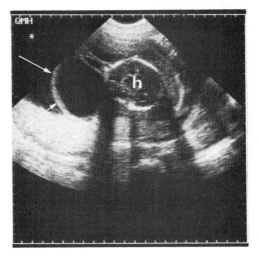

Figure 14. Transverse section through head (h) showing cystic hygroma (boundaries arrowed).

the right and left ventricles and the right and left atria, to be seen. The apex of the fetal heart points to the left and the right ventricle lies anteriorly opposite the spine; this allows orientation.

The lungs are poorly imaged in early pregnancy but are more readily appreciated in later gestation as their tissue density increases. It is not possible to predict lung maturity from an ultrasound image.

The ribs often cause shadowing which makes visualization of the heart particularly difficult. The 12 pairs may be counted if a syndrome is suspected in which absence of ribs is a feature. The twelfth rib is also a useful landmark for the lowest thoracic vertebra when determining the exact position of a spinal defect.

The diaphragm should always be identified and care should be taken not to misinterpret rib shadowing for the diaphragm. The usual clue to a diaphragmatic hernia is the presence of the stomach or bowel shadows in the chest.

5.1.5 Abdomen

The gastro-intestinal tract, the urogenital tract and the adominal wall will be considered separately.

(i) *The gastro-intestinal tract.* The stomach is most easily seen as a fluid-filled organ usually situated in the left upper quadrant of the abdomen. It is not consistently filled since swallowing and emptying occur physiologically. It may be over-distended in pathological states such as duodenal atresia when a 'double bubble' appearance is noted resulting from the distended stomach and the first part of the duodenum (*Figure 15*). In contrast the stomach will never fill in oesophageal atresia unless there is a fistula present to allow amniotic fluid to pass into the stomach. Small bowel shadows are rarely noted unless there is pathological distension. The large bowel becomes more easily seen as pregnancy progresses and the typical pattern of haustration of the colon and the increasingly dense echoes from the meconium contained within the colon is a sign of late gestation. Meconium ileus is suspected when a cluster of very dense echoes is seen. This has been reported in association with cystic fibrosis.

(ii) *The urogenital tract.* The bladder is the most easily recognized part of this system and, like the stomach, fills and empties physiologically. It may be confined to the pelvis but as it fills it is seen in the lower part of the abdomen. The kidneys are not readily identified until 18–20 weeks and their characteristic appearance of cortex, medulla and renal pelvis only becomes apparent in later gestation. Their position on either side of the spine produces in a transverse scan the appearance of a pair of spectacles. The ureters are not normally identified but, in cases of obstruction, whether unilateral (*Figure 16*) or bilateral, the bladder, ureters and renal pelvis may all show evidence of dilatation. Cysts may also be noted in the kidneys or, if microcystic disease is present in infantile polycystic

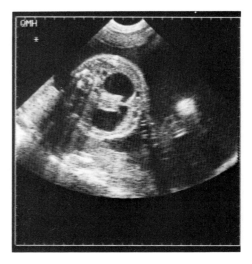

Figure 15. Transverse section through upper abdomen showing 'double bubble' — dilated stomach and duodenum. Note also polyhydramnios. The fetus had trisomy 21.

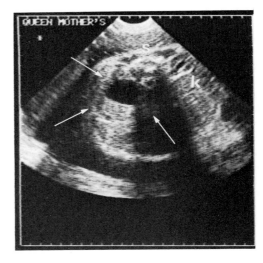

Figure 16. Unilateral hydronephrosis (margins of kidney arrowed). The other kidney (k) is normal. This is the optimal position for inspection of fetal kidneys with the spine (s) anterior.

kidney disease, the kidneys will be enlarged and more dense in appearance due to the increased number of tissue interfaces present.

Absence of the kidneys, renal agenesis, is not an easy diagnosis to make. It is always easier to make a positive diagnosis about the presence of a tumour

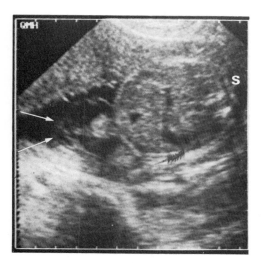

Figure 17. Transverse section of fetal abdomen showing spine (s) and exomphalos (arrowed).

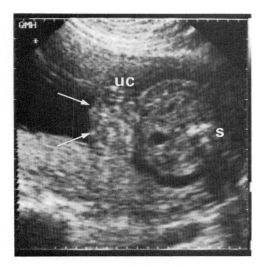

Figure 18. Gastroschisis (arrowed) — the umbilical cord (uc) has separate attachment to the anterior abdominal wall.

or a defect rather than to make a negative diagnosis that an organ is absent. This problem is accentuated in renal agenesis by oligohydramnios. From approximately 20 weeks onwards almost all the liquor is the result of fetal urine production. Oligohydramnios results in increased fetal flexion and the lack of any ultrasound window makes imaging of the fetal anatomy extremely difficult.

The fetal sex may be of importance in inherited disease or in obstructive uropathies where bilateral obstruction is more common in the male resulting from posterior urethral valves. It is also useful in the differential diagnosis of a cystic lesion within the abdomen when an ovarian cyst can be excluded if the fetus is male!

Recognition of the normal gastro-intestinal and urogenital tracts is required to help differentiate various intra-abdominal pathologies. Two cystic areas, the stomach and the bladder, are normally present. A third could represent the second bubble in the 'double bubble', or a dilated segment of the urinary tract, or an ovarian cyst or a mesenteric cyst. Clues as to which of these diagnoses is most likely are firstly in the amount of liquor—polydramnios being associated with upper gastro-intestinal obstruction and oligohydramnios with urinary tract obstruction or poor renal function, and secondly in checking for the known features, for example the presence of meconium filled transverse colon excludes an upper intestinal obstruction.

(iii) *The abdominal wall.* In early embryonic life the loops of intestine lie outside the abdomen before being returned within the abdominal cavity. Failure to be replaced will result in an abdominal wall defect of which there are two types. An exomphalos (*Figure 17*) is a defect involving the umbilical cord, and the abdominal contents, which are extruded, are contained within a sac. About one third of these defects are associated with other abnormalities, often chromosomal. Gastroschisis (*Figure 18*) on the other hand is not usually associated with other defects, does not involve the umbilical cord but arises usually to the right of the cord, and the extruded bowel lies free in the amniotic fluid. Both these abdominal wall defects are surgically correctable. They are often identified as the result of a raised maternal serum α-fetoprotein.

A lethal form of abdominal wall defect is the body stalk anomaly which arises at an even earlier stage of embryonic development and results in gross distortion of the lower half of the body including the lower limbs which may be absent.

5.1.6 *The skeleton*

The head, spine and ribs have already been considered. The long bones should all be identified, absence of for example the radius being associated with various syndromes. The position of the hands and feet may also be significant, club foot may occur in isolation but its presence would be an indication to look carefully for spina bifida. Fingers and toes may be counted, not just to impress the patient, but because the presence of polydactyly may be associated with other abnormalities.

Routine measurement of the femur for gestational ageing has already been described and all the long bones may be similarly measured. This is of particular importance in cases at high risk of a skeletal dysplasia. Serial measurement is necessary since shortening may not become apparent until late in the second trimester. Polyhydramnios is a common association.

Hydrops fetalis comprises oedema, ascites and pericardial effusion. A decade ago it was virtually pathognomonic of rhesus isoimmunization. Today only a small percentage of hydrops is associated with rhesus disease, the majority being non-immune in origin. Hydrops may result from structural heart disease or cardiac arrhythmias, viral infections such as the recently described human parvovirus infection, or be associated with placental tumours, chromosomal abnormalities or multiple anomaly syndromes. Many cases are of unknown aetiology.

5.1.7 Tumours

Cystic hygromata have been described. The other commonly encountered tumour is the teratoma, usually in the sacro-coccygeal area. These may reach a large size presenting as an abnormal lie and necessitating abdominal delivery. They are a mixture of solid and cystic elements.

5.1.8 The recognition of fetal abnormalities

There is a great deal of truth in the saying that one only sees what one looks for and this is particularly relevant in the field of pre-natal diagnosis of fetal abnormality. If one is asked to assess the gestational age of a fetus and only makes a BPD measurement, concentrating on that to the exclusion of all else, then inevitably there will be times when an anomaly will be missed. Ultrasound examination is an art rather than a science and the keen observer will be aware of subtle alterations in liquor volume which are so often the clue to the presence of pathology. Difficulty in obtaining a correct section for measurement is another trigger to question why? Is it simply the fetal position or is there another reason? The majority of pregnancies progress uneventfully and the demonstration of a

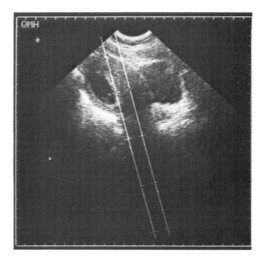

Figure 19. Transabdominal chorion villus sampling. The 'tramlines' show the path of the needle.

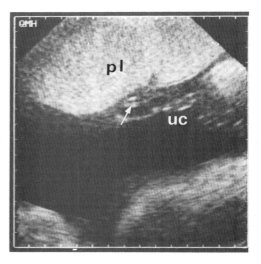

Figure 20. The arrow shows the tip of the needle in the umbilical vein at the insertion of the umbilical cord (uc) into the placenta (pl).

normal fetus is reassuring. However, as technology develops enabling a wider range of pre-natal diagnoses to be made and medical and surgical advances provide for their treatment, it is encumbent on those providing ultrasound services to develop their skills and recognize the potential of pre-natal diagnosis.

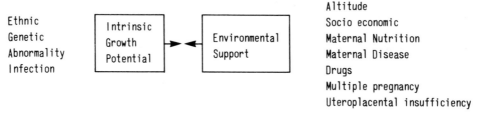

```
Ethnic          ┌──────────────┐      ┌──────────────┐      Altitude
Genetic         │ Intrinsic    │      │              │      Socio economic
Abnormality     │ Growth       │ ──►◄── │ Environmental│      Maternal Nutrition
Infection       │ Potential    │      │ Support      │      Maternal Disease
                └──────────────┘      └──────────────┘      Drugs
                                                            Multiple pregnancy
                                                            Uteroplacental insufficiency
```

Figure 21. Factors influencing fetal growth rate.

6. Interventional procedures

Ultrasound alone may be used to diagnose structural defects but it also has an important role to play in various other diagnostic procedures. It is possible to visualize the path of a needle through maternal and fetal tissues (*Figure 19*) and to accurately place a needle tip in a chosen site. Use is made of this in obtaining blood, aspirating fluid or tissue and in therapeutic procedures such as intravascular (*Figure 20*) or intra-abdominal transfusion, or placing a catheter in the fetal bladder to relieve the back pressure in an obstructive uropathy.

Amniocentesis has been the accepted method of obtaining amniotic fluid for cell culture or biochemical analysis. The risk of abortion following the procedure is 1% and is minimized by avoiding placental puncture, fetal trauma, more than one needle insertion and 'bloody taps'. Direct aspiration of fluid under ultrasound guidance provides the safest method.

Chorionic villus sampling is a newer method of obtaining tissue for DNA, biochemical or chromosomal analysis. This involves aspiration of villi either by the transcervical or transabdominal route and may only be undertaken with safety under ultrasound guidance.

Fetal blood sampling was first attempted using a fetoscope which allowed direct visualization of the umbilical cord from which blood could be aspirated. With improved image quality it is now possible to pass a needle with the precision required to obtain blood using ultrasound guidance alone. This results in a safer procedure since a finer needle may be used compared with the fetoscope. Blood may be withdrawn for diagnostic purposes and, if necessary, transfused directly knowing the haematocrit and eliminating the guesswork previously used in judging the volume to be transfused. Drugs may also be administered in this way.

As the diagnostic potential increases so do the therapeutic possibilities. These are not restricted to medical intervention but include the surgical relief of, for example, obstruction of the urinary tract and the drainage of cysts. Ultrasound has a vital role to play in all these developments.

7. Fetal growth

In later pregnancy, fetal measurement is performed to assess fetal size at a point in time, or monitor fetal growth over a period of time, or both. An estimation of fetal weight may be useful to the clinician in managing pre-term labour, breech presentation, diabetic patients and those who have had a previous Caesarean section, but usually fetal measurement is employed where there is concern about actual or potential retardation of fetal growth. This may occur in association with maternal disease such as hypertension or may be of unknown aetiology. The terms 'small-for-dates' and 'growth retarded' are often (but wrongly) used as synonymous although there is overlap. The usefulness of different ultrasound measurements is usually assessed by their ability to predict babies that are small-for-dates (SFD) at birth, although it must be stressed that cessation of growth may occur (and be detected by ultrasound) without the baby being SFD. The main clinical risks facing the SFD group are intrauterine death, asphyxia during labour, hypoglycaemia after birth and, possibly, long-term handicap. The main risks facing individual SFD fetuses will vary with aetiology which may be complex (*Figure 21*). Less than half of SFD fetuses are detected by abdominal palpation during routine antenatal care although around 65% will be identified by tape measurement of symphysial—fundal height: these are the clinical standards against which ultrasound should be assessed.

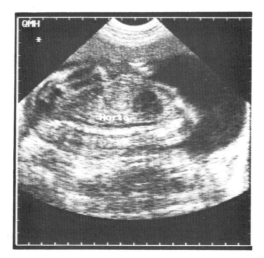

Figure 22. Fetal aorta.

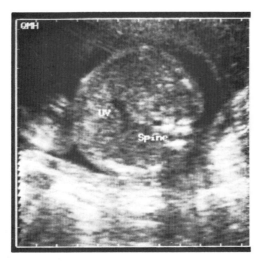

Figure 23. Correct section for abdominal measurements displaying umbilical vein (uv).

Table 4. Prediction of small-for-dates fetuses (7).

Measurement	Sensitivity (%)	Specificity (%)
Biparietal diameter	58	90
Head area	59	90
Head circumference	.56	92
Trunk area	81	89
Trunk circumference	83	90
Transverse trunk diameter	61	88

longitudinal section of the fetus [optimally by imaging the whole length of the abdominal aorta (*Figure 22*)], rotating the transducer through 90°, and freezing a transverse section which includes the umbilical vein (*Figure 23*). In the correct plane there is usually only a short segment of umbilical vein, although when the fetus is flexed, for example in a transverse lie, a longer segment is seen. Most modern equipment allows electronic measurement of area or circumference which is better than simply measuring diameters. Although area measurement is theoretically better than circumference, this makes no difference in practice. As was expected from knowledge of the 'brain sparing effect', trunk measurements give better prediction of SFD fetuses than head measurements (7) (*Table 4*).

Abdominal circumference measurements may be converted to estimates of fetal weight (8) said to be accurate to within ±10%. Other formulae exist to include BPD or both BPD and femur length measurements although this seems to make little difference to the accuracy of estimation.

The ratio of head to abdominal circumferences, whilst not a useful method of identifying SFD fetuses, may be of some value in assessing the fetus already identified as SFD (9). The theory is that asymmetry (larger head, smaller trunk) represents a fetus malnourished because of uteroplacental insufficiency and therefore vulnerable to the classical hazards of the SFD group. The symmetrically SFD fetus, on the other hand, is said to represent the genetically small baby or those with decreased intrinsic growth potential for other reasons (e.g. associated with abnormality). As will be discussed later, inspection of placental 'texture' and of amniotic fluid volume should be included in the ultrasound examination of the SFD fetus as they may provide additional clues to pathogenesis and to prognosis.

A number of other fetal measurements are being evaluated at present, as indices of abnormal growth, including femur length, thigh circumference and liver

For many years, the standard ultrasound method of assessing fetal growth was serial measurement of the BPD. However, the size of the brains of SFD infants is relatively less affected than that of other organs including the liver whose stores of glycogen become depleted by intrauterine malnutrition. It therefore seemed logical to perform measurement of the abdomen at liver level and the standard section for such measurement is identified by firstly obtaining a volume but these are not yet standard measurements and may not become so (10). Measurement of total intrauterine volume (TIUV) was popular in North America for some time and was based on the argument that not only is the SFD fetus small but it also tends to be associated with decreased amniotic fluid and with a small placenta. However, TIUV is associated with an unacceptable rate of interoperator variability and in any case, is not any better than simple abdominal area measurements in predicting SFD babies, even when performed in a painstaking and precise fashion.

Although abdominal measurement undoubtedly provides the best single measurement to predict SFD babies, the exact role of repeated abdominal measurements has yet to be explored in detail. Because of measurement error, examinations at less than 2 weekly intervals are unlikely to be of value in assessing growth. The concept of using each fetus as its own control by evaluating its intrinsic growth potential by two measurements before 28 weeks is attractive (11). Subsequent measurements would allow comparison of actual growth rate with that projected, although the fact that growth retardation may occur before 28 weeks raises obvious difficulties.

8. Multiple pregnancy

Theoretically, no multiple pregnancy should go undetected at an ultrasound examination provided the ultrasonographer makes a conscious effort to scan the entire uterine cavity in both longitudinal and transverse planes. Especial care should be taken with patients in whom conception followed pharmacologically-induced ovulation or *in vitro* fertilization or when the uterus is large for dates. Diagnosis is more difficult in very early and in late pregnancies. Undue concentration on counting heads rather than making an overall inspection of uterine contents misses discordant anencephaly.

Fetal abnormality and growth retardation (12) are more common in multiple than single pregnancies, and it is our practice to carefully inspect fetal anatomy in twin pregnancies at around 17 weeks and to perform abdominal measurements at 4-weekly intervals thereafter. When there is a single monochorionic placenta, vascular anastomoses between the twins' circulations may result in fetal polycythaemia with excess amniotic fluid in one sac and anaemia, growth retardation and diminished fluid in the other. These features, although uncommon, constitute twin-to-twin transfusion syn-

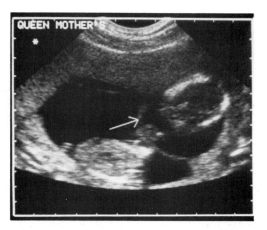

Figure 24. Twin pregnancy at 17 weeks with thick dividing membrane arrowed.

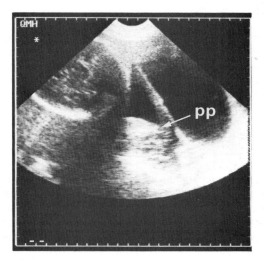

Figure 25. Posterior placenta praevia (pp).

drome and are recognizable with ultrasound. If the dividing membrane between sacs (*Figure 24*) is either not visible or is thin, almost certainly the placenta is monochorionic.

9. The placenta

Ultrasound has a well-established role as the optimal method of locating placental position—this is of particular importance prior to chorion villus sampling

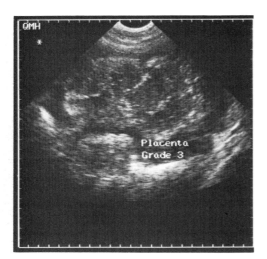

Figure 26. Grade 3 placental 'texture'.

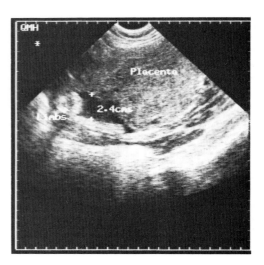

Figure 27. Measurement of largest pool of amniotic fluid in a case of oligohydramnios.

and amniocentesis, and in cases of antepartum haemorrhage, abnormal fetal presentation or other clinical situations which suggest the presence of placenta praevia. A full bladder is necessary for satisfactory inspection of the lower pole of the uterus. The lower margin of posteriorly sited placentas is more difficult to identify than those of anterior placentas because of acoustic shadowing from the fetus (*Figure 25*). The exact margins of the lower uterine segment cannot be identified ultrasonically although the uterovesical angle is a convenient landmark for reporting the position of anterior placentas. The extent of the lower segment increases as pregnancy advances so that a placenta that appears low earlier in pregnancy may appear to have risen on subsequent examination and thus no longer seems praevia. Approximately 5% of patients have ultrasound evidence of a low placenta at 16−18 weeks although only 10% of these actually have placenta praevia at delivery (13). Because of these variables, there remains a small group of patients in whom ultrasound findings are equivocal and in whom ultimate management relies entirely on clinical assessment. The other major cause of antepartum haemorrhage is placental abruption in which the ultrasound appearances are often curiously undramatic.

The 'texture' of the placenta on ultrasound examination has been a source of interest. The changes that occur may be graded (14), and advanced grades (*Figure 26*) may appear at earlier gestational ages in pregnancies complicated by hypertensive diseases or

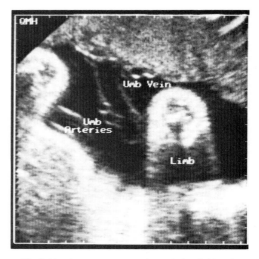

Figure 28. Subjective assessment of amniotic fluid volume by inspecting the limbs.

fetal growth retardation. A suggested correlation between placental texture grading and fetal lung maturation has not proved useful in our experience.

10. Amniotic fluid

Subjective assessment of amniotic fluid volume should be part of any ultrasound examination during pregnancy. Excessive amniotic fluid (polyhydramnios) may be associated with many fetal abnormalities (as has been discussed) and also with multiple pregnancy and diabetes mellitus although in the majority of cases no cause is found. Diminished amniotic fluid (oligohydramnios) may also be associated with fetal abnormality, fetal growth retardation and spontaneous rupture of the membranes. When associated with growth retardation, oligohydramnios indicates an even less favourable prognosis. When spontaneous rupture of membranes occurs in early pregnancy, but abortion does not ensue, the fetus is at risk of developing pulmonary hypoplasia—this may be predictable by measurement of chest dimensions.

The best method of assessing amniotic fluid volume, with ultrasound, is not certain. Some use a semi-quantitative method in measuring the dimensions of the largest pool of fluid (*Figure 27*) but, because liquor is not evenly distributed, and because fetal movement will alter the dimensions of fluid collections, an overall subjective impression may well be of more value. Inspection of fluid around the limbs is recommended (*Figure 28*).

11. Miscellaneous applications

Fetal presentation may be difficult to determine on occasions by clinical examinations, but this is easily established with ultrasound. Rarely, the umbilical cord is seen to present at the cervix and thus early warning of possible cord prolapse may be given.

Ultrasound is useful in the investigation of pelvic masses in pregnancy. An ovarian cyst (other than the functional cysts of <5 cm in diameter) should be removed surgically because of the risk of torsion and the possibility that it may be malignant. A fibroid, which is the main differential diagnosis, is readily distinguished with ultrasound (see Chapter 10) and requires no surgical intervention.

It has been suggested that ultrasonography is a useful adjunct to the successful performance of external cephalic version. Whether or not this is true, ultrasound examination may assist detection of predisposing causes of breech presentation such as fetal malformation, placenta praevia and fetal growth retardation.

Uterine abnormality, a cause of second trimester abortion and abnormal fetus lie, may occasionally be identified although this is difficult to diagnose with ultrasound.

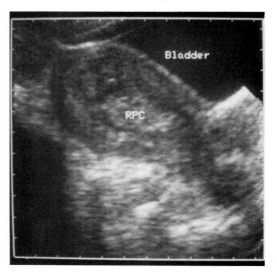

Figure 29. Retained products of conception (RPC) within the uterine cavity.

Locating an intrauterine contraceptive device within the pregnant uterus may not be of much practical value. The consensus of opinion is that such a device should be removed and this is only possible if the threads of the device can be identified on vaginal examination.

Ultrasound examination during the puerperium may be performed to search for retained products of conception which, if present, appear as echogenic tissue within the uterine cavity (*Figure 29*). Sometimes only blood is present within the cavity and gives an echo-free appearance. There is a 'grey area' in reporting in which it is difficult to make a judgement as to whether significant products remain or endometrial echoes are unusually prominent.

12. Routine early pregnancy ultrasonography

There has been considerable controversy generated by the question of whether routine ultrasonography, that is the screening of all patients, is desirable or not. In 1984, the consensus panel of the National Institutes of Health in the USA concluded that routine ultrasonography could not be justified from available evidence, but they did cite such an extensive list of indications for ultrasound examination that would have included virtually all pregnancies. During the same year, a working party of the Royal College of Obstetricians and Gynaecologists in Britain reviewed the same evidence and reached the conclusion that routine early ultrasonography at 16 – 18 weeks was justified.

Possible rationales for routine early pregnancy ultrasonography are:

(i) to establish accurate gestational age in all pregnancies;
(ii) to detect multiple pregnancies;
(iii) to detect certain fetal malformations;
(iv) to detect some non-viable pregnancies;
(v) to predict placenta praevia;
(vi) to encourage 'pre-natal bonding'.

The most important of these is assessment of gestational age. The advantages of accurate knowledge of gestational age include:

(i) a necessary reference point with which the SFD fetus may be identified;
(ii) optimal planning of the time of delivery when necessary;
(iii) optimal interpretation of tests of fetal wellbeing and of pre-natal diagnosis.

As discussed earlier, it is frequently not evident early in pregnancy in which case a precise knowledge of gestational age will be vital later when this cannot be determined by ultrasound or any other technique.

Multiple pregnancy is associated with a high perinatal mortality rate mainly due to pre-term labour. To what degree this may be modified by early identification of all multiple pregnancies is open to question, but this does seem desirable.

Our own practice is to perform a routine ultrasound scan on all patients when they first attend the antenatal clinic to optimize the assessment of gestational age. This preempts useful inspection of fetal anatomy in many patients because of the early gestational age. Others time the routine early scan for 16 – 18 weeks with the specific intention to identify possible fetal malformation. If this is done, parents should be made aware of this purpose as there are those who would not consider termination of pregnancy in the event of major abnormality. Some of the human quandaries and difficulties have been discussed elsewhere (15). Because we perform (often) very early routine ultrasonography, we detect some non-viable pregnancies before any vaginal bleeding has occurred. Whether this is more safe and kind than awaiting spontaneous miscarriage is open to debate.

Prediction of placenta praevia could not *per se* justify routine early ultrasonography, nor could 'pre-natal bonding' although a sympathetically structured real time scan can encourage the mother to have a more positive view of the pregnancy and help her to take health-promoting steps such as stopping smoking.

Whilst there is a good theoretical case for routine early ultrasonography, it remains to receive scientific justification. This would require a large controlled trial in which patients would be randomly allocated to either undergo routine ultrasonography or not. Those controlled trials which have been performed have not provided a conclusive answer although there is some evidence that routine scanning in early pregnancy does, by establishing gestational age, decrease the necessity of induction of labour of apparent post-term pregnancies.

12.1 *Routine late ultrasonography*

The rationale for routine late ultrasonography is primarily to detect clinically unsuspected SFD fetuses although occasionally fetal malformations, abnormal fetal lie and unsuspected placenta praevia may be identified. The timing of such an examination will be a compromise between increasing sensitivity of identification of SFD fetuses as pregnancy advances with progressively decreasing clinical usefulness. It is important that a sensitive ultrasound measurement is used and some trials have failed to do this. Our own study (16) used a measurement (product of crown – rump length and trunk area at 34 – 36 weeks) which successfully identified more than 90% of SFD babies, but we were unable to show that this improved pregnancy management or fetal outcome. The theoretical case for routine late ultrasonography is less convincing than for early scanning and available evidence does not support its use.

13. Fetal activity

There continues to be interest in assessing fetal breathing and overall fetal activity as indices of wellbeing. One problem that became apparent during research investigations of fetal breathing was the large number of extraneous factors that could diminish the frequency of breathing, including drugs, cigarette smoking and even abdominal palpation. While the absence of breathing and movement during the course of an ultrasound examination may mean nothing, their presence is generally a favourable feature and this should be reported by the ultrasonographer investigating the compromised pregnancy.

14. Doppler ultrasound

Although Doppler ultrasound has been used to study blood flow in adult vessels for a number of years (Chaper 11), its application to obstetrics has been recent. The measurement of flow is difficult because of the inherent difficulties in accurately measuring such elements as vessel diameter and most workers use alternative methods of waveform analysis. There is no uniformity of opinion as to the optimal vessel for study, although both fetal and maternal vessels have been studied and the fetal umbilical artery has proved most popular. While the reasons for diminished blood velocities during diastole in the maternal uterine arterial system are usually relatively easy to explain, the pathophysiological processes underlying similar waveform changes in the umbilical artery are not clear.

At present, Doppler ultrasound is a research technique in obstetrics and a careful consideration of available evidence does not justify its use in clinical practice (17). It may find a role in assessment of fetal wellbeing and predicting adverse outcome, in identifying fetal growth retardation, in establishing aetiology when a fetus is identified as SFD, or in evaluating twin pregnancies. It is to be hoped that this interesting new technique of fetal surveillance does not follow many others into premature clinical practice before it has been adequately evaluated.

15. References

1. Wells,P.N.T. (ed.) (1987) The safety of diagnostic ultrasound. *Br. J. Radiol.*, Suppl. 20.
2. Campbell,S., Reading,A.E., Cox,S.N., Sledmere,C.M., Mooney,R., Chudleigh,P., Beadle,J. and Ruddick,H. (1982) Ultrasound scanning in pregnancy. *J. Psychol. Obstet. Gynaecol.*, **1**, 57−61.
3. Robinson,H.P. (1975) The diagnosis of early pregnancy failure. *Br. J. Obstet. Gynaecol.*, **82**, 849−857.
4. Robinson,H.P. and Fleming,J.E.E. (1975) A critical evaluation of sonar 'crown−rump length' measurements. *Br. J. Obstet. Gynaecol.*, **82**, 702−710.
5. Campbell,S. (1976) Fetal growth. In Beard,R.W. and Nathanielsz,P.W. (eds), *Fetal Physiology and Medicine*. Saunders, London, pp. 271−301.
6. O'Brien,G.D., Queenan,J.T. and Campbell,S. (1981) Assessment of gestational age in the second trimester by real-time ultrasound measurement of the femur length. *Am. J. Obstet. Gynecol.*, **139**, 540−545.
7. Neilson,J.P., Whitfield,C.R. and Aitchison,T.C. (1980) Screening for the small-for-dates fetus: a two-stage ultrasound examination schedule. *Br. Med. J.*, **280**, 1203−1206.
8. Campbell,S. and Wilkin,D. (1975) Ultrasonic measurement of fetal abdomen circumference in the estimation of fetal weight. *Br. J. Obstet. Gynaecol.*, **82**, 689−697.
9. Campbell,S. and Thoms,A. (1977) Ultrasonic measurement of the fetal head to abdomen circumference ratio in the assessment of growth retardation. *Br. J. Obstet. Gynaecol.*, **84**, 165−174.
10. Gennser,G. and Persson,P.-H. (1986) Biophysical assessment of placental function. *Clin. Obstet. Gynaecol.*, **13**, 521−552.
11. Deter,R.L., Rossavik,I.K., Harrist,R.B. and Hadlock,F.P. (1986) Mathematic modeling of fetal growth: development of individual growth curve standards. *Obstet. Gynecol.*, **68**, 156−161.
12. Neilson,J.P. (1981) Detection of the small-for-dates twin fetus by ultrasound. *Br. J. Obstet. Gynaecol.*, **88**, 27−32.
13. Rizos,N., Miskin,M., Benzie,R.J. and Ford,J.A. (1979) Natural history of placenta praevia ascertained by diagnostic ultrasound. *Am. J. Obstet. Gynecol.*, **133**, 287−291.
14. Grannum,P.A., Berkowitz,R.L. and Hobbins,J.E. (1979) The ultrasonic changes in the maturing placenta and their relation to fetal pulmonary maturity. *Am. J. Obstet. Gynecol.*, **133**, 915−922.
15. Hutson,J.M., McNay,M.B., MacKenzie,J.R., Whittle,M.J., Young,D.G. and Raine,P.A. (1985) Antenatal diagnosis of surgical disorders by ultrasonography. *Lancet*, **1**, 621−623.
16. Neilson,J.P., Munjanja,S.P. and Whitfield,C.R. (1984) Screening for small-for-dates fetuses: a controlled trial. *Br. Med. J.*, **289**, 1179−1182.
17. Neilson,J.P. (1987) Commentary: Doppler ultrasound. *Br. J. Obstet. Gynaecol.*, **94**, 929−934.

16. Recommended further reading

Smith,D.W. (1982) *Recognizable Patterns of Human Malformation*. 3rd Edition. W.B.Saunders, Philadelphia.

Ferguson-Smith,M.A. (ed.) (1983) Early Prenatal Diagnosis. *Br. Med. Bull.*, **39**, No. 4.

Clinics in Obstetrics and Gynaecology (1983) **10**, No. 3: *Ultrasound in Obstetrics and Gynaecology: Recent Advances*. W.B.Saunders, London.

Chudleigh,P. and Pearce,J.M. (1986) *Obstetric Ultrasound: How, Why and When?* Churchill Livingstone, Edinburgh.

Allan,L.D. (1986) *Manual of Fetal Echocardiography*. MTP Press, Lancaster.

Young,D.G. (ed.) (1988) *Perinatal Practice—Treatable Congenital Malformations*. John Wiley, Chichester, in press.

Chapter 7

Abdominal scanning

P.A.Dubbins

1. Introduction

Cross-sectional imaging techniques rely heavily on a knowledge of anatomy and anatomical relations for their correct interpretation. This is particularly so in ultrasound imaging. There is almost no limit to the possible planes of scan and a knowledge of anatomy both surface and sectional is not only vital for interpretation but also for proper scan technique. An exhaustive description of the anatomy of the relevant structures within the peritoneum, retroperitoneum and gut is not feasible in a volume of this nature and therefore only a brief review of the salient features relevant to scan performance and interpretation can be considered here. The reader is referred to many excellent works on topographic, sectional and ultrasound anatomy (1−4).

2. General anatomy

Although the anatomy of individual structures will be discussed further below, general considerations of the peritoneal and retroperitoneal compartments are important for proper understanding of anatomical relations and the ideal scan planes for best demonstration of intra-abdominal anatomy.

The topography of the posterior abdominal wall is in large part responsible for the position and anatomical relations of many of the intra-abdominal organs and not just those in the retroperitoneum. In transverse section the posterior wall of the abdomino−pelvic cavity can be considered as a figure 3 placed on its back (3) (*Figure 1*). This shape is augmented moving more caudally by the lumbar lordosis and the increasing size of the lumbar vertebrae, although the slopes of the para-colic gutters are modified to become flatter by the large iliac wings. Structures situated in the mid-line will be relatively anteriorly placed while those lying in the para-colic gutters will have a more posterior location. This factor is important in the approach to ultrasound imaging of the retroperitoneal structures. Those placed in the mid-line, for instance the aorta, may seem unexpectedly close to the anterior abdominal wall to the beginner when considering that this structure is in the same anatomical compartment as the kidney whose upper pole lies deep within the abdominal cavity.

The *retroperitoneum* can be largely considered to contain those structures on the posterior abdominal wall covered on their anterior surface by peritoneal membrane. They include the pancreas, duodenal loop, ascending and descending colon, both kidneys and adrenals, the aorta and the inferior vena cava and their branches and tributaries, lymph nodes, posterior wall muscles, fat and nerves.

The kidneys and pancreas are considered elsewhere in this book and will not be discussed further except where anatomical relations are important.

The normal peritoneum is not visualized on ultrasound. It is a thin membrane lining the abdominal cavity. Although structures are described as intra-peritoneal (such as the small bowel) or retroperitoneal (kidneys and pancreas) all abdominal structures are lined at least on one surface by peritoneum. Their position related to the peritoneal cavity is a consequence of invagination of the peritoneum by the gut and organs derived from the gut during embryological development. Structures considered to lie within the peritoneal cavity are therefore lined on all sides by a peritoneal membrane but remain attached to the peritoneal covering of the abdominal wall by mesenteries which consist of two layers of the peritoneal reflection thrown up by the developmental invagination. These contain the vessels of supply of

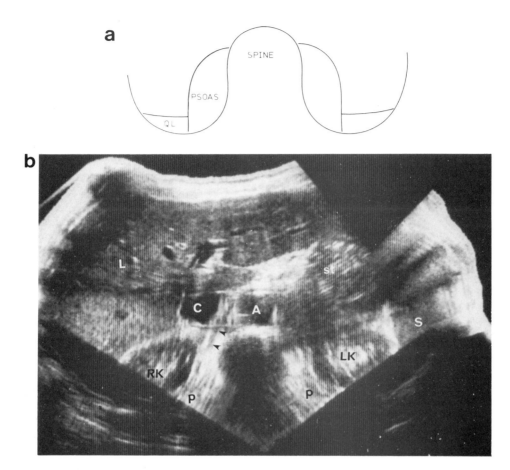

Figure 1. (a) Stylized drawing of the posterior abdominal wall in transverse section. The '3' shape determines the position of the various retroperitoneal structures. QL = quadratus lumborum muscle. (b) Transverse section through the upper abdomen (compounded). L = liver; C = inferior vena cava; A = aorta; st = stomach; S = spleen; LK = left kidney; RK = right kidney; p = psoas muscle. The arrowheads point to the right crus of the diaphragm. The relative positions of the various organs determined by the shape of the posterior abdominal wall are well shown.

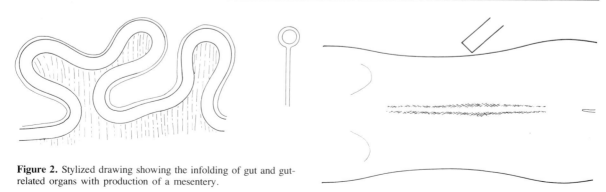

Figure 2. Stylized drawing showing the infolding of gut and gut-related organs with production of a mesentery.

Figure 3. Coronal plane of scan with patient in decubitus position.

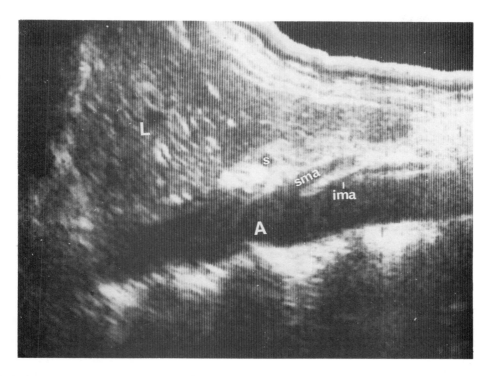

Figure 4. Sagittal scan just to the left of mid line demonstrating the aorta (A) and its anterior branches, the superior mesenteric artery (sma) and the inferior mesenteric artery (ima). Although the origin of the coeliac axis is not seen in this plane, its left branch, the splenic artery (s) is demonstrated. L = liver. The anterior branches of the aorta do not necessarily arise in linear fashion. Frequently the mesenteric vessel arises to the left of the aorta compared with the origin of the coeliac axis.

the organ (*Figure 2*). These mesenteries include the greater and lesser omentum, small bowel mesentery and indeed the falciform ligament which can be demonstrated on routine ultrasound imaging of the abdomen.

The division therefore between intra- and extra-peritoneal spaces may seem somewhat false since all these organs are covered by peritoneum. Indeed certain pathology in the retroperitoneum may involve the mesentery and vice versa but the potential space, the peritoneal cavity, into which these organs project is separate anatomically from the retroperitoneum. Its anatomy is inconstant since this is determined by the position of bowel loops, the size of the constituent lobes of the liver, etc. The potential space is further sub-divided by the ligaments and mesenteries of its resident organs. Pathological processes of the retro-peritoneal space tend not to spread intraperitoneally although the resident organs may influence intra-peritoneal pathology.

3. Ultrasound techniques, general considerations

Techniques for evaluation of the peritoneal cavity and the retroperitoneal space generally follow the same criteria as for ultrasound imaging of other organs. While special techniques are required for complete demonstration of some structures, certain general principles apply.

3.1 *Patient position*

Although the supine position is the starting point for most ultrasound examinations the patient may be examined in both decubitus positions, as well as prone or erect: longitudinal and transverse scan planes are used to provide orthodox anatomical sections but other planes of scan are useful in certain circumstances. The coronal plane, with the transducer placed on either flank (*Figure 3*) is particularly useful for the evalu-

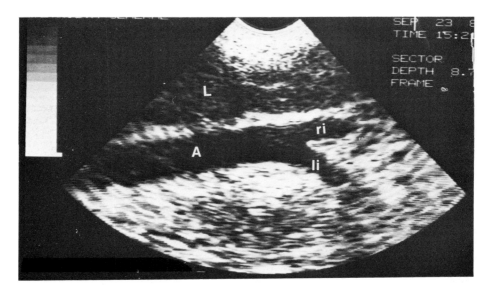

Figure 5. Coronal scan of the aorta (A) produced by scanning from the flank using the liver (L) as an acoustic window demonstrating some of the lateral branches. The right and left iliac vessels are shown (ri and li) straddling the superior rectal artery. Multiple small lateral branches are also demonstrated which probably represent lumbar arteries arising from the main aorta.

ation of retroperitoneal structures which may be combined with a slightly oblique coronal scan for visualization of the inferior vena cava and aorta. Oblique scanning planes through the ribs are useful for the visualization of the adrenal glands.

3.2 The major vessels (5)

The aorta and inferior vena cava course in almost true sagittal fashion on the anterior surface of the bodies of the lumbar spine. The aorta enters the abdomen at the aortic hiatus at the level of T12 and extends to its bifurcation at the level of L3/4 into the right and left common iliac arteries. Its course through the abdomen tends to bring it more anteriorly consequent upon the lumbar lordosis and increasing size of the lumbar vertebrae. It has predominantly anterior single branches and paired lateral branches (*Figure 4*). Scan technique has to take account of the course and position of the aorta, anatomical relations, aortic branches as well as the pathology of the aorta and periaortic regions. Sagittal scans just to the left of midline utilizing either sector or linear array real-time scanners will demonstrate the long axis of the aorta. Anterior branches are well demonstrated by this means as well as in true transverse plane which is achieved by rotating the transducer 90°. Lateral branches of the

aorta will be demonstrated on the transverse scan and also on a modified coronal scan achieved by placing the transducer on the right flank (*Figure 5*). The coronal scan plane is slightly modified by angling the transducer posteriorly. Visualization from the left flank is less reliable. Use of compression with the transducer is useful for displacement of bowel gas although care must be exercised in the presence of an aortic aneurysm. Because the course of the aorta may occasionally be very markedly anterior particularly in thin patients, it may be necessary to use a high frequency short focused transducer for adequate visualization, although in general terms a 3.5−5 MHz medium to long focus transducer is used in the larger patient. Modification of the scan planes becomes necessary when the aorta is tortuous when simple sagittal scans will not show the full extent of the aorta and a three-dimensional image must be compiled by a series of short scans each one sagittal to that portion of the aorta.

Using this technique the aorta can be visualized as an echo-free tube with highly reflective walls and regular pulsations. Antero-posterior measurements of the diameter of the aorta are up to 2.5 cm at the aortic hiatus and 1.8 cm at the bifurcation. This diameter varies slightly throughout the cardiac cycle.

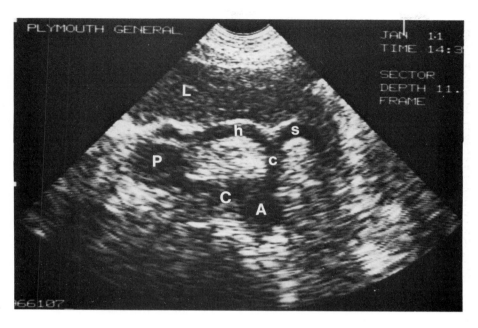

Figure 6. Transverse section through the origin of the coeliac axis (c) demonstrating its right and left branches, hepatic artery (h) and splenic artery (s) forming a 'Y' shape. L = liver; P = portal vein; C = cava; A = aorta.

3.3 *Aortic branches*

Almost immediately upon entering the abdomen the aorta gives off the coeliac artery (*Figure 6*). This is an anterior branch which normally divides into the splenic, hepatic and left gastric arteries. [However the branching pattern of the coeliac axis and the superior mesenteric arteries is very variable. For instance branches to the liver arise from the superior mesenteric artery and the left gastric artery in a number of cases (4).] The splenic artery follows a somewhat undulating course to the left related to the posterior superior surface of the pancreas while the hepatic artery crosses to the right moving somewhat cephalad to occupy the left side of the porta hepatis in the free edge of the lesser sac related to the anterior and left surface of the portal vein. The left gastric artery is a smaller vessel whose course cannot always be followed by ultrasound but supplies the lesser curvature of the stomach. The appearances of the coeliac axis in transverse section may be those of a 'Y' or a 'T' depending upon the course followed by the two major branches.

The superior mesenteric artery (SMA) arises from the anterior surface of the aorta at the level of L1

(*Figure 4*), is the vessel of supply to the small bowel and to the right side of the colon and enters the small bowel mesentery just below the neck of the pancreas. It therefore serves as a useful landmark for the pancreas lying posterior to the neck and anterior to the uncinate process. The pancreatic head lies to the right.

The inferior mesenteric artery is not always seen on ultrasound imaging but represents the third anterior branch (*Figure 4*).

The angle subtended by these vessels with the aorta is variable and does not necessarily indicate pathology.

The paired lateral branches of the aorta are many and include multiple lumbar arteries. These can only rarely be seen on ultrasound and thus the only important paired lateral branches are the renal arteries (*Figures 5, 7* and *8*). Although there may be more than one renal artery on either side it is not common to visualize multiple arteries perhaps because of the small size of most accessory renal arteries. The left renal artery commonly arises from the lateral or postero-lateral aspect of the aorta to travel posteriorly, inferiorly and laterally to the hilum of the left kidney. The right renal artery arises from the lateral or

Figure 7. Coronal oblique scan demonstrating the lateral branches of the aorta (A), the renal arteries both right and left (r and l). The left renal vein (lrv) is also demonstrated. L = liver.

antero-lateral aspect of the aorta to travel posteriorly, inferiorly and laterally to the hilum of the right kidney coursing behind the inferior vena cava (IVC).

3.4 The inferior vena cava

The IVC enters the abdomen at the level of T8 and its course follows a gentle sickle-shaped curve in longitudinal axis through the liver to lie to the right of the vertebral bodies. In the upper abdomen it lies slightly anterior to the plane of the aorta but its course through the abdomen is to move gradually posteriorly and this relative position to the aorta is reflected in the anatomical relations of its tributaries; the left renal vein courses anterior to the aorta while the left iliac vein lies posterior to the left iliac artery.

The IVC receives several tributaries but those visible on ultrasound are the left, right and middle hepatic veins entering the anterior aspect just before the caval hiatus (*Figure 9*), the paired lateral renal veins and the common iliac veins which join to form the IVC at lumbar vertebra 3. The left is the longer of the two renal veins coursing anterior to the aorta but posterior to the SMA to enter the antero-lateral surface of the IVC (*Figure 8*). This relationship is important because the left renal vein acts as a landmark for the inferior

margin of the pancreas but also because the SMA and aorta will produce a 'nut-cracker' effect in some people with functional narrowing of the left renal vein and dilatation of the proximal portion. The walls of the IVC and of its tributaries are thinner and less highly reflective than those of the arteries. They are therefore more potentially compressible by adjacent structures. However, in the normal patient there are only two impressions on the IVC; that produced by the head of the pancreas which may cause a very slight, smooth impression on the anterior surface of the IVC and the right renal artery which may indent the posterior aspect of the IVC (*Figure 10*). All other impressions and compressions of the IVC must be considered abnormal and indeed if these two here described are unusually marked they may indicate pathology either in the pancreatic head or the renal artery.

The IVC and its related anatomical structures serve as a useful guide for determining retroperitoneal pathology. Masses distorting the IVC above the level of the porta hepatis are normally of hepatic or adrenal origin, those between the porta hepatis and the lower border of the pancreas of renal or vascular origin, whereas primary retroperitoneal tumours and lymphadenopathy may occur anywhere along the course of the vessel.

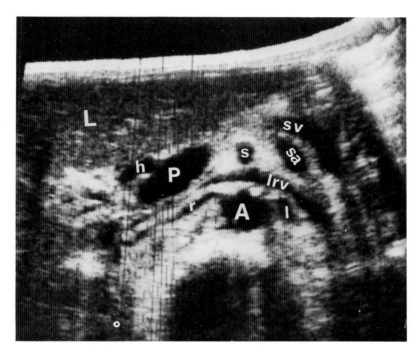

Figure 8. Transverse section through the upper abdomen demonstrating the lateral branches of the aorta (A) and inferior vena cava. The right and left renal arteries (r and l) the left renal vein (lrv). Other arterial branches are also demonstrated. The hepatic artery (h) situated superior to the confluence of superior mesenteric vein and splenic vein (P), the superior mesenteric artery situated anteriorly to the left renal vein (s), the splenic artery and splenic veins (sa, sv). L = liver.

Imaging the vessels is necessarily of a dynamic nature, the aorta in the young patient is a pulsatile structure moving slightly with each systole and changing diameter by as much as 20% between systole and diastole. These changes are of course less marked with age as the aorta becomes less elastic as a result of atheroma. The IVC is similarly dynamic changing in diameter from up to 2 cm to almost complete occlusion consequent upon the respiratory cycle and the A,C,V waves corresponding to phases of the cardiac cycle. Thus there is no absolute maximum or minimum dimension of the IVC although absence of normal respiratory or cardiac variation may be taken as evidence of disease.

4. Vessel pathology

This may be considered under three main headings: alteration of diameter; luminal contents; abnormality of course.

Although atheroma and tortuosity of the aorta are perhaps a normal concomitant of ageing, other changes in course may be due to extra-vascular disease within the retroperitoneum and will be considered under this heading.

4.1 Alteration of vessel diameter

Although it is difficult to set upper limits for caval dimensions the diameter does vary in response to certain disease processes. Increases in blood flow through veins such as occurs in arterio-venous malformations and A−V shunting in vascular tumours will cause venous engorgement and enlargement (renal carcinomas may cause enlargement of the renal vein and IVC). However more common is the enlargement of the IVC secondary to congestion due to congestive cardiac failure or constrictive pericarditis. In congestive cardiac failure respiratory and cardiac modulation of caval size is diminished with the cava remaining dilated throughout respiratory and cardiac cycles. In constrictive pericarditis there may initially

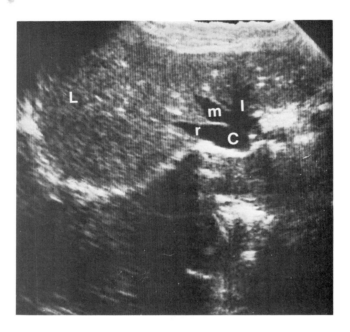

Figure 9. Transverse scan of the inferior vena cava (C), just proximal to the caval hiatus. The right, middle and left (r,m,l) branches of the inferior vena cava are demonstrated draining from the liver (L) into the inferior vena cava.

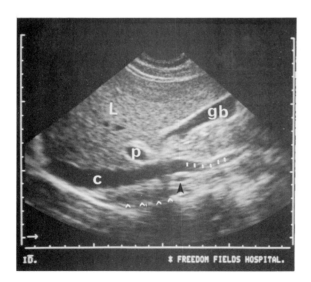

Figure 10. Sagittal scan through the inferior vena cava (c) demonstrating the anterior impression of the pancreas (short lines) and the posterior impression by the right renal artery (closed arrowhead). The anatomical relations of the inferior vena cava are well shown. The right lobe of the liver (L), the portal vein (p), gall bladder (gb) and the right crus of the diaphragm (open arrowheads).

be augmentation of the cardiac modulation with almost 'bouncing' motion of the IVC. These diameter changes in veins are secondary and not due to primary disease of the vessel. The aorta however, may undergo marked changes in diameter as a result of atheroma.

4.2 Aortic aneurysm

This is dilatation of any part of the aorta and is usually diagnosed when the antero-posterior (AP) diameter exceeds 3 cm. The dilatation is usually fusiform (*Figure 11*) but may be saccular and, rarely, an aortic dissection may extend from the thoracic aorta into the abdomen. Most fusiform aneurysms arise below the level of the renal arteries but it is important to determine whether the renal vessels are involved since this will alter management and, if operation is considered, the operative approach and the operative procedure. The position of the renal arteries is best determined by scanning in the transverse plane or the coronal plane with the patient in the lateral decubitus position (*Figure 12*). In this latter plane the IVC and aorta are imaged in long axis simultaneously and the right renal artery is seen coursing beneath the IVC. In the transverse plane the renal arteries may be seen

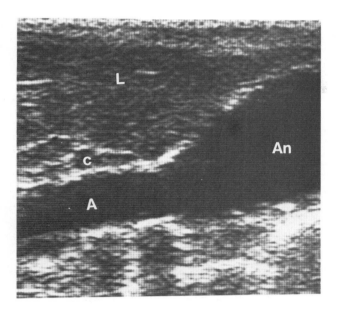

Figure 11. Fusiform aortic aneurysm (An) demonstrates gradual dilatation of the aorta (A). The relationship of the right crus of the diaphragm (c) as it decussates over the surface of the aorta is shown. L = liver.

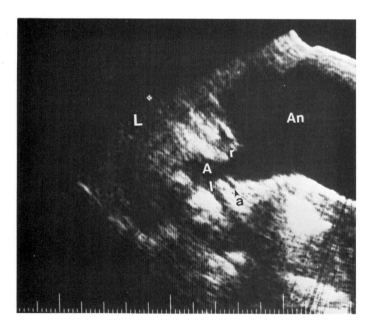

Figure 12. Coronal scan of the aorta (A) through the right flank showing a very large aortic aneurysm (An). This scan plane shows the relationship of the renal arteries (r and l) as well as an accessory renal artery on the left (a). L = liver.

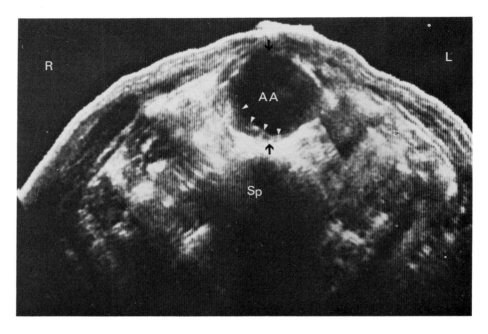

Figure 13. Transverse scan through large abdominal aortic aneurysm (AA). The measurement end points for assessment of aneurysm size are demonstrated by the arrows. The closed arrowheads demonstrate the position of mural clot. Sp = spine.

coursing laterally from the aorta in a plane slightly inferior to the SMA. Occasionally the size and course of the aneurysm are such that the anatomy of origin of the renal vessels is distorted and in these cases their position must be inferred from the relationship of the aneurysm to the SMA; the renal arteries usually arising within 2 cm of this origin.

The appearances of aortic aneurysms are variable. There may be different amounts of lamellated mural clot which narrows the true lumen, as well as differing amounts of calcification of the aortic wall (*Figure 13*). The size of the aneurysm is of great importance in planning management. The likelihood of aortic rupture is greater the larger the size of the aneurysm. Furthermore a progressive increase in aneurysm size also places the patient at higher risk for aortic rupture. Accurate measurement of aneurysm size is important both in the immediate assessment and for regular follow up. Measurements of the aneurysm should represent the maximum dimension usually in the anterior posterior plane, and should be from outer to outer wall if they are to correlate with surgical findings. Standardization of technique between examinations is most important to allow accurate

follow up. Measurement of the clot-free or true lumen may also be performed which correlates well with arteriographic findings. The longitudinal extent of the aneurysm is best documented descriptively relating the upper and lower limits to branch vessels rather than by measurement. Ultrasound is an insensitive indicator of aortic leak.

Aneurysms of branch vessels may occur and splenic artery aneurysms are particularly common. These may similarly be demonstrated by ultrasound as fusiform or saccular dilatation of the vessel with varying amounts of intraluminal echoes usually adjacent to the vessel wall representing clot. Narrowing of vessel diameter may occur focally as in atheromatous stenosis, in response to extrinsic pressure, or generally in the arteritides such as Takayasu's arteritis. Demonstration of arterial narrowing as a result of intrinsic disease is seen by reduction of luminal diameter, the presence of mural plaques which may be relatively echo poor (soft plaque) or highly echogenic often with acoustic shadowing as a result of calcification. However, imaging is a poor predictor of the haemodynamic importance of the narrowing. Doppler studies of the abdominal vessels, major and

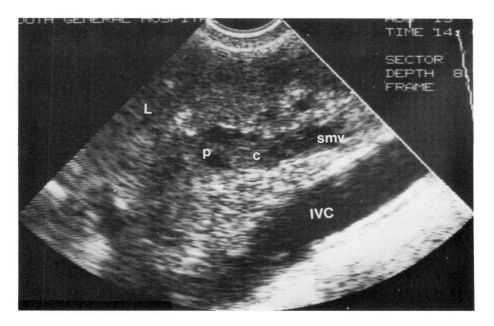

Figure 14. Longitudinal scan demonstrating the superior mesenteric vein (smv) as it travels to the portal confluence (p). The veins are distended by intraluminal clot (c). IVC = inferior vena cava; L = liver.

branch, have been performed and have been found to be useful in the detection of such stenoses but full consideration of their findings is beyond the scope of this chapter (6,7). Extra-luminal causes of vascular narrowing also occur but usually also involve alteration of vessel course and will be considered later.

4.3 Luminal contents

The contents of the vessels in normal situations are fluid and although intraluminal echoes may be demonstrated in the veins by real time these are transient phenomena. Echoes within the vessels may be seen when there is intravascular clot or tumour. Both are more commonly seen in the veins and are usually associated with venous distention (*Figure 14*). Clearly demonstration of intravenous clot and venous occlusion are important but special significance is attached to the demonstration of intraluminal content in the presence of tumour since this may indicate tumour spread along the vein. For example in the renal vein this is evidence of spread of renal carcinoma. Intraluminal clot is rarely seen in arteries except when there is aneurysm formation and intraluminal tumour is only seen in the very rare primary haemangio-endothelial tumours.

4.4 Alterations in course

The aorta is a thick walled muscular vessel whose course just to the left of mid-line and anterior to the spine is less subject to displacement than that of the adjacent thin walled IVC. Atheromatous change and the resultant lengthening and tortuosity of the aorta may produce a change in course and the aorta in this situation may weave its way down the abdomen occasionally coming to lie in front of or behind the IVC. Such a course is of no clinical significance but is of great imaging significance. Although real time facilitates the demonstration of the vessel in these cases it is still difficult to produce images representing the long axis of the aorta and to provide accurate measurements because of the obliquity of scan plane to the aortic lumen.

Both vessels may be distorted and displaced by adjacent pathology although the IVC is most susceptible. Indeed scanning the IVC in long axis may provide subtle signs of the presence of a retroperitoneal tumour simply by displacement of the vessel anteriorly away from the spine and distortion of its normally smooth curve. Although many of the pathologies producing this displacement may well be obvious if a large mass is present, smaller masses, and those with

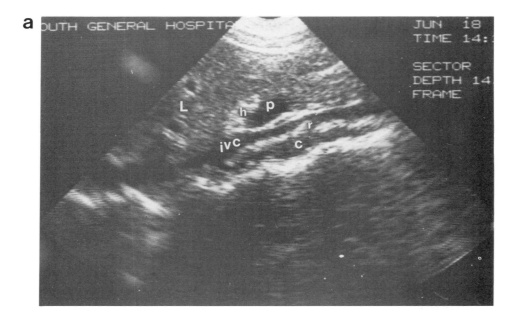

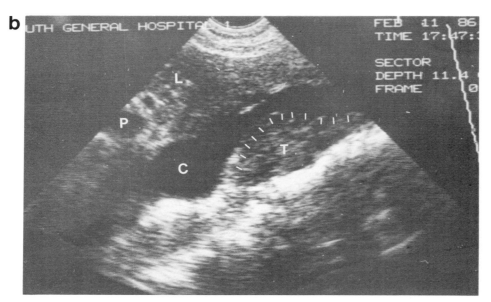

Figure 15. (a) Longitudinal scan through a normal inferior vena cava (ivc) with a prominent right crus of the diaphragm (c). The relationship of this and of the inferior vena cava to other structures is shown. L = liver; h = hepatic artery; p = portal vein; r = right renal artery. (b) Longitudinal scan through the inferior vena cava (C) demonstrating a tumour mass (T). This was a recurrence of a renal carcinoma which produces an abnormal impression on the posterior aspect of the inferior vena cava. P = portal vein; L = liver.

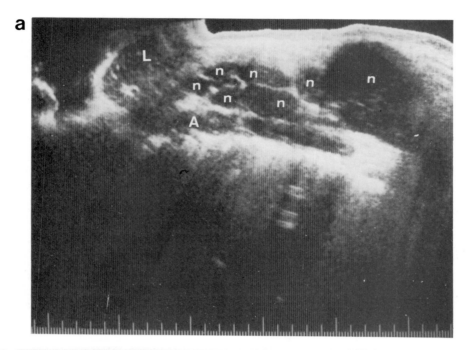

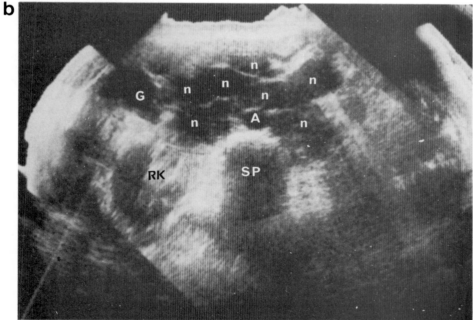

Figure 16. (**a** and **b**) Longitudinal transverse scans of the upper abdomen in a patient with lymphoma. Multiple confluent echo-poor nodal masses are shown surrounding the aorta (A), the inferior vena cava is not visualized separately because of its similar echo pattern to the nodal masses (n). G = gall bladder; RK = right kidney; SP = spine; L = liver.

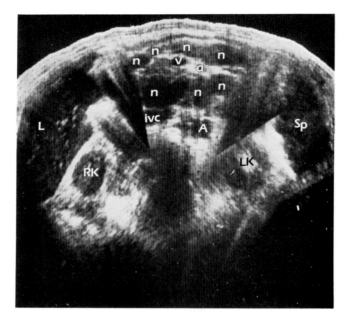

Figure 17. Transverse scan through the upper abdomen in a patient with lymphoma. There is extension of lymphomatous tissue into the mesentery with nodal masses (n) surrounding the superior mesenteric artery and vein (a and v). There is also splenomegaly (Sp). A = aorta; L = liver; RK, LK = right and left kidney.

similar echogenicity to retroperitoneal fat, may go unnoticed. Care must be taken however, not to confuse normal retrocaval structures for pathology. The right crus of the diaphragm and the right renal artery are normal posterior relations (*Figure 15*).

Displacement of the branch vessels of the aorta may similarly be used as evidence of pathology. The alteration of the angle subtended by the SMA with the aorta may be greater than 60° in normal patients although this is unusual and may instead be the result of adenopathy.

Occasionally, such is the distortion that adjacent pathology causes, the lumen of the vessel may be altered. The IVC may be draped over larger retrocaval masses and its lumen consequently narrowed to slit-like proportions thus putting at risk the venous drainage from the legs. Arteries are less likely to become narrowed by tumour although tumour encasement of vessels may occasionally produce significant stenoses.

5. The retroperitoneum (8,9)

Structures resident within the retroperitoneal space include the pancreas, kidneys, parts of the gastro-intestinal tract and the major and branch vessels. The retroperitoneum also contains fat, muscle, lymph nodes and neural tissue. Pathology may affect any of these constituent structures but pathology of the kidney and pancreas is considered elsewhere.

The ultrasound demonstration of pathology of the lymph nodes depends upon increase in size, since resolution by ultrasound is not sufficient to detect pathological but non-enlarged nodes. Nodal pathology is imaged by the same techniques as are used for the aorta and IVC. The transverse and coronal scan planes are particularly useful for demonstration of the relationship of the aorta to the IVC and whether there are intervening enlarged nodes. Extending the scan to the iliac fossae, iliac lymphadenopathy may be demonstrated by scan planes following the long and short axes of the iliac vessels, directing the longitudinal plane of scan along a line connecting the umbilicus and the femoral pulse. Here again compression with the transducer may displace bowel gas and a linear array probe may be more effective than a sector scanner. A full bladder is not necessary for examination of this part of the pelvis and may be a positive hindrance since it will prevent proper compression.

Enlarged lymph nodes are seen as echo-poor oval

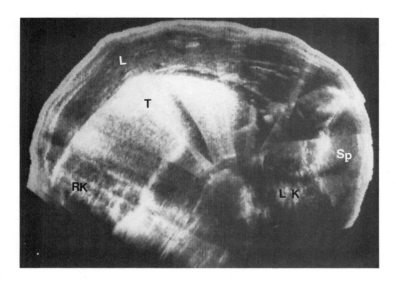

Figure 18. Transverse scan through the upper abdomen in a patient with a retroperitoneal liposarcoma (T). This is a highly reflective tumour due to the fat content. The mass has displaced the right kidney laterally (RK) and has displaced and flattened the right lobe of the liver (L). Sp = spleen; LK = left kidney.

or rounded structures adjacent to or even surrounding the major and branch vessels. They are of varying size and may be confluent (*Figure 16*). There is no way to distinguish between metastatic lymphadenopathy and that due to lymphoma. Extension of lymph node involvement into the mesentery represents an important factor in adequate staging of malignant disease, particularly the lymphomas. Such extension can be detected by ultrasound by the demonstration of the 'sandwich sign' (*Figure 17*). Here the mesenteric vessels are sandwiched by echo-poor lymph node masses in the mesentery.

Primary tumours of the retroperitoneum are uncommon but derive from muscle, fat or nervous tissue. Although for the most part tumours cannot be distinguished by their echogenic appearance, liposarcomas with their high fat content tend to be highly reflective (*Figure 18*). The demonstration of a high fat content is important and is better performed by computed tomography or by magnetic resonance imaging since the greater the ratio of soft tissue to fat within the tumour the greater the malignant potential.

6. Muscle pathology

Apart from the sarcomas arising from muscle tissue the only significant pathology involves the psoas muscle. The normal psoas can be demonstrated on transverse scans as a round or oval structure lying lateral to the spine. The rather thinner quadratus lumborum muscle lines the posterior abdominal wall more laterally. The psoas muscle forms the postero-medial relationship of the kidney and may be involved in pathology arising from the kidney, from the spine or from the colon to which it is also closely related. The normal psoas muscle has a parenchymal texture of mid-range echoes in contrast to its more echo-poor neighbour the quadratus lumborum (*Figure 1*). Furthermore its parenchymal texture appears to be arranged in fibres and this may be appreciated on long axis scans. Both psoas muscles are usually symmetrical in size and echo pattern and this is important in assessing pathology which normally produces muscle enlargement. However, care must be taken when assessing sportsmen since over development of one or other of the muscles may occur particularly in footballers.

The ultrasound appearance of psoas abscess and psoas haematoma are essentially the same. The muscle is enlarged, echo poor and tender. Eventually echo-free or fluid areas may be identified within the muscle but this is frequently a late finding. Abscess formation is associated with pathology in adjacent organs and haematoma with bleeding diatheses.

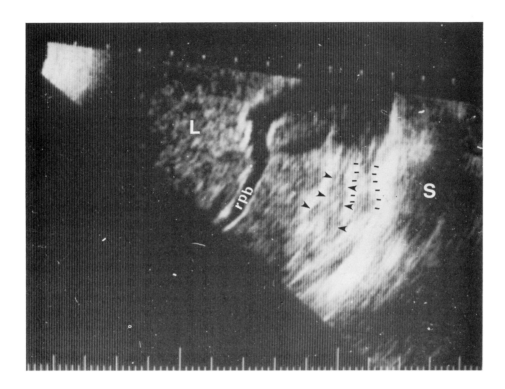

Figure 19. Transverse scan through the right side of the upper abdomen via an intercostal space, demonstrating a normal right adrenal gland (arrowheads). The adjacent right crus of the diaphragm is also marked. It is sometimes difficult to differentiate between the crus and the right adrenal gland. The echo pattern of the adrenal gland is noted with echo-poor cortex and more reflective medulla. The relationship of the gland to the right lobe of the liver (L) is shown; the right posterior branch (rpb) of the portal vein is also shown within the liver. S = spine.

7. The adrenal gland (10,11)

Visualization of the adrenal gland with ultrasound is inconstant with most real-time equipment. The gland is a small structure shaped either like an inverted 'V' or 'Y' whose component limbs are thin and have a central echogenic medulla and an echo-poor peripheral cortex (*Figure 19*). However, adrenal reflectivity is not markedly dissimilar from adjacent structures, particularly the crura of the diaphragm and retroperitoneal fat. Its location medial and superior to the upper pole of the kidney within the perirenal fat makes for difficult visualization. Special scan planes developed for use with static B-mode equipment involved the use of oblique scan planes along the intercostal spaces with the patient in both decubitus positions. Visualization of the adrenals was claimed to be 92% for the right and as high as 75% for the left. Although multiple scan planes can be used with real-time imaging the lateral resolution of most sector scanners and the contrast resolution of most linear arrays result in much poorer adrenal visualization in practice. While newer methods of beam focusing (for example Acuson) appear to allow visualization of the adrenal with similar reliability to that originally described for static machines, such equipment is not yet widely available in this country. While the reliability of the visualization of the normal adrenal in the adult is less than ideal, the neonatal adrenal is relatively larger in all dimensions and in this situation ultrasound is the imaging method of first choice. Furthermore although ultrasound may lack sensitivity in the demonstration of the normal adult adrenal, the adrenal bed can usually be visualized and the sensitivity of detection of significant adrenal pathology is much higher.

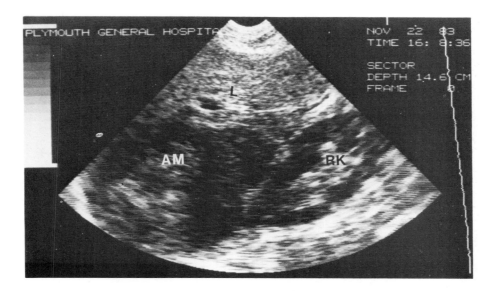

Figure 20. A large adrenal mass (AM) situated cephalad to the right kidney. This was a large adrenal metastasis from a bronchogenic carcinoma which also had invaded the liver (L). Note that there is no demonstrable capsule between the liver and the adrenal metastasis. RK = right kidney.

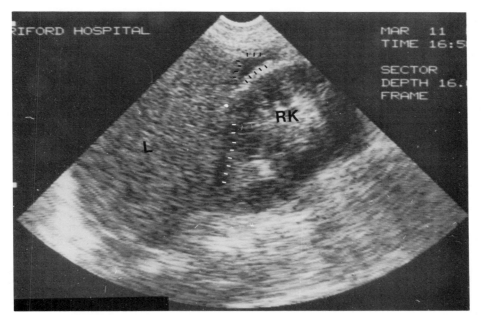

Figure 21. Early ascites, small amounts of ascitic fluid are seen between the right kidney (RK) and the liver (L) migrating over the surface of these organs by capillary effect.

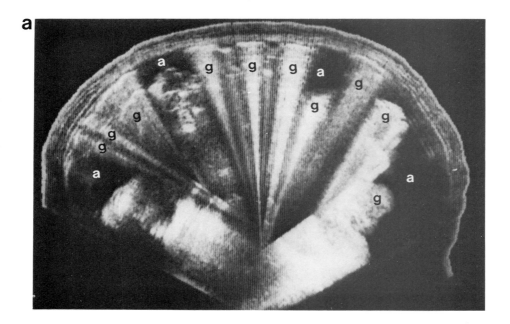

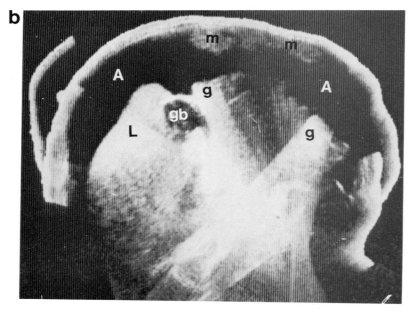

Figure 22. (a) Transverse scan of the abdomen in a patient with loculated ascites (a), although many of the bowel loops containing gas (g) are situated at the peritoneal surface as would happen in a benign ascites; there is loculation of fluid and some of the bowel loops are abnormally tethered. **(b)** Malignant ascites, matted bowel loops (g) are held close to the posterior abdominal wall by malignant infiltration. The ascitic fluid (A) appears to contain no echoes but there are peritoneal deposits (m) on the anterior surface of the abdominal wall. L = liver; gb = gall bladder.

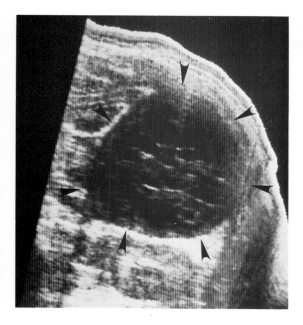

Figure 23. Large intra-abdominal abscess. This abscess contains multiple thick and thin septa but this appearance is not specific for an abscess and may be seen in lymphocoeles and haematomas.

7.1 *Adrenal pathology*

The adrenal glands may be diffusely enlarged or may be involved by neoplastic processes.

In the neonate adrenal enlargement may be seen in congenital adrenal hyperplasia and enlargement and alteration of echo pattern (usually a decrease in echogenicity) may be seen in adrenal haemorrhage.

The adult adrenal may be involved by many different neoplastic processes; adrenal cysts occur but care must be taken in the diagnosis of an adrenal cyst because phaeochromocytomas in particular may be cystic. Other primary tumours of the adrenal occur but apart from phaeochromocytoma the most common lesion in the adult is an adrenal metastasis commonly from a primary bronchogenic carcinoma. These are irregular masses with mid-range echoes and may frequently be bilateral (*Figure 20*).

8. The peritoneal cavity (12,13)

The peritoneal cavity is divided into several different spaces by the bowel mesenteries and other peritoneal reflections. The major spaces are the supramesocolic space and the intramesocolic space: the upper and lower part of the abdomen being separated by the mesentery of the transverse colon, the transverse mesocolon. The supramesocolic space is further divided by the liver into the subphrenic and subhepatic spaces and by the falciform ligament into the right and left subphrenic space and the perisplenic space. The lesser sac is a further space bounded by the stomach and lesser omentum anteriorly and the pancreas posteriorly. It communicates with the remainder of the peritoneal cavity through its free edge in which lies the portal vein, common bile duct and hepatic artery.

Although none of these spaces can be visualized by ultrasound in the normal person, their importance lies in the distribution of fluid collections within the abdomen. Thus these tend to be confined to certain potential spaces and to spread to others via certain specific pathways, viz. the pelvis to the paracolic gutters, the paracolic gutters to the subhepatic and pericolic spaces and thus to the subphrenic spaces or vice versa. The lesser sac remains largely separate and fluid tends to remain confined here, for example pancreatic pseudocysts; or excluded from the lesser sac: large volume of ascites may be present without entering the lesser sac.

Normally the peritoneal cavity contains a very small amount of fluid sufficient only to line the peritoneal surface of the bowel loops and lubricate their motion. Abnormal collections of fluid are ideally suited for evaluation by ultrasound and their distribution and echo pattern may help to determine their nature.

Ascites represents free fluid within the peritoneal cavity. Because it tends to gravitate to the most dependent portion of the abdomen it will collect in the pelvis and in the paracolic gutters when the patient is supine. Fluid may be detected by ultrasound examination of the more dependent parts of the patient. For example the patient may lie in the decubitus position with the transducer on the dependent flank, may lie prone (or on hands and knees) with the transducer on the dependent abdomen or may stand erect with the transducer applied to the pelvis. Small amounts of fluid may migrate over the surface of a solid viscus such as the liver by a capillary effect (*Figure 21*) and may be detected as a very thin echo-free film over the liver surface. Simple ascites is usually freely mobile and bowel loops float freely within it. Complicated ascites which may be infective or neoplastic may contain reflective particulate matter, demonstrate septa and/or

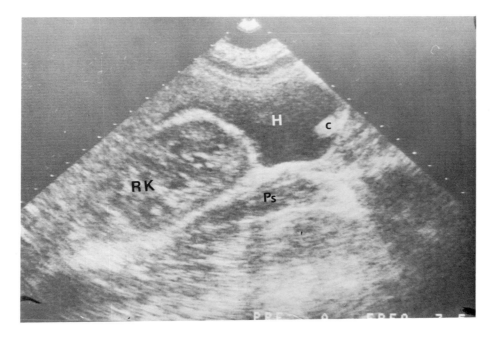

Figure 24. Peri renal haematoma (H) as a result of abdominal trauma involving the right kidney (RK). There is a small focus of dense mural clot (c). These clot aggregations are much more commonly seen in intra abdominal haematomas rather than other fluid collections. The relationship of the haematoma and kidney to the retroperitoneal muscles is seen. Ps = psoas muscle.

be loculated. Bowel loops may not float in malignant ascites but are often matted and stuck to the posterior abdominal wall (*Figure 22*).

Other collections of fluid within the peritoneal cavity are usually either pus or blood. Although lymphocoeles and urinomas also occur there is no way on ultrasound to differentiate between these fluids or between these and the rather rare mesenteric cyst. All are variously echo free with differing amounts of reflective particulate matter of varying size and differing degree of irregularity of containing wall (*Figures 23* and *24*). While an abscess may have an irregular wall with fine particulate matter within the fluid and a haematoma may demonstrate larger reflective particles of aggregated clot, quite often both can demonstrate all the characteristics of a simple cyst. Septa appear to be more common in lymphocoeles.

The technique of evaluation of a patient to detect an abnormal intra-abdominal fluid collection bears special consideration. Although the basic principles remain the same, certain factors demand a change in technique. Perhaps the most important of these is the fact that the single most common cause today of an intra-abdominal abscess is the surgeon's knife. Thus many patients presenting for ultrasound in this situation will be unwell, perhaps unable to be moved from their bed, swathed in dressings with tubes emanating from most natural orifices and many unnatural ones. Resourcefulness on the part of the Sonographer is required here—moving the patient where possible, removing preferably all the dressings and often resorting to non-standard scan planes where necessary. For the most part the examination is directed according to the site of the operation, for example the left upper quadrant following a gastrectomy; or according to symptoms—the right subphrenic space in the presence of hiccoughs and a right pleural effusion on chest X-ray. If however the history provides no guide, a scanning routine to evaluate all the likely sites for fluid collection is necessary. The author's preference is to scan the right subphrenic space using intercostal and subcostal scans angled cephalad to ensure adequate visualization of the diaphragm (*Figure 25*). Formal scan technique to examine the liver allows evaluation of the perihepatic spaces and Morrison's Pouch. The left subphrenic and perisplenic regions follow,

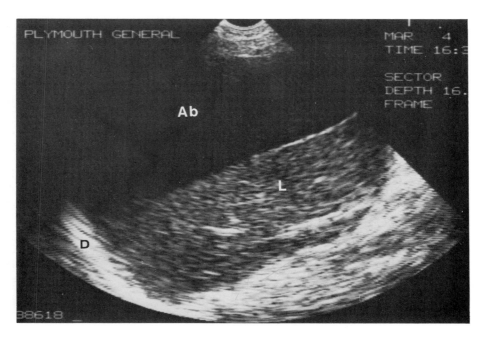

Figure 25. Intercostal scan demonstrating large subphrenic abscess (Ab) just beneath the diaphragm (D) displacing the liver (L) downwards and medially.

scanning in the intercostal spaces for adequate visualization of the diaphragm through the spleen. The use of the decubitus or oblique position where possible is usually of great value, although the flexibility of real-time scanners has made this non-essential in large collections and when patients are difficult to move. The Sonographer's attention is then drawn to the pelvis. The full bladder technique is not always possible but compression by firm pressure with the transducer will usually allow visualization of the muscle layers of the iliac wings (the iliacus and psoas) and similar pressure in the mid-line will often identify large pelvic abscesses. The paracolic gutters are then examined from the anterior approach, in coronal plane and if possible from the posterior aspect. Finally the mid-abdomen is scanned since this is the least likely site for an abscess to collect and indeed the most difficult to demonstrate with ultrasound.

Attention must be paid throughout the examination not to confuse fluid-filled bowel for an abnormal extra-intestinal fluid collection. This is a particular problem with the stomach in the left upper quadrant and with small bowel loops in the mid-abdomen and pelvis. Although recognition of bowel structure (vide infra)

and the demonstration of peristalsis on real time has to some degree facilitated this differentiation, the paralytic ileus which occurs post-operatively and in response to inflammation may result in significantly distended loops of bowel which show no peristalsis. Careful evaluation is necessary, sometimes repeating the examination after a day or two to avoid confusing bowel for an abscess.

Once the abnormal fluid collection is diagnosed ultrasound may be used for guidance of needle aspiration for bacteriology and cytology or for percutaneous catheter drainage (14).

9. The gastrointestinal tract (15)

The gastrointestinal tract has been blamed and continues to be blamed for much of the problems associated with imaging the upper abdominal organs with ultrasound. Attempts at reducing bowel gas in its deleterious effects on ultrasound image quality have occupied the minds of many. However, high resolution real time and a growing appreciation that understanding ultrasonic patterns of bowel depends

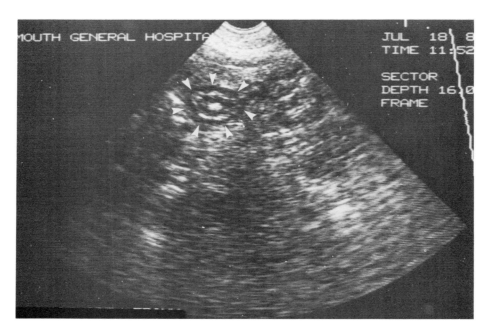

Figure 26. 'Bullseye' pattern of collapsed bowel loop seen in cross-section. This scan is a transverse section through the antrum of the stomach which contains a mixture of gas and fluid producing this rather complex 'Bullseye' within the echo-poor gastric mucosa.

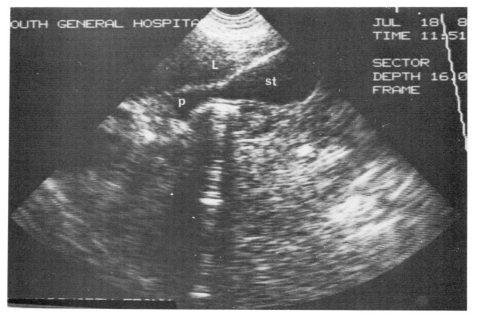

Figure 27. Fluid containing stomach (st) adjacent to the left lobe of the liver (L). The shape of the stomach as it narrows towards the pylorus (p) is seen.

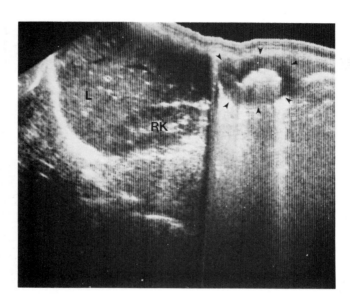

Figure 28. Carcinoma of the hepatic flexure. A bowel-associated mass is shown by the arrowheads with evidence of bowel wall thickening of an irregular nature. The location of the mass to the hepatic flexure is made purely by the anatomical relations to the liver and right kidney (L and RK).

Table 1. Lesions producing atypical target configuration.

Tumours	Carcinoma
	Sarcoma, e.g. Leiomyosarcoma
	Lymphoma
	Serosal metastases
Inflammatory disease	Crohn's disease
	Diverticular disease
	Pseudomembranous colitis
	Gastritis
	Amyloidosis
	Whipples disease
	Granulomatous disease
	Appendicitis
Other	Hypertrophic pyloric stenosis
	Intramural haemorrhage
	Lymphangiectasia
	Intussusception
	Menetrier's disease

upon recognition of the contribution of bowel contents, has meant that while bowel remains an occasional impediment to adequate visualization of other abdominal structures ultrasound may be used to image the bowel itself.

Three patterns of bowel have been recognized; the mucus (*Figure 26*), fluid and gas patterns. The first is the appearance of a 'target' representing a transverse section of a loop of bowel containing only mucus and air which produce a highly reflective core and the bowel wall which is responsible for the echo-poor halo. Visualization of this same bowel loop in long axis has an almost tramline appearance. Bowel containing fluid may give the appearance of a tubular structure or a cyst depending upon whether it is imaged along its long or short axis (*Figure 27*). This pattern may be further modified by the mucosal structure. Haustra of large bowel and valvulae conniventes of small bowel may be recognized, the latter providing a step-ladder appearance. The pattern of gas-containing bowel is well known for its bright reflections, shadowing and reverberation or 'ring down' artefact sometimes tapering in the form of a comet tail (*Figure 22a*).

Bowel is usually examined by ultrasound during the examination of other structures in the abdomen and pelvis although there are specific technical manouevres which allow the better visualization of bowel pathology. Real-time examination allows the demonstration of motility of bowel loops, the motion of bowel contents and the change in size and configuration of segments of bowel. Higher frequency transducers may be used to assess bowel close to the skin surface and particularly to evaluate abnormal bowel in superficial

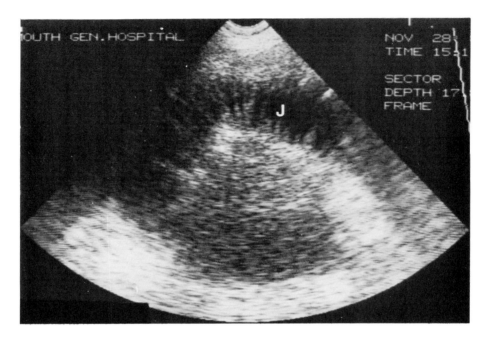

Figure 29. Scan in the mid abdomen demonstrating a single loop of jejenum (J). The jejenum is recognized by the transverse echogenic mucosal folds or valvulae conniventes.

locations, for example in hernias. Compression may be used not only to displace bowel loops to reveal deeply placed pathology but also to change the shape of bowel and validate diagnosis of bowel abnormalities.

9.2 Bowel pathology

The appearances of bowel in pathological conditions may be considered under the three main patterns described for normal bowel and these appearances are again dependent upon bowel contents. Abnormalities involving thickening of the bowel wall produce what is described as the atypical target or pseudo kidney pattern (*Figure 28*); this feature is seen in many different pathologies both inflammatory and neoplastic and these are summarized in *Table 1*.

Abnormalities of the fluid pattern are seen in bowel obstruction, particularly when this is a closed loop. In this situation air is excluded from the bowel and the plain abdominal X-ray may be unhelpful. Ultrasound may demonstrate dilated fluid-filled loops of bowel. Recognition of mucosal structures such as valvulae conniventes may enable not only the diagnosis

of obstruction to be made but the level of obstruction assessed (*Figure 29*).

Abnormalities of the nature and distribution of bowel gas are more difficult to evaluate. The pattern is used in differentiation of benign from malignant ascites because of tethering of gas-filled bowel loops. In certain types of malabsorption abnormal distribution of bowel gas produces a characteristic pattern of widespread ring down and comet tail artefact.

10. Summary

Evaluation of peritoneal and retroperitoneal structures with ultrasound requires a knowledge of surface and sectional anatomy. Careful attention to scan techniques with the use of multiple scan planes allows the demonstration of abdominal anatomy and pathology.

11. References

1. Bo,W.J., Meschen,I. and Krueger,W.A. (1980) *Basic Atlas of Cross Sectional Anatomy. A Clinical Approach.* W.B. Saunders, Philadelphia.

2. Wagner,M. and Lawson,T.L. (1982) *Segmental Anatomy. Applications to Clinical Medicine.* Collier MacMillan.

3. Schneck,C.D. (1983) The anatomical basis of abdomino pelvic sectional imaging. In *Ultrasound in Inflammatory Disease. Vol. 11 Clinics in Diagnostic Ultrasound.* Joseph,A.E.A. and Cosgrove,D.O. (eds), Churchill Livingstone, pp. 13−41.

4. Vogler,J.B., Helms,C.A. and Callen,D.W. (1986) *Normal Variants and Pitfalls in Imaging.* W.B. Saunders, Philadelphia.

5. Dubbins,P.A. and Goldberg,B.B. (1984) Abdominal vasculature. In *Abdominal Ultrasonography.* Goldberg,B.B. (ed.), John Wiley and Sons, 2nd edition, pp. 19−79.

6. Dubbins,P.A. (1986) Sound on, vision on. *Lancet,* **II,** 137−138.

7. Hricak,H., Taylor,K.J.W., Marich,K., Lue,T.S. and Burns,T. (1986) Doppler in urology. In *Genito Urinary Ultrasound. Vol. 18 Clinics in Diagnostic Ultrasound.* Hricak,H. (ed.), Churchill Livingstone, pp. 241−270.

8. Goldberg,B.B. and Pollack,H.M. (1984) Retroperitoneum. In *Abdominal Ultrasonography.* Goldberg,B.B. (ed.), John Wiley and Sons, 2nd edition, pp. 393−425.

9. Love,L., Meyers,M.A., Churchill,R.J., Reynes,C., Moncada,R. and Gibson,D. (1981) Computed tomography of extra peritoneal spaces. *Am. J. Radiol.,* **136,** 781−789.

10. Yeh,H.C. (1984) Adrenal sonography. In *Ultrasound in Breast and Endocrine Disease. Vol. 12 Clinics in Diagnostic Ultrasound.* Leopold,G.R. (ed.), Churchill Livingstone, pp. 101−130.

11. Marchal,G., Gelin,J., Verbecken,E., Baert,A. and Lauwerijans,S. (1986) High resolution real time sonography of the adrenal glands: a routine examination. *J.U.M.,* **5,** 65−68.

12. Dubbins,P.A. and Kurtz,A.B. (1984) Intra abdominal fluid collections. In *Abdominal Ultrasonography.* Goldberg,B.B. (ed.), John Wiley and Sons, 2nd edition, pp. 241−285.

13. Meyers,M.A. (1970) The spread and localisation of acute peritoneal effusion. *Radiology,* **95,** 547−554.

14. Gerzof,S.G., Spira,R. and Robbins,A.H. (1981) Percutaneous abscess drainage. *Seminars Roentgenol.,* **16,** 62−71.

15. Dubbins,P.A. (1988) Ultrasound evaluation of the gastro-intestinal tract. In *Gastrointestinal Ultrasonography. Clinics in Diagnostic Ultrasound.* Kurtz,A.B. and Goldberg,B.B. (eds), Churchill Livingstone, New York.

Chapter 8

Liver, spleen, pancreas and kidneys

Maria Chiara Bossi and David O.Cosgrove

1. Introduction

The application of ultrasound to the regions that are the subject of this chapter constitutes the core of the 'radiology' use of the technique: these are problems where conventional radiology methods have important limitations, many of which can be solved by ultrasound or by computerized tomography (CT) scanning. They form the major workload of ultrasound departments working alongside X-ray units − indeed, the two are often integrated as a diagnostic imaging unit for this reason.

2. Technique

In general, for these applications, high resolution scanners are preferred and a sector configuration is most useful because all of these structures are difficult to access, being obscured by bone (such as ribs) or by gas (in the gastrointestinal tract). The key to technical success is flexibility of approach, so that whatever windows are available can be maximally exploited. Thus, for the liver, spleen and pancreas, intercostal views in oblique and coronal planes are used to supplement the more conventional sagittal and transverse subcostal views. Shifting the patient's position is often most helpful: the left decubitus is invaluable for the biliary tree, while scanning with the patient standing is often helpful for the pancreas. For the most part preparation is not required, but fasting is essential for reliable interpretation of gall bladder images as is bladder filling for the pelvis.

The addition of pulsed Doppler to conventional imaging is proving to be a powerful tool for answering anatomical questions regarding the vascular nature of tubular structures. It can confidently be predicted that more sophisticated uses will also emerge.

3. Liver

The liver is the largest organ in the abdomen and is divided into lobes and segments by ligaments and vascular structures (1). The arterial and venous blood enters the liver through the porta hepatis, the artery lying anterior and medially. Also lying in the porta hepatis is the bile duct which lies antero-lateral to the portal vein. Blood leaves the liver through three main hepatic veins which empty directly into the upper inferior vena cava.

3.1 Diffuse disease

The most common and important liver diseases fall in this group. Ultrasound can be used to assess the liver size and its texture as well as changes in the absorption of ultrasound. Three main patterns are recognized. In fatty change the texture is abnormally fine with small, closely packed dots and loss of contrast between the liver parenchyma and the strongly reflective margins of the portal vein radicals. Ultrasound attenuation is increased in most cases. This is the so called 'bright liver' (2) (*Figure 1*). Fatty change is caused by a number of metabolic alterations, for example obesity and diabetes, as well as toxic insult such as alcohol abuse and ulcerative colitis.

The second pattern, classified as the 'dark liver' (*Figure 2*), is seen in acute inflammatory changes, especially viral hepatitis (3). The liver texture and attenuation are unremarkable, but the echo levels are low so that the echogenic peri-portal tissues are abnormally prominent.

Cirrhosis represents the third pattern (4) (*Figure 3*). Here the echo levels are increased, the texture is coarse with large, often confluent, dots. The attenuation is usually high. Large regenerative nodules may produce lobulation on the surface of the liver and in the late stage the liver is scarred and small with a relatively

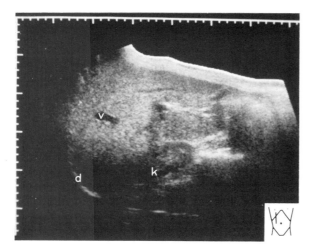

Figure 1. Bright liver pattern. The high intensity echoes and the fine pattern that are seen on this scan on a patient with fatty change. There is high attenuation to the sound beam so that the deeper parts of the liver are not well seen. d = diaphragm; k = kidney; v = hepatic vein.

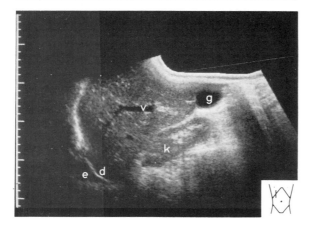

Figure 2. Dark liver pattern. In this patient with congestive cardiac failure, the echo intensity from the liver is reduced so that the renal parenchymal echoes are at about the same level. Note the pleural effusion. e = effusion; d = diaphragm; g = gall bladder; k = kidney; v = portal vein.

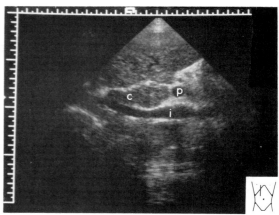

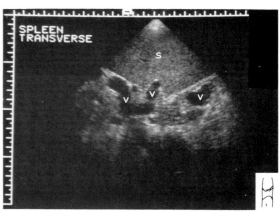

Figure 3. Cirrhosis. In this patient with cirrhosis the liver echo intensity is normal, but the texture is somewhat coarsened. The caudate lobe is prominent and has a lobulated contour because of nodular regeneration. There are prominent varices around the splenic hilum. c = caudate lobe; i = inferior vena cava; p = portal vein; s = spleen; v = varices.

prominent caudate lobe. In clinical practice it is difficult to establish a diagnosis of cirrhosis on the ultrasound alone and many cases have a normal appearance. The severity of the cirrhosis also correlates poorly with the ultrasound appearances. Associated findings may be more useful.

Ascites has the usual pattern of fluid spaces that follow the peritoneal contours; when marked, bowel loops are seen to float within it. In portal hypertension, the portal vein initially dilates (normally < 15 mm in diameter) and then the spleen enlarges due to congestion (5). Later, porto-systemic anastomotic channels open up producing oesophageal and splenic varices and recanalization of the umbilical vein (6).

The limitations of ultrasound in diffuse liver diseases are its inability to establish a precise histological

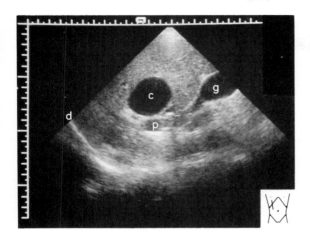

Figure 4. Hepatic cyst. A very well-defined, smooth walled, echo-free space is typical of a hepatic cyst. It shows distal enhancement manifested as a band of apparent increased intensity distal to the lesion. c = cyst; d = diaphragm; g = gall bladder; p = portal vein.

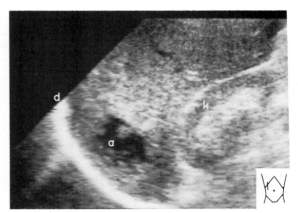

Figure 5. Liver abscess. A rather ill-defined lesion is seen in the posterior part of the liver in this patient with an amoebic abscess. It has irregular shaggy walls and low level echoes within the fluid. a = abscess; d = diaphragm; k = kidney.

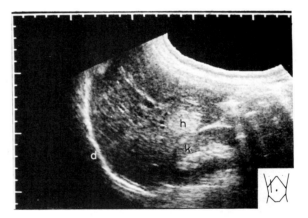

Figure 6. Haemangioma. A well-defined, rounded, echogenic lesion is seen in a subcapsular position in the right lobe of the liver. It was an incidental finding. The pattern suggests the haemangioma; if clinically critical, confirmation by dynamic CT or arteriography may be required. d = diaphragm; h = haemangioma; k = kidney.

diagnosis and the poor correlation between the portal pressure and the degree of portal vein distension. In practice, the role of ultrasound is restricted to confirming abnormalities, excluding focal disease and indicating the need for liver biopsy.

3.2 Focal benign lesions

Simple cysts are common and are readily detected on ultrasound as thin walled, echo-free, enhancing cavities (7) (*Figure 4*). Occasionally they are multi-loculated. Usually they are an incidental finding, although when large they can cause discomfort. Very large cysts distort the local anatomy so the site of origin may be difficult to assess. The presence of internal echoes suggests alternative diagnoses.

Hydatid cysts have a characteristic pattern with multiple daughter cysts partly, or completely, filling the main cyst (8). They are often calcific. However, when no daughter cysts are formed the pattern is indistinguishable from a simple cyst.

Abscesses are easily detected as fluid spaces with thick, shaggy walls, containing debris and sometimes intense internal echoes due to gas (9) (*Figure 5*). Before a fluid cavity is formed the pattern is subtle since it appears as an area of slightly reduced echo level due to the inflammatory process. Fine needle aspiration to provide a bacteriological diagnosis is invaluable in management and serial studies are useful to follow the evolution of the abscess.

Haemangiomata have a wide range of appearances: they may be solitary or multiple and may have high or low intensity echoes (10) (*Figure 6*). There are two typical patterns. Bright lesions with ill-defined margins, impossible to differentiate from liver metastases, are the commonest. Echo-free, lobulated areas, sometimes containing internal echoes and without distal enhancement, are also sometimes encountered. Haemangiomata are often detected incidentally and

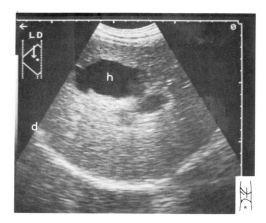

Figure 7. Hepatic trauma. Ill-defined echo-poor regions were found in the liver of this patient who suffered a road traffic accident. They represent haematomas. There is no sign of extra hepatic fluid which would indicate liver rupture. d = diaphragm; h = haematoma.

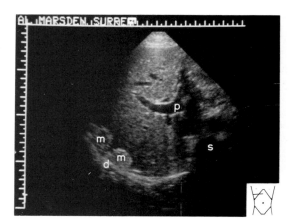

Figure 8. Liver metastases. Malignancy in the liver has a variety of appearances, a common one being focal echogenic lesions. They are usually rounded and fairly well defined. Echo-poor lesions are also common. These were secondaries from a carcinoma of colon. d = diaphragm; m = metastases; p = portal vein; s = spine.

fine needle aspiration biopsy is required for a definitive diagnosis.

3.3 Trauma

The typical pattern of liver trauma is of an echo-poor mass containing a variable amount of internal echoes; they are often completely echo-poor initially, but will become more echogenic as clot and fibrosis develop (11) (*Figure 7*). Ultrasound is useful in localizing and determining the extent of the lesion and in evaluating peritoneal bleeding which is seen as peritoneal fluid that is indistinguishable from ascites.

3.4 Hepatocellular carcinoma

Primary liver carcinoma is well visualized by ultrasound with an overall accuracy of approximately 90%. The echo patterns are variable with lesions of high, low or mixed echoes (12). Although the lesions are readily detected in either a normal or a cirrhotic liver the distinction between hepatoma and metastasis is impossible and a fine needle aspiration biopsy is required in case of doubt.

3.5 Metastases

Ultrasound can detect metastases down to a size of 1−2 cm with a 90% accuracy. They usually appear as solid, rounded or lobulated masses (13) (*Figure 8*). The echogenicity varies, with echo-poor lesions being the commonest and echogenic deposits particularly associated with gastrointestinal and urogenital primary

tumours. Cystic deposits are also encountered, especially from the ovary, stomach and pancreas (all mucin-secreting organs). Calcification and necrosis may occur. The liver is often enlarged and shows surface nodularity. The space-occupying lesions may displace the liver vessels and cause biliary tree obstruction. In some cases the liver involvement is widespread with extensive replacement so that no focal lesion is seen. The texture becomes diffusely irregular and biopsy is required to distinguish between tumour and inflammatory disease.

4. Biliary tree

The biliary tree consists of the gall bladder with its cystic duct, the hepatic ducts draining the liver and the common bile duct. The cystic duct is rarely visualized on ultrasound, but the main right and left hepatic ducts are seen at the level of the porta hepatis; they measure 1−2 mm in diameter as they join to form the main hepatic duct which is usually seen where it crosses anterior to the right portal vein. Here it measures 4 mm in internal diameter. Below the confluence of the cystic duct (the common bile duct), the calibre increases. The mid portion of the common bile duct lies posterior to the duodenum and so is usually obscured by gas, but the distal portion within the head of the pancreas is usually detectable where it measures up to 7 mm in diameter.

4.1 *Gall bladder*

The gall bladder is a thin walled structure with a pear shape which is sometimes altered by folding and septa. Its position is variable; although it usually lies in the gall bladder fossa on the visceral surface of the liver, it is sometimes attached by a long mesentery or even embedded within the liver. Gall bladder function may be assessed by its contraction after a fatty meal; the wall of the contracted gall bladder is 3−4 mm in thickness. Thickening is also seen in ascites where the wall becomes oedematous. In normal subjects the bile in the gall bladder is echo-free, but when non-functioning, for example in patients on parenteral nutrition, it may contain fine echogenic debris which sediments to a gravitationally dependent position.

Gall stones are readily detected as strongly reflected foci (*Figure 9*). They are mobile and absorb the ultrasound beam, thus casting an acoustic shadow (4). Ultrasound is highly sensitive in the detection of stones, but problems arise with very small stones which can be missed, with patients where the gall bladder neck is very tortuous and casts an acoustic shadow that mimics an impacted stone, or with an empty gall bladder that escapes detection.

In acute cholecystitis the gall bladder wall becomes thickened and there is tenderness that can be localized using the ultrasound probe to the region of the gall bladder itself ('ultrasound Murphy's sign') (15) (*Figure 10*). There are often associated gall stones, one of which has impacted in the neck. The gall bladder lumen may be echo-free or contain debris. There is usually an echo-poor halo around the gall bladder wall due to inflammatory oedema.

Empyema is a complication of acute cholecystitis, characterized by echoes due to pus within the gall bladder lumen and by pericholecystic collections due to perforation.

Tumours appear as soft tissue masses attached to the gall bladder wall; in the early stages carcinoma cannot be distinguished from small polyps. More advanced tumours penetrate into the gall bladder wall and invade the liver and porta hepatis often causing jaundice.

5. Jaundice

Dilated bile ducts are easily detected as fluid-filled, tubular structures lying anterior to the portal vein branches within the liver (16) (*Figure 11*). Normal

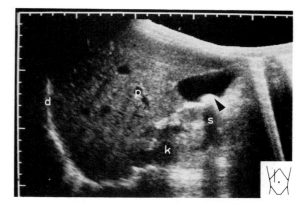

Figure 9. Gallstone. Intense reflectivity and marked distal shadowing are the typical features of gall stones. In this patient they were an incidental finding and the gall bladder is otherwise normal. d = diaphragm; k = kidney; p = portal vein; s = acoustic shadowing; arrowhead = gall stone.

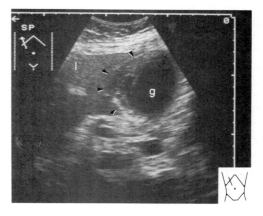

Figure 10. Acute cholecystitis. In the patient with right upper quadrant pain and tenderness the gall bladder wall is diffusely thickened and, in addition, there is a collection between the gall bladder and the liver (arrows). A peri-cholecystitic abscess complicating acute cholecystitis was the diagnosis. There were gall stones seen in other sections. g = gall bladder; l = liver.

intrahepatic bile ducts are not visualized, but when they dilate 'too many tubes' are seen in the periphery of the liver. A typical parallel channel pattern is produced by the dilated duct lying alongside the normal portal vein branch. The ducts form a stellate pattern near the porta hepatis and there is often enhancement of the echoes distal to the lumen. Overall, ultrasound has an accuracy of about 95% in distinguishing between duct obstruction and the medical causes for jaundice.

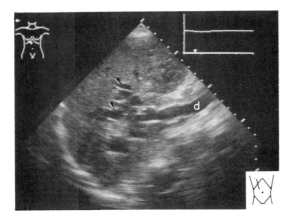

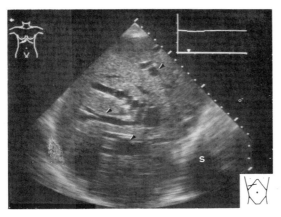

Figure 11. Obstructive jaundice. An increase in the number and size of the intrahepatic vessels in this jaundiced patient was due to dilatation of the intrahepatic biliary tree (arrows). The dilated common bile duct could be traced leaving the porta hepatis and lymphadenopathy was demonstrated at its lower end. The patient had lymphoma. d = common bile duct; s = spine.

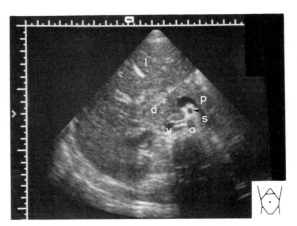

Figure 12. Normal pancreas. The curved shape of the pancreas overlying the splenic vein is shown in this transverse section. In this young person its echo intensity is about the same as that of the liver; in older subjects the pancreas is more reflective. a = aorta; d = duodenum; l = liver; p = pancreas; s = splenic vein; v = vena cava; arrowhead points to superior mesenteric artery.

6. Pancreas

Under optimal conditions the pancreas can be imaged in its entirety and with great detail so that ultrasound has become the method of choice for initial evaluation (17) (*Figure 12*). The appearances of the normal pancreas change with age because of progressive fibrosis and fatty infiltration so that, whereas in the child it returns a lower level of echoes than the liver, in the adult and elderly it is more intensely reflective.

6.1 *Acute pancreatitis*

Because of the paralytic ileus that is often associated with severe pancreatitis, visualization of the pancreas by ultrasound can be difficult or impossible. When adequate images can be obtained the gland appears diffusely enlarged and echo-poor with ill-defined margins (18) (*Figure 13*). The biliary tree needs to be examined carefully to detect gall stones since they are a major cause of pancreatic disease. The pseudocysts which commonly develop a few weeks after the initial attack of pancreatitis are easily evaluated by ultrasound (19) (*Figure 14*). They need to be distinguished from haemorrhagic pancreatitis where the shape of the contours of the pancreas are retained and from pancreatic abscesses which have an identical appearance and, therefore, need diagnostic aspiration for their distinction. Many pseudocysts disappear

If the lesion causing the obstruction is large enough (1−2 cm) to be resolved by ultrasound the level and cause of obstruction can also be defined. Large stones in the common bile duct, tumours in the head of pancreas and nodes in the porta hepatis are examples. On the other hand small impacted stones, ampullary carcinoma and bile duct strictures are difficult or impossible to detect.

Cholangiocarcinoma of the extra hepatic biliary tree has a non-specific appearance presenting with obstructive jaundice, the obstructing lesion being difficult to detect. Intra-hepatic cholangiocarcinoma appears as a mass in the periphery of the liver and cannot easily be distinguished from other tumours.

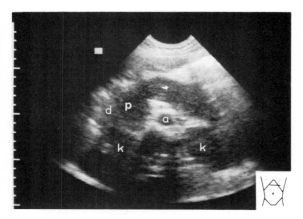

Figure 13. Acute pancreatitis. In this patient with severe upper abdominal pain the pancreas is enlarged and echo-poor. The prominent pancreatic duct (arrowhead) suggests the aetiology: a gall stone was impacted in the lower end of the common bile duct. a = aorta; d = duodenum; p = pancreas; k = kidney.

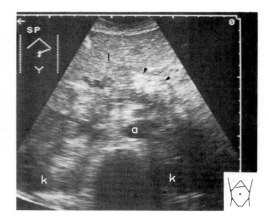

Figure 15. Chronic pancreatitis. High level, clumped, echoes are seen in the body of the pancreas in this patient with chronic calcific pancreatitis. a = aorta; k = kidney; l = liver.

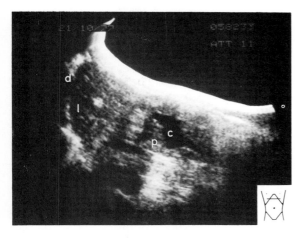

Figure 14. Pancreatic pseudocyst. Within the region of the enlarged pancreas a fluid region is seen representing pseudocyst formation. c = cyst; d = diaphragm; p = pancreas; l = liver.

spontaneously, but progressive enlargement is easily documented on ultrasound which, therefore, forms a useful guide to the surgeon.

6.2 Chronic pancreatitis

The common appearance of chronic pancreatitis is of an irregular pancreas with heterogeneous internal echoes; in advanced cases the pancreas shrinks and acquires calcification (20) (*Figure 15*). Duct dilatation

can sometimes be demonstrated. However, it is important to bear in mind that even in severe chronic pancreatitis the ultrasound appearances may be normal. Pseudocysts are a frequent association and sometimes there are regions of low level echoes due to superimposed focal pancreatitis. In these cases the differentiation from carcinoma is impossible and fine needle aspiration biopsy is required (21).

6.3 Carcinoma

Ultrasound can detect small focal textural abnormalities before there is enlargement of the pancreas. Carcinomas are usually echo-poor lesions, but the commonly associated haemorrhage produces a heterogeneous texture (22) (*Figure 16*). When the tumour is large or superficially placed it produces distortion of the contour of the gland with displacement of adjacent vessels and, if located in the head of the pancreas, obstruction of the pancreatic and common bile ducts. Focal pancreatitis is often associated with the tumour, either in the distal part of the gland (due to duct obstruction) or surrounding the mass. Unfortunately, by the time patients present with symptoms, the tumour is large so that the impact of ultrasound diagnosis is not to increase survival, but merely to expedite the diagnosis and staging (local and liver invasion) and obtaining biopsies in order to avoid unnecessary surgery. Where there is lymphadenopathy in the upper retroperitoneum, the possibility of lymphoma should be borne in mind; this can be established by biopsy with fine cutting needles to obtain a histological sample.

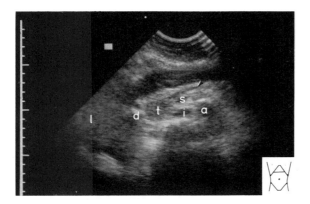

Figure 16. Carcinoma of the pancreas. An echo-poor mass is seen in the posterior part of the head of the pancreas. The pancreatic duct (arrowhead) is somewhat prominent and may be obstructed by the lesion. a = aorta; d = duodenum; i = inferior vena cava; l = liver; s = superior mesenteric artery; t = tumour.

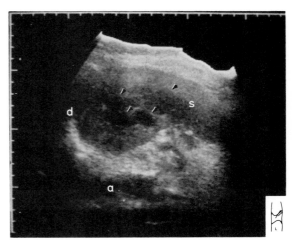

Figure 17. Splenic lymphoma. Several echo-poor foci were seen in the spleen in this patient with a non-Hodgkins lymphoma. Positive demonstration of focal lesions in the lymphoma is unequivocal evidence of involvement, but negative findings are unreliable. a = aorta; d = diaphragm; s = spleen.

Cystadenocarcinoma is a rare pancreatic tumour found in the tail of the pancreas and predominantly affecting middle-aged women. The ultrasound appearance may be similar to a cyst with a fluid collection containing internal septa and small calcifications.

Islet cell tumours may be functioning or non-functioning; the former are usually very small and, therefore, difficult to visualize even at surgery, though intra-operative ultrasound is promising in this application (23). The non-functioning tumours are often large and more easily detected on ultrasound.

Where major surgery is planned for pancreatic tumours, CT is usually also needed to assess the tumour extent and arteriography is often also required to evaluate vascularity.

7. Spleen

The subcostal location of the spleen means that it is more easily imaged using a sector probe with the patient rotated into the right decubitus position (24). It has a triangular shape with a rather variable size, the maximum longitudinal length being 14 cm.

7.1 Benign lesions

Cysts and abscesses in the spleen are relatively uncommon and their appearances are identical to those in the liver. Splenic infarction occurs in patients with endocarditis, pancreatitis, leukaemia, sickle cell anaemia and polyarteritis nodosa. The ultrasound features are of a localized, peripheral, wedge-shaped echo-poor lesion which becomes increasingly reflective as scarring develops (25). Benign tumours of the spleen (cavernous haemangioma, hamartoma, cystic lymphangioma) are rare and are seen as a non-specific inhomogeneous mass.

7.2 Malignant lesions

While primary tumours of the spleen (e.g. haemangiosarcoma) are rare, metastatic involvement is common in advanced malignancy, especially in melanoma, carcinoma of the breast, ovary, lung, prostate and stomach. On ultrasound intra-splenic masses are seen which may be echo-poor or echo-rich. Splenic involvement with lymphoma is common at presentation; focal lesions are seen as echo-poor masses, but ultrasound is not able to detect minute foci or diffuse involvement (26) (*Figure 17*).

7.3 Splenomegaly

The spleen is often affected by systemic disease, especially haematological and inflammatory disorders and in portal congestion. The spleen becomes enlarged to a moderate degree in acute conditions and on ultrasound is generally echo-poor, whereas in chronic processes it can be greatly enlarged and is usually echogenic.

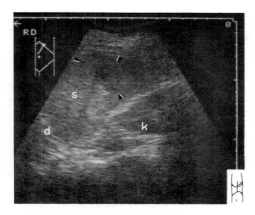

Figure 18. Splenic trauma. This young man suffered left lower rib fractures in a road traffic accident. An echo-poor lesion is seen within the spleen: it contains irregular echoes. The pattern is consistent with an intra-splenic haematoma. d = diaphragm; k = kidney; s = spleen.

7.4 Trauma

The appearances of splenic trauma are similar to those of the ruptured liver and the degree of echogenicity of the haematoma is influenced by its age (11) (*Figure 18*). The trend to more conservative management of splenic trauma makes ultrasound particularly useful, especially with portable real-time scanners, to provide a bedside follow-up service. Fresh haematomata are echo-poor while subcapsular collections and peri-splenic fluid are echo-free. With time, the haema-tomata become more fluid and eventually resolve without trace.

8. Lymphatic system

Because normal sized nodes are too small to be detected on ultrasound, only gross pathology can be evaluated. In the upper abdomen, the acoustic window (provided mainly by the liver) and the vascular landmarks allow nodes of 1 cm or so in size to be detected, but in the mid-abdomen and in obese or gassy patients even large masses of nodes escape detection. In the pelvis a full bladder is required to visualize the external iliac nodes.

Enlarged nodes appear as echo-poor, rounded or lobulated masses without distal acoustic enhancement (*Figure 19*). They lie along the line of the major blood vessels which are often distorted by the enlarged nodes. There are no specific features which allow for differ-

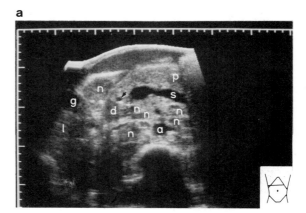

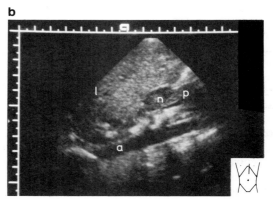

Figure 19. (a) Retroperitoneal lymphadenopathy. Numerous rounded echo-poor masses are seen in the peri-aortic and peri-pancreatic region in this patient with a lymphoma. a = aorta; d = duodenum; g = gall bladder; l = liver, n = nodes; p = pancreas; s = splenic vein; arrowhead points to the lower end of the common bile duct. **(b)** Pre-aortic lymphadenopathy. A prominent echo-poor lymph node is seen lying superior to the pancreas in this patient with a lymphoma. a = aorta; l = liver; n = node; p = pancreas.

entiation between malignant and non-malignant node enlargement.

9. Renal tract

Very simple uses of ultrasound are important in clinical practice (27). The renal length can be assessed (9 – 11 cm in the normal adult) and in children periodic reassessment can be used to check for proper growth. Localization of the lower pole of the kidney is invaluable in planning a renal biopsy: the success rate is improved and there are fewer complications. Ultrasound of the kidneys is important in evaluating obstruction of the ureter and for mass lesions within the renal parenchyma.

On ultrasound in longitudinal section the kidney has an oval shape, while in transverse section it is more circular. The relatively echo-poor parenchyma is arranged around the central sinus tissue which is intensely reflective because of the vessels and collecting system embedded in fat.

The sinus communicates with the retroperitoneum at the renal hilum. This is situated in the mid portion of the kidney and faces anterio-medially. Through it pass the renal vein and artery and the renal pelvis (from anterior to posterior). In the parenchyma the two components of the cortex and medulla can be made out, the medulla forming echo-poor pyramidal shapes that are completely enclosed by cortex except centrally where they contact the renal sinus. At this point the collecting ducts empty the formed urine into the calyces of the collecting system. The relatively echogenic cortex surrounds the pyramid.

Renal masses are often discerned on urography either incidentally or during the investigation of haematuria. Ultrasound can reliably distinguish the clean-walled echo-free space of a simple cyst from the sinister, solid, often irregular mass, of a renal carcinoma (28) (*Figures 20* and *21*). Most cases clearly fall into one of these categories, but occasionally a calcified cyst or one that has undergone haemorrhage will have an irregular wall or internal echoes. In this case, the lesion cannot be distinguished from a necrotic tumour on ultrasound and needle aspiration is required. Where a carcinoma is discovered, ultrasound can be used to evaluate local, nodal and hepatic spread as well as invasion into the renal vein and inferior vena cava.

The sensitivity of ultrasound to fluid spaces makes it the best way of confirming polycystic disease where both kidneys are replaced by numerous, closely packed cysts of 0.5−5 cm in diameter (29). It is useful in evaluating the families of this dominantly inherited disease, the cysts appearing in late childhood or early adult life. Complications of advanced cystic disease, such as infection or haemorrhage, are very difficult to assess because of the chaotic architecture that the cysts produce. Other rarer forms of cystic disease, such as infantile and dysplastic cysts, can also be evaluated by ultrasound, as may some other focal processes such as abscesses, haematomas due to trauma and benign tumours.

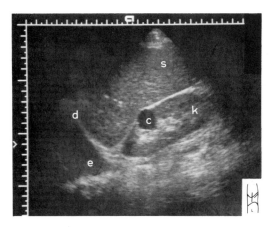

Figure 20. Renal cyst. This well-defined echo-free lesion in the kidney has the typical appearances of a renal cyst. The patient also had a left pleural effusion. c = cyst; d = diaphragm; e = effusion; k = kidney; s = spleen.

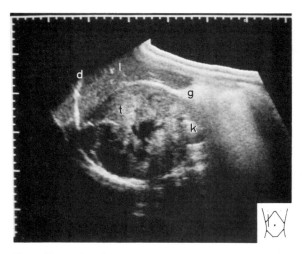

Figure 21. Renal carcinoma. A large heterogeneous mass replacing the upper pole of the kidney is typical of a renal carcinoma. The lower pole of the kidney can just be discerned. d = diaphragm; g = gas in bowel; k = kidney; l = liver; t = tumour.

The commonest diffuse change to be detected on ultrasound is an increase in the reflectivity of the renal cortex so that it returns higher level echoes than the adjacent liver (30). In extreme cases the cortex becomes as intense as the renal sinus. A wide range of disease may produce the 'bright' kidney, especially those where there is a cellular interstitial infiltrate (e.g. glomerulonephritis) or fibrosis (e.g. hypertension or

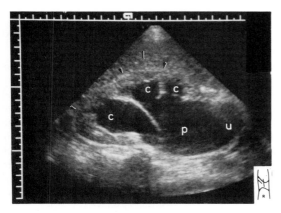

Figure 22. Hydronephrosis. The dilated calyces, renal pelvis and upper ureter can be made out in this patient with hydronephrosis. The outline of the thinned renal parenchyma is marked by arrows. c = calyces; l = liver; p = pelvis; u = ureter.

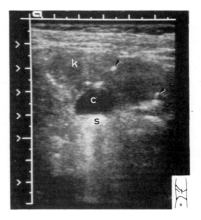

Figure 23. Renal stone. In this close up view of the kidney in a patient with renal stones the dilated calyx above the stone can be made out. There are several calcific foci also within the renal substance (arrows). c = calyx; k = kidney parenchyma; s = stone.

diabetes) and they cannot be distinguished. It is important to note also that in some cases there is no ultrasonic abnormality. The main role of ultrasound here is in determining renal size because acute diffuse disorders cause enlargement while in chronic disorders the kidneys shrink.

Ultrasound plays a major role in the evaluation of dilatation of the pelvi-calyceal system (31). Separation of the walls of the collecting system produces echo-free spaces in the renal sinus (*Figure 22*). From their position and shape the particular parts that are affected can be distinguished. The most important cause of collecting system dilatation is obstruction, but it is important to note that physiological dilatation occurs during an intense diuresis and that other pathology, such as reflux, may be responsible for dilatation so that its demonstration must not be equated with obstruction. The degree of dilatation is more closely related to the length of time that the obstruction has been present than to its severity, so that mild, long-standing obstruction may produce gross dilatation whereas with severe obstruction there may be little or no dilatation because renal shutdown may occur. Once dilatation has occurred it tends to persist even after relief of the obstruction. When only the pelvis is dilated a central fluid bag is seen in the renal sinus, but as the calyces also dilate these form a series of smaller fluid spaces around the pelvis and communicating with it. The upper ureter can usually be defined when it is dilated.

Renal stones can be detected on ultrasound as intensely reflective foci which cast an acoustic shadow if they are large enough to obstruct the ultrasound beam (32) (*Figure 23*). These features apply even if the stone is not calcified so that non-radio opaque stones are equally detectable on ultrasound. Stones that are too small to produce shadowing are very difficult to detect because they have the same echo properties as the surrounding sinus fat. Stones in the ureter are usually too small for ultrasonic detection.

In the lower urinary tract, ultrasound offers a simple and reliable method of assessing the bladder volume, for example to check for residual urine after micturition, to evaluate low abdominal mass that might be a distended bladder or to localize the bladder for a supra-pubic puncture (33). It can evaluate the bladder wall thickness in a qualitative way, demonstrating the trabeculation and diverticulum formation typical of bladder neck obstruction. Focal thickening suggests carcinoma: the lesion may be sessile and thus seen as a flat plaque or be predunculated, appearing as an ingrowth, sometimes with fronding (34) (*Figure 24*). Evaluation of the penetration of the bladder is more difficult though sometimes an extra-vesical mass can be documented. It is also possible to demonstrate some malformations of the urinary tract, especially wide dilatation of the lower ureters and the presence of ureteroceles.

Since the normal prostate can regularly be imaged at the bladder base, ultrasound can be used to assess enlargement (35). Carcinoma can only be diagnosed if extension beyond the prostatic capsule has occurred;

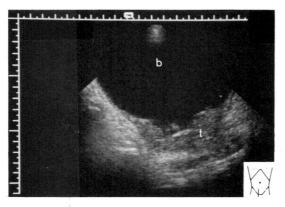

Figure 24. Carcinoma of the bladder. An extensive carcinoma of the bladder is seen to be destroying the bladder wall. b = bladder; t = tumour.

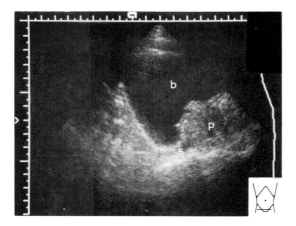

Figure 25. Carcinoma of the prostate. The prostate is enlarged with irregular echoes and an ill-defined margin. b = bladder; p = prostate; s = seminal vesicle.

a confined carcinoma is difficult to differentiate from benign prostatic hyperplasia (*Figure 25*). This can be improved by the use of endoprobe scanning via the rectum. The essential advantage is that a high frequency transducer can be used to give better resolution because the probe is placed close to the region of interest.

9.1 *Renal transplants*

Because of their superficial position, renal transplants are especially suitable for ultrasound imaging (36). The

applications can be divided into two categories:

(i) the demonstration of surgical complications; and
(ii) the evaluation of medical disorders.

Fluid collections around the transplant such as haematomas, urinomas and lymphoceles are easily detected on ultrasound. The appearances overlap and further evaluation is often required, but if conservative management is adopted serial ultrasound scans can be useful to evaluate progress. Doppler ultrasound is required to assess the patency of the arterial and venous anastomoses. There are usually no changes in acute tubular necrosis whereas in rejection the kidney enlarges rapidly and there may be either accentuation or effacement of the internal architecture. Where there is focal necrosis echo-poor patches appear. The Doppler signal may alter indicating an increase in the impedance in the renal arterial bed.

10. References

1. Niederau,C., Sonnenberg,A. and Mueller,J.E. (1983) Sonographic measurements of the normal liver, spleen, pancreas and portal vein. *Radiology*, **149**, 537–540.
2. Joseph,A.E., Dewbury,K.C. and McGuire,P.E. (1979) Ultrasound in the detection of chronic liver disease. *Br. J. Radiol.*, **52**, 184–186.
3. Kurtz,A.B., Rubin,C.S., Cooper,H.S., Nisenbaum,H.L. and Cole-Beuglet,C. (1980) Ultrasound findings in hepatitis. *Radiology*, **136**, 717–720.
4. Dewbury,K.C. and Clark,K.B. (1979) The accuracy of ultrasound in the detection of cirrhosis of the liver. *Br. J. Radiol.*, **52**, 945–950.
5. Kane,R.A. and Katz,S.G. (1982) The spectrum of sonographic findings in portal hypertension. *Radiology*, **142**, 453–458.
6. Saddekni,S., Hutchinson,D.E. and Cooperberg,P.L. (1982) The sonographically patent umbilical vein in portal hypertension. *Radiology*, **145**, 441–443.
7. Weaver,R.M., Goldstein,H.M. and Green,B. (1978) Grey scale ultrasound evaluation of hepatic cystic disease. *M.J. Runtginol.*, **130**, 849–852.
8. Hadidi,A. (1979) Ultrasound findings of hydatid disease in the liver. *J. Clin. Ultrasound*, **7**, 365–368.
9. Terrier,F., Becker,C.H.D. and Triller,J.K. (1983) Morphological aspects of hepatic abscesses at computed tomography and ultrasound. *Acta Radiol.*, **24**, 129–134.
10. Bree,L.M., Schwab,R.E. and Neiman,H.L. (1983) Solitary echogenic spot in the liver: is it diagnostic of a haemangioma. *Am. J. Runtganol.*, **140**, 41–45.
11. Viscomi,G.M., Gonzales,R., Taylor,K.J.W. and Crade,M. (1980) Ultrasonic evaluation of hepatic and splenic trauma. *Arch-Surg.*, **115**, 320–321.
12. Tanaka,S., Kitamura,T. and Imaaka,S. (1983) Hepatocellular carcinoma: sonographic and histologic correlation. *Am. J. Runtganol.*, **140**, 700–707.

13. Hillman,B.J., Smith,E.H. and Gammelgaard,J. (1979) Ultrasonographic pathological correlation of malignant hepatic masses. *Gastrointest. Radiol.*, **4**, 361−365.

14. Cooperberg,P.L. and Burhenne,H.J. (1980) Real time ultrasound: diagnostic technique of choice in calculus gall bladder disease. *New England J. Med.*, **302**, 1277−1279.

15. Ralls,P.W. (1985) Real time sonography in suspected acute cholecystitis. *Radiology*, **155**, 767−770.

16. Dewbury,K.C., Joseph,A.E., Mayes,S. and Murray,C. (1979) Ultrasound in the evaluation and diagnosis of jaundice. *Br. J. Radiol.*, **52**, 276−280.

17. Clark,L.R., Jaffe,M.H., Choyke,P.L., Grant,E.G. and Zeman,R.K. (1985) Pancreatic imaging. *Radiol. Clin. North. Am.*, **23**, 489−501.

18. Irving,H.C. and Mitchell,C.J. (1983) *The Pancreas. Clinics in Diagnostic Ultrasound. Vol. 11.* Churchill-Livingstone.

19. Hashimoto,B.E., Laing,F.C., Jeffrey,R.B. and Federle,M.P. (1984) Haemorrhagic pancreatic fluid collection examined by ultrasound. *Radiology*, **150**, 803−808.

20. Alpern,M.B., Sandler,M.A., Kellman,G.M. and Madrazo,B.L. (1985) Chronic pancreatitis: ultrasonic features. *Radiology*, **155**, 215−219.

21. Neff,C.C., Simeone,J.F., Wittenberg,J., Mueller,P.R. and Ferrucci,J.T. (1984) Inflammatory pancreatic masses − problems in differentiating focal pancreatitis from carcinoma. *Radiology*, **150**, 35−38.

22. White,M. and Wittenberg,J. (1984) Pancreatic neoplasia. *Semin. Ultrasound, CT and MR*, **5**, 401−413.

23. Rifkin,M.D. and Weiss,S.M .(1984) Intraoperative sonographic identification of non-palpable pancreatic masses. *J. Ultrasound Med.*, **3**, 409−411.

24. Mittelstaedt,C.A. (1981) Ultrasound of the spleen. *Semin. Ultrasound, CT and MR*, **3**, 233−240.

25. Yeh,H.C., Eacks,J. and Jurado,R.A. (1981) Ultrasonography of splenic infarction. *Mount Sinai J. Med.*, **48**, 5−10.

26. Kande,J.V. and Joyce,P.H. (1980) Evaluation of abdominal lymphoma by ultrasound. *Gastrointest. Radiol.*, **5**, 249−254.

27. Hricak,H. (ed.) (1986) *Genito-urinary Ultrasound. Clinics in Diagnostic Ultrasound. Vol. 18.* Churchill-Livingstone.

28. Joseph,N., Heiman,H.G. and Vogelzang,R.L. (1986) Renal masses. In *Genito-urinary Ultrasound. Clinics in Diagnostic Ultrasound. Vol. 18.* Hricak,H. (ed.), Churchill-Livingstone, pp. 135−160.

29. Pollack,H.M., Banner,M.P. and Harger,P.H. (1982) Grey scale ultrasound in the differentiation of cystic neoplasms from benign cysts. *Radiology*, **143**, 741−745.

30. Hricak,H., Crus,C. and Romanski,B. (1982) Renal parenchyma disease. *Radiology*, **144**, 141−147.

31. Ellenbogen,E., Scheibel,F. and Talner,B.B. (1978) Sensitivity of grey scale ultrasound in urinary tract obstruction. *Am. J. Runtganol.*, **131**, 731−735.

32. Erwin,B., Carroll,B. and Summer,F. (1982) Renal colic: the role of ultrasound. *Radiology*, **152**, 147−150.

33. Resnick,M.I., Willard,J.N. and Boyce,W.H. (1977) Ultrasound of the bladder and prostate. *J. Urol.*, **117**, 444−446.

34. Abu Yousef,M., Narayana,S. and Franken,K. (1984) Urinary bladder tumours studied by cystostenography. *Radiology*, **153**, 223−226.

35. Rifkin,M.D. and Kurtz,A.B. (1986) Prostatic ultrasound. In *Genito-urinary Ultrasound. Clinics in Diagnostic Ultrasound. Vol. 18.* Hricak,H. (ed.), Churchill-Livingstone, pp. 195−218.

36. Hricak,H. and Hoddick,P. (1986) Ultrasound in renal transplants. In *Genito-urinary Ultrasound. Clinics in Diagnostic Ultrasound. Vol. 18.* Hricak,H. (ed.), Churchill-Livingstone, pp. 161−180.

11. Further reading

Sarti,D.A. (1987) *Diagnostic Ultrasound*. 2nd edition (Year Book).

Goldberg,B.B. (ed.) (1985) *Abdominal Grey Scale Ultrasonography*. Wiley.

Cosgrove,D.O. and McCready,V.R. (1982) *Ultrasound Imaging: Liver, Spleen, Pancreas*. Wiley.

Netter,F.H. (1964) *The Ciba Collection of Medical Illustrations. Vol. 3/III. Digestive System; Vol. 6. Kidneys, Ureter and Urinary Bladder*. Ciba-Geigy.

Chapter 9

Echocardiography

J.C.Rodger

1. Introduction

This chapter presumes a basic knowledge of the technical aspects of conventional and Doppler echocardiography. It is intended for beginners who are not cardiologists and aims to help them to 'problem-orientate' their approach to the echocardiographic examination.

Echocardiography has a place in the investigation of all forms of heart disease. There are thus many reasons for requesting an echocardiographic examination. Only a few of the more common are dealt with here and the emphasis is on adult cardiology.

2. Mitral murmur—cause, haemodynamic significance?

Remember that although the murmur has been judged to originate from the mitral valve, its source may, in fact, be elsewhere, for example the tricuspid or the aortic valve; be ready to redirect your examination accordingly.

First, using the parasternal long axis view and an M-mode through the leaflets, decide whether or not the mitral valve is rheumatic. Anterior diastolic motion of the posterior leaflet is pathognomonic of rheumatic involvement (*Figure 8*). For further investigation of the rheumatic valve see Section 2.2.

2.1 Non-rheumatic causes of a mitral murmur—differential diagnosis and assessment

The disorders which should be considered are floppy mitral valve, calcific degeneration of the mitral ring and functional mitral regurgitation: all are common causes of a mitral systolic murmur. In the context of acute myocardial infarction (see Section 7) a flail valve

resulting from papillary muscle necrosis is an additional possibility. All are well displayed by cross-sectional imaging in the parasternal long axis view and this should always be the starting point.

2.1.1 *The floppy mitral valve*

This is the result of so-called myxomatous degeneration of the leaflets and their chordae. It is recognized by its billowing motion, the apparent thickening of its

a

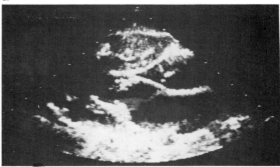

b

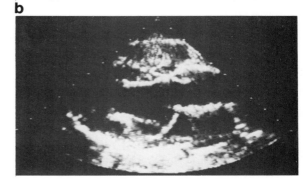

Figure 1. Parasternal long axis images of floppy mitral valve. In diastole (**a**), note the voluminous leaflets. In late systole (**b**), both leaflets arch into the left atrium; the configuration of the chordae is in keeping with laxity of the subvalvar apparatus.

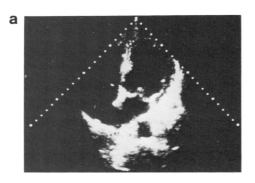

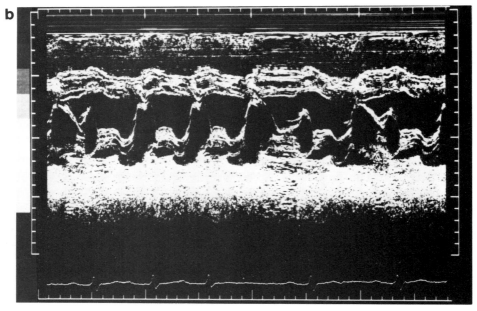

Figure 2. (a) Apical four-chamber image of floppy mitral valve (late systole) showing arching of the posterior leaflet into the left atrium. (b) M-mode record from same patient showing late systolic posterior motion of the leaflets.

voluminous cusps and by the laxity of its subvalvar apparatus (*Figure 1*). With a myxomatous valve, significant systolic prolapse is almost invariable. It is identified as systolic arching of the leaflets above the plane of the atrio−ventricular ring; when looking for it, it is advisable to use both the parasternal long axis and apical four-chamber views (*Figures 1* and *2a*). Remember that a normal valve can exhibit a minor degree of prolapse and that trivial mitral regurgitation is a normal Doppler finding. It is the combination of myxomatous degeneration and significant prolapse that characterizes the floppy mitral. Although a floppy mitral can be diagnosed on cross-sectional images, the diagnosis (particularly of significant prolapse) is more reliably made by M-mode echocardiography. Ideally the M-mode should be directed through the tip of the open anterior leaflet towards the posterior wall of the left atrium, just above the atrio−ventricular junction. Sometimes it is easier to select an appropriate M-mode from short axis scans just below the level of the aortic root than from parasternal long axis views. The M-mode features (*Figure 2b*) are increased diastolic excursion, additional echoes due to folding of the leaflets and holosystolic sagging or late systolic

Figure 3. M-mode record showing flail posterior mitral leaflet; its motion at the onset of diastole is paradoxical and there is coarse diastolic fluttering.

posterior motion of the leaflets reflecting significant prolapse. A floppy mitral can become acutely regurgitant due to chordal rupture; the chaotic motion of the flail leaflet is best demonstrated on M-mode records (*Figure 3*). When the mitral is classically floppy look for myxomatous degeneration of the remaining valves; this is rarely of haemodynamic consequence.

2.1.2 Degenerative calcification of the mitral ring

In this situation the leaflets are unaffected and the calcification is best seen in the parasternal long axis view. It is a common cause of mitral regurgitation in the elderly; associated calcification of the aortic valve is common and calcification may extend to the interventricular septum (*Figure 4*).

2.1.3 Functional mitral regurgitation

This is usually caused by left ventricular dilatation and very rarely by hypertrophic cardiomyopathy (see Section 10); its haemodynamic significance is subsidiary to that of its cause and is not usefully assessed separately.

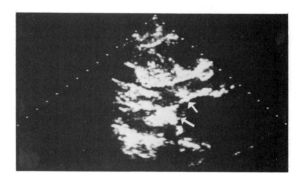

Figure 4. Parasternal long axis image from an elderly patient with left ventricular hypertrophy and dense calcification of the mitral ring (arrows). Calcification also involves the upper part of the interventricular septum.

Although obviously it can provide information about left atrial and left ventricular size, conventional echocardiography is of little value in *assessing the severity of mitral regurgitation*. This is best assessed by Doppler and, if simultaneous imaging is feasible, the method of choice is to map how far the regurgitant flow extends into the left atrium. It is worth noting,

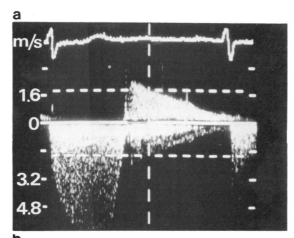

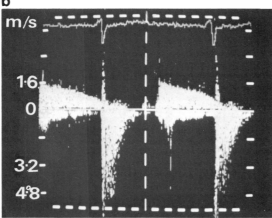

Figure 5. (a) CW Doppler (from the apex) showing mitral stenosis and regurgitation; peak velocity diastolic flow 2.3 m s^{-1}, maximum diastolic gradient 23 mmHg, pressure half-time 130 ms. The mitral regurgitation is not severe. (b) CW Doppler (from the apex) from a severely regurgitant mitral valve; the velocity of the regurgitant jet decreases in later systole.

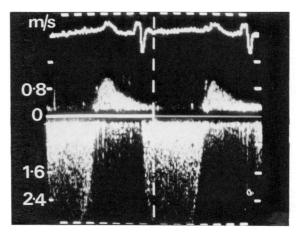

Figure 6. CW Doppler showing tricuspid regurgitation. The peak velocity of the regurgitant jet is 2.96 m s^{-1}. The right ventricular−right atrial systolic gradient is thus 35 mmHg. Assessing the right atrial systolic pressure as 5 mmHg, the pulmonary artery systolic pressure is 40 mmHg.

2.2 Assessment of rheumatic mitral valve

There are two basic considerations. Firstly, is surgical treatment necessary? Secondly, if surgery is indicated, is the valve suitable for an operation other than replacement? For economy of effort, start with a Doppler study. If this indicates a haemodynamically insignificant lesion, the rest of the echocardiographic examination can be confined to a search for left atrial thrombus.

In mitral stenosis the peak velocity of forward flow (normally around 1 m s^{-1}) is increased, and velocity decreases only slowly throughout diastole (*Figure 5a*). It is generally accepted that the pressure half-time (normal < 60 m s^{-1}) is a reliable index of stenosis. It increases as the valve area decreases; values in excess of 220 m s^{-1} are associated with a valve area of less than 1 cm^2 and are thus surgically important. The quantification of mitral regurgitation is dealt with above (see note under 2.1.3). The Doppler study should include an assessment of pulmonary artery pressure. This is conveniently calculated from the tricuspid regurgitant jet (tricuspid regurgitation is a normal finding). Thus, the peak systolic gradient between the right ventricle and right atrium can be calculated from the peak velocity of regurgitant flow. Addition of right atrial pressure (estimated from the

however, that without colour flow imaging (not yet widely available), the mapping process can be time consuming. The signs of severe regurgitation that can be obtained with non-imaging Doppler are increased intensity of the regurgitant jet (compared with the intensity of the forward flow signal), decreased velocity (*Figure 5b*) of regurgitant flow in late systole (unreliable if there is left ventricular dysfunction) and increased velocity of forward flow in early diastole.

a

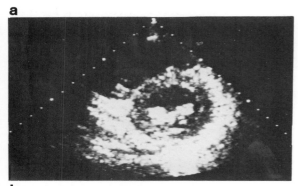

b

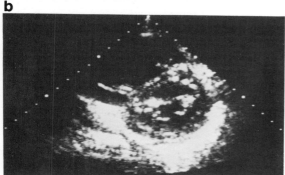

Figure 7. Parasternal short axis images of rheumatic mitral valve in (**a**) systole and (**b**) diastole. Calcification predominantly involves the anterior leaflet and is heaviest close to the commissures.

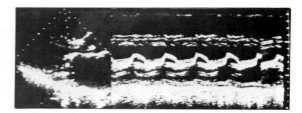

Figure 8. Parasternal long axis image (diastole) and M-mode record of rheumatic mitral valve. Both leaflets are calcified and the posterior leaflet moves anteriorly in diastole.

If operation is necessary, look in detail at valve structure: the information obtained can be of value to a surgeon considering open valvotomy or reconstruction of the valve as alternatives to replacement. Examine the subvalvar apparatus in both the parasternal and apical long axis views; look for thickening, shortening and fusion of the chordae tendineae and for fusion of the papillary muscles with the leaflets. Identify calcification of the leaflets and note its distribution [short axis scans are best for this (*Figure 7*)]. Assess whether the restricted mobility of the anterior leaflet is generalized or confined to the commissures. Measure left atrial size; this is most conveniently done by measuring the left atrial dimension (normal <4 cm) on an M-mode record derived from the parasternal long axis view. Look for rheumatic involvement of the aortic valve and assess its severity. Make an assessment of left ventricular function; remember that the patient may have co-existing coronary heart disease.

3. Aortic murmur—cause, haemodynamic significance?

Remember that although the murmur has been judged to originate from the aortic valve, its source may be elsewhere: mitral and, less often, pulmonary murmurs can cause confusion.

Start with cross-sectional echocardiography and a short axis view of the aortic root; angulate the probe for optimal display of the leaflets (*Figure 9*). Note that non-calcified bicuspid valves are rarely seen in general clinical practice and that, in the absence of heavy calcification, failure to identify three cusps usually reflects technical inadequacies. Look for

jugular venous pulse) to this gradient provides an estimate of right ventricular systolic pressure and, assuming that there is no obstruction to right ventricular outflow, this in turn reflects pulmonary artery systolic pressure (*Figure 6*).

Without Doppler, assessment of the severity of rheumatic mitral regurgitation is impossible. However, some indication of the severity of mitral stenosis can be obtained by visual assessment of orifice size on short axis scans (*Figure 7*) and by measurement of cusp separation on an M-mode derived from them (measurements of mitral valve area are inevitably inaccurate and are not worth attempting). Further, the presence of right ventricular hypertrophy and dilatation is indirect evidence of pulmonary hypertension.

Whatever the conclusions regarding the severity of the lesion, it is essential to look for left atrial thrombus and for co-existing rheumatic involvement of the aortic valve (see Section 4 and *Figure 8*).

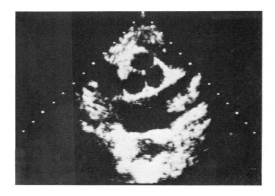

Figure 9. Parasternal short axis image (diastole) of normal tricuspid aortic valve.

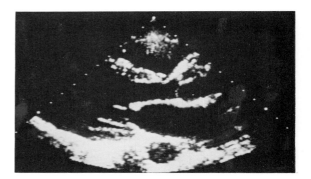

Figure 11. Parasternal long axis image (diastole) showing prolapse of the non-coronary cusp of the aortic valve.

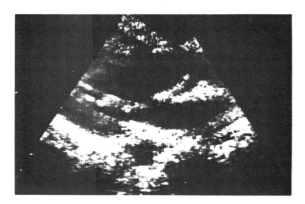

Figure 10. Parasternal short axis image showing systolic doming of congenitally stenotic aortic valve.

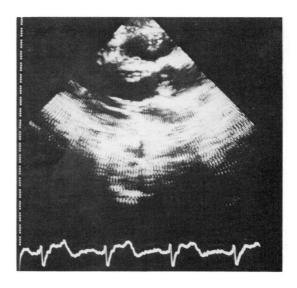

Figure 12. Parasternal long axis image (systole) showing bright echoes from the voluminous cusps of a floppy aortic valve (cusp mobility was unrestricted).

calcification [bright echoes and restricted cusp mobility (*Figure 14*)]. For further investigation of the significantly calcified aortic valve see Section 4.

3.1 The non-calcified aortic valve— differential diagnosis and assessment

Although the disorders that need to be considered can all be identified by conventional echocardiography, it is useful to start with a Doppler study. If there is evidence of stenosis (see Section 4), look for a non-calcified rheumatic or congenitally stenotic valve. The first is invariably associated with rheumatic involvement of the mitral valve. The second is characterized by systolic doming of the leaflets within the aortic root (*Figure 10*) and can be diagnosed only from para-

sternal long axis images (M-modes are unhelpful). If aortic regurgitation is present (see Section 4), look in the parasternal long axis view for dilatation of the aortic root with or without aortic valve prolapse and for destruction of the valve by endocarditis (see Section 6). Aortic valve prolapse (*Figure 11*) is recognized as diastolic sagging of the cusps below their points of attachment. Assessment can be difficult because, as with mitral prolapse, a trivial degree of aortic valve prolapse is not abnormal. Aortic valve prolapse can

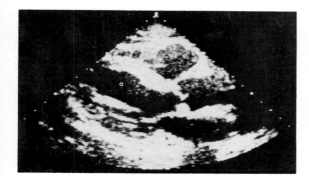

Figure 13. Parasternal long axis image (early systole) showing calcified aortic valve. Calcification particularly affects the non-coronary cusp which was immobile.

involve any or all of the cusps; it occurs with bicuspid valves, myxomatous valves and perimembranous ventricular septal defects. Myxomatous degeneration of the aortic valves should be considered when abnormally bright echoes are recorded from normally mobile leaflets (*Figure 12*) and particularly in patients who have a classically floppy mitral valve.

4. Assessment of the calcified aortic valve

If surgery is indicated, there is no alternative to valve replacement. The nature of the valve underlying the calcification (normal, bicuspid, rheumatic, etc.) is thus irrelevant, as are details of the distribution of the calcium (*Figure 13*). The only important consideration is the haemodynamic significance of the lesion and this can be reliably determined by Doppler but not by conventional echocardiography.

For Doppler quantification of aortic stenosis (*Figure 14*), use the continuous wave (CW) mode and, to ensure that peak velocity is identified, record from apical, right parasternal and suprasternal probe positions. Doppler quantification of aortic regurgitation is best done by mapping the regurgitant flow within the left ventricle; this requires simultaneous imaging (preferably colour flow). The signs of severe regurgitation that can be obtained without imaging are a high intensity of the regurgitant jet (compared with the intensity of the forward flow signal) and an early diastolic decrease in its velocity (*Figure 14a*).

With conventional echocardiography, the presence

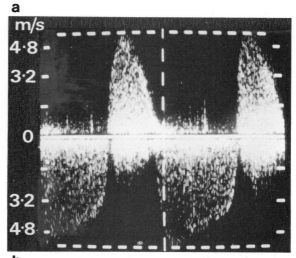

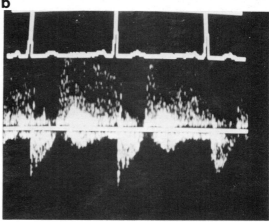

Figure 14. (a) CW Doppler (from the right sternal edge) showing aortic stenosis and regurgitation. Peak velocity systolic flow 5.3 m s^{-1}, maximum systolic gradient 113 mmHg. There is a moderate decrease in the velocity of the regurgitant flow during diastole; this is usual and not evidence of severe regurgitation. (b) CW Doppler record (from the apex) of a floppy aortic valve; there is aortic regurgitation but no significant systolic gradient.

of aortic regurgitation can be inferred if there is diastolic oscillation of the mitral apparatus. Acute severe regurgitation can be diagnosed by its effect on mitral valve closure (see Section 6). With this exception, aortic regurgitation cannot be reliably quantified although obviously, provided that other possible causes (for example co-existing coronary heart disease) can be excluded, left ventricular dilatation is evidence that it is severe.

a

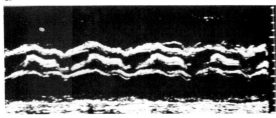

b

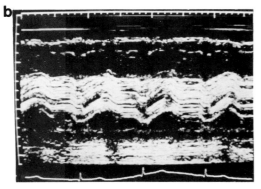

Figure 15. M-mode records. (**a**) From a modestly calcified aortic valve; mobility of the right coronary cusp is normal and that of a non-coronary cusp is only slightly restricted. (**b**) From a heavily calcified aortic valve; there is little evidence of cusp separation and cusp and root echoes move together.

Conventional echocardiography has more to offer in the assessment of aortic stenosis. Thus, visual assessment of orifice size and/or measurement of cusp separation on M-modes derived from the parasternal long axis view (normal 20 mm) will distinguish between trivially and grossly stenotic valves (*Figure 15*), and the presence of left ventricular hypertrophy (provided systemic hypertension is absent) is additional evidence of severe stenosis.

5. Prosthetic valve—malfunction?

Ideally all prosthetic valves should be fully assessed echocardiographically within a month of operation. Without this baseline information for comparison, the echocardiographic diagnosis of malfunction can be difficult.

5.1 Mechanical prostheses

Conventional echocardiography has a limited role (*Figure 16*). Thus a rocking motion of the prosthesis

is reliable evidence of valve dehiscence and it is occasionally possible to identify thrombus formation and abnormal disc motion on cross-sectional images. M-mode records can be confusing: reverberations from the disc can be confused with thrombus and, although the amplitude of disc motion and its opening and closing velocities decrease with thrombus formation, no normal range for these measurements has been established.

In the assessment of a mechanical prosthesis, the emphasis should be on Doppler (although preferable, imaging Doppler is not essential).

5.1.1 Aortic prostheses

Measure the peak velocity of systolic flow and calculate the gradient across the prosthesis. There is normally a gradient across all prosthetic valves; it varies with the type and the size of the prosthesis and with left ventricular function (a visual assessment of function and measurements of cavity size should always be included in the assessment). The gradient across an aortic Bjork−Shiley prosthesis is typically around 20 mmHg; it has been suggested that values in excess of 40 mmHg should be regarded as abnormal.

Look for aortic regurgitation and quantify it by mapping regurgitant flow within the left ventricle. Doppler will detect mild regurgitation in up to 50% of normally functioning mechanical prostheses; more severe regurgitation should be regarded as evidence of malfunction.

5.1.2 Mitral prostheses

Measure the peak velocity of forward flow and calculate the peak gradient. A typical peak gradient for a mitral Bjork−Shiley prosthesis is around 10 mmHg (*Figure 17a*). It has been suggested that peak gradients in excess of 22 mmHg should be regarded as abnormal. Calculate the pressure half-time. This is typically around 100 ms (*Figure 17a*) but values up to 200 ms are not definitely abnormal. Look for mitral regurgitation and assess its severity. Mild mitral regurgitation occurs in about 10% of normally functioning prostheses; severe regurgitation should be regarded as evidence of malfunction.

5.2 Tissue prostheses

Both cross-sectional and M-mode echocardiography are of value. Look for thickening of the leaflet (these are normally less echo-dense than the struts; *Figure*

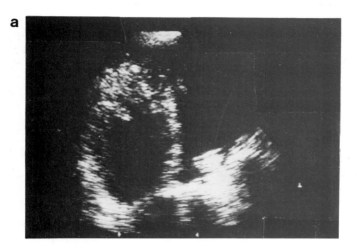

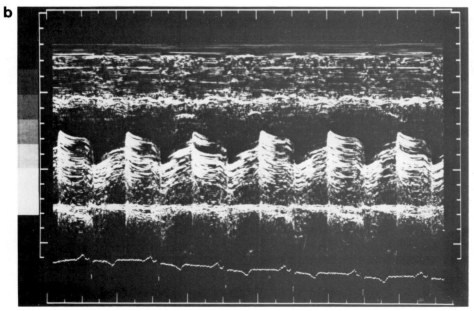

Figure 16. Bjork−Shiley prosthesis in mitral position. (**a**) Apical four-chamber image. (**b**) M-mode record (see text), from a different patient.

18), for leaflet prolapse and abnormal valve motion on cross-sectional scans, and use M-mode echocardiography to look for chaotic leaflet motion suggesting that there is a tear and that part of the leaflet is thus flail. The Doppler assessment is as for mechanical prostheses (*Figure 17b*). Typical values for peak gradients across tissue prostheses in the aortic and mitral positions are similar to those recorded across mechanical prostheses. The incidence of mild

regurgitation in normally functioning tissue prostheses is possibly slightly less than in their mechanical counterparts.

6. Infective endocarditis?

The echocardiographic diagnosis depends on the identification of vegetations. These appear on cross-

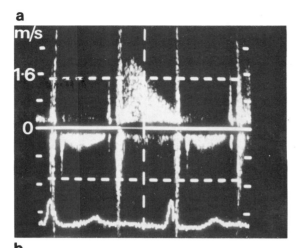

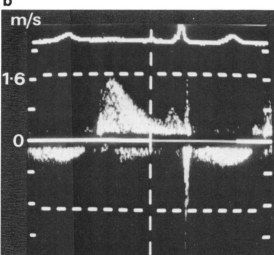

Figure 17. CW Doppler (from apex). (**a**) Bjork−Shiley prosthesis in the mitral position; peak velocity diastolic flow 1.96 m s^{-1}, maximum diastolic gradient 15 mmHg, pressure half-time 140 m s. (**b**) Carpentier Edwards prosthesis in mitral position; peak velocity diastolic flow 1.6 m s^{-1}, maximum diastolic gradient 11 mmHg.

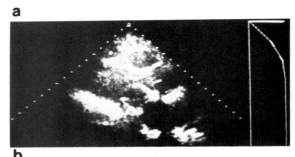

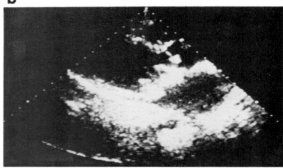

Figure 18. Parasternal long axis images. (**a**) Normal tissue prosthesis in mitral position showing thin leaflet echoes. (**b**) Abnormal tissue prosthesis (previous endocarditis) showing thick cusp echoes.

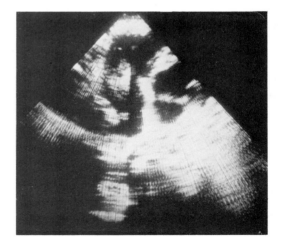

Figure 19. Apical four-chamber image showing vegetations on the tricuspid valve of a heroin addict.

sectional scans as mobile, rounded or elongated echo-dense masses (*Figure 19*) and on M-mode records as characteristically 'shaggy' echoes. Vegetations most commonly occur in valves (natural or prosthetic) but may also be found on the endocardium surrounding a septal defect. Note that as vegetations can only be visualized if they are larger than 3 mm, failure to identify them does not exclude the diagnosis of infective endocarditis and that vegetations of active and previous endocarditis cannot be differentiated echocardiographically.

The basic aims of the echocardiographic examination should be first to look for vegetations and then, if these are present, to look for evidence of valve disruption and haemodynamic upset. All the valves should be examined, but pay particular attention to the right heart valves in intravenous drug abusers, to prosthetic valves and to valves previously known to be abnormal. There may be abscess formation particularly around prosthetic and aortic valves but this is difficult to visualize. Look for aortic cusp prolapse and flail mitral or tricuspid valve leaflets. Use Doppler to identify and quantify regurgitation, make a visual assessment of ventricular function and use M-mode echocardiography to measure cardiac dimensions; this information is important in assessing progress. Use M-mode too, to look for premature closure of the mitral valve which is an important sign of acute gross aortic regurgitation.

7. Acute myocardial infarction

Be careful to confine the use of echocardiography to what it does best. Echocardiography does not reliably determine the site or size of an infarct; this is best done by nuclear imaging using an infarct avid tracer. Echocardiographic estimates of ejection fraction are inaccurate and measurement of ejection fraction by nuclear angiography is the method of choice for serial assessment of left ventricular performance following myocardial infarction. Echocardiography provides more information about wall motion than does any other technique. However, visual assessment of all but gross abnormalities is unreliable and attempts to monitor the effects of interventions, such as thrombolysis, on wall motion are not advised.

7.1 Left ventricular thrombus?

Thrombi (*Figure 20*) are usually associated with acute anterior infarcts. They may be apical or less often septal and are invariably associated with hypokinesis of the underlying myocardium. Fresh clot usually protrudes well into the left ventricular cavity; it can be frond-like and is often ominously mobile. Organized thrombi can be less obvious and particularly in the case of small apical clots, differentiation from trabeculae and artefact can be difficult.

7.2 Murmur—cause?

Usually the murmur will be long-standing and unrelated to the infarct or it will be due to functional

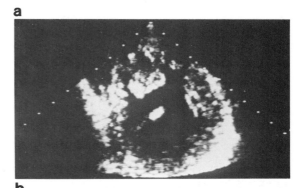

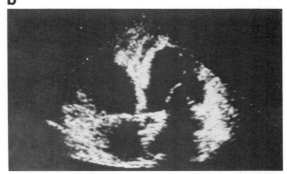

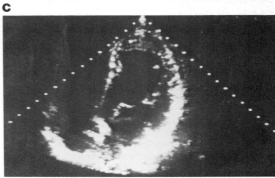

Figure 20. Apical four-chamber images. (**a**) Large, fresh (mobile) apical thrombus. (**b**) Organized apical thrombus extending onto the intraventricular septum. (**c**) Organized thrombus within apical aneurysm.

mitral regurgitation. Rarely the cause will be ventricular septal rupture or papillary muscle rupture, both of which produce a major haemodynamic disturbance. To differentiate the two, use Doppler (simultaneous imaging is unnecessary). With papillary muscle rupture, there is evidence of severe mitral regurgitation (see Section 2) and with a ventricular septal defect

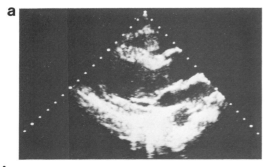

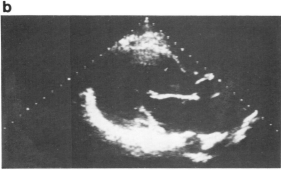

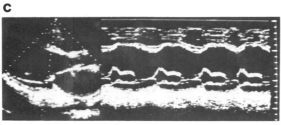

Figure 21. Parasternal long axis images. (a) Normal left ventricle. (b) Grossly dilated left ventricle with thin septum from patient with previous resection of LV aneurysm. (c) Dilated left ventricle with diminished posterior wall motion and a compensatory increase in septal motion (see M-mode).

(see also Section 12), a high velocity jet (directed towards the transducer) is recorded from the sternal edge. The flail mitral valve resulting from papillary muscle rupture can be identified in parasternal long axis images and on M-modes derived from them; it is often not possible to image the site of septal rupture.

7.3 Left ventricular aneurysm?

Although most aneurysms are apical, posterior wall aneurysms do occur. Use cross-sectional echocardiography in the apical four-chamber and the parasternal

long axis views. Look for dilatation confined to the apex and clearly demarcated from the rest of the ventricle (*Figure 20c*). The dilated area may exhibit systolic expansion or it may be akinetic; both are consistent with a diagnosis of aneurysm. The examination should be completed with a search for thrombus within the aneurysm and an assessment of the function of the remainder of the ventricle.

7.4 Persisting failure—cause?

The causes that should be considered are left ventricular aneurysm, global left ventricular dysfunction, major segmental wall motion abnormalities (*Figure 21*), ventricular septal rupture and papillary muscle rupture. All can be diagnosed by echocardiography. Minor segmental wall motion abnormalities are rarely relevant in this particular context and time should not be spent looking for them.

8. Cerebrovascular accident—cause?

There is an increasing emphasis on the investigation of strokes, particularly of transient ischaemic episodes. Conventional echocardiography can identify potential sources of embolism (see below). With Doppler, the investigation can be extended to the identification of stenotic lesions in the extracranial arteries; simultaneous imaging is necessary for precise anatomical information.

Cardiac sources of thromboembolism are the left ventricle (in myocardial infarction, left ventricular aneurysm and dilated cardiomyopathy), the left atrium (in rheumatic mitral valve disease), floppy valves and thrombosed prostheses. Non-thrombotic embolism may occur in infective endocarditis (vegetations and portions of cusp tissue), left atrial myxoma and calcific aortic valve disease. These disorders are all discussed elsewhere in this chapter and all should be considered when you investigate the cause of a stroke.

9. Assessment of left ventricular function

Use cross-sectional echocardiography to look at the size and shape of the left ventricular cavity and to make a visual assessment of wall motion and wall thickness. Use parasternal long axis images to look at the septum and posterior wall, apical four-chamber images to look

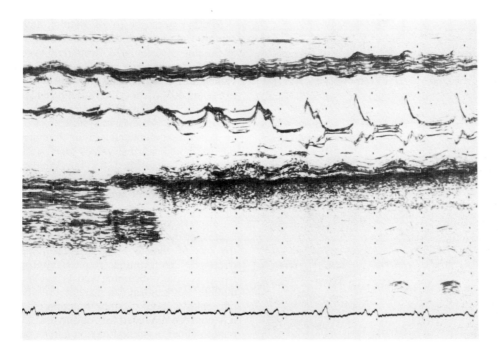

Figure 22. M-mode record from patient with dilated cardiomyopathy, showing dilatation of both ventricles, reduced amplitude of motion of the ·septum and the left ventricular posterior wall and late diastolic notching of the mitral echogram. Wall thickness is normal.

at the septum and postero-lateral walls and apical two-chamber images to examine the anterior and inferior walls. Is the left ventricular cavity normal, is it generally dilated or is there localized bulging? Does wall motion appear uniform and, if so, is it of normal amplitude? Is there an obvious segmental abnormality? Remember when assessing septal pulsatility that the septum is normally relatively immobile in its upper third. Is the myocardium at the site of a localized bulge akinetic or dyskinetic? Although the qualitative information obtained in this way is imprecise, it is often of considerable practical value. Thus, it will reliably differentiate between a normal ventricle, a ventricle with an aneurysm (*Figures 20c* and *21*) or other major segmental abnormality (*Figure 21c*), and a globally dilated, hypokinetic left ventricle. The last is characteristic of dilated cardiomyopathy and of the heart muscle disorders associated with, for example, thyroid disease and cancer chemotherapy with adriamycin and daunorubicin.

Dilated cardiomyopathy is a disease of unknown cause characterized by impaired systolic function, dilatation of the ventricles and normal wall thickness (*Figure 22*). Although identical echocardiographic features occur in some patients with coronary heart disease and the diagnosis of dilated cardiomyopathy ultimately depends on the angiographic demonstration of normal coronary arteries, most ischaemic patients with major left ventricular dilatation also have some evidence of segmental disease. Note that the presence of clot within the left ventricle is not a distinguishing feature as it occurs in up to 50% of patients with dilated cardiomyopathy.

For routine clinical purposes, quantification of left ventricular systolic function by cross-sectional echocardiography should not be attempted. There are formulae for calculating left ventricular volumes (and from them, ejection fraction) but the results are inevitably inaccurate partly because poor endocardial detail makes measurement difficult and partly because it is impossible to make simultaneous measurements in more than one plane. At present nuclear angiography is the method of choice for measuring the left ventricular ejection fraction.

It is convenient to measure the internal (antero−posterior) dimension of the left ventricular cavity on

a

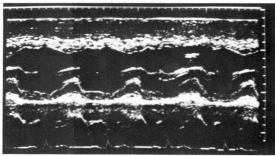

b

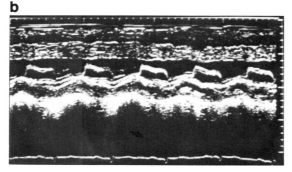

Figure 23. M-modes with vertical axis marked off in centimetres.
(**a**) Normal left ventricle; posterior wall and septal thickness 9 mm.
(**b**) Hypertrophied left ventricle from patient with non-rheumatic
aortic stenosis showing narrow cavity and reduced mitral diastolic
closure rate; posterior wall and septal thickness 18 mm.

Routine measurement of systolic thickening of the
septum and posterior wall is unnecessary. Percentage
thickening is calculated from (end-systolic − end-
diastolic thickness) ÷ end-diastolic thickness; normal
values for the septum and posterior wall are in excess
of 30 and 40%, respectively. Failure to achieve these
values indicates that the segment under investigation
is hypokinetic. Absence of systolic thickening indicates
an akinetic segment and systolic thinning a dyskinetic
one. Although normal values for systolic thickening
have been established for the standard recording
position, the criteria for identifying an akinetic or
dyskinetic segment are applicable throughout the
ventricle.

M-mode records can be digitized and instantaneous
values for internal dimension, wall thickness, etc. and
the rate of change of these measurements, can be
obtained. The technique requires appropriate equipment, is time consuming and is not reliably reproducible. Although it can provide useful information,
particularly about diastolic function of the ventricle,
it is not recommended for routine use. Limited information about diastolic function can be obtained from
standard M-mode analysis. Thus, the diastolic closure
rate of the mitral valve is slowed when a non-compliant
ventricle presents resistance to filling (*Figures 23b* and
26b) and late diastolic notching of the mitral valve
(*Figure 22*) is evidence of a raised left ventricular end-
diastolic pressure.

M-mode records, but M-mode quantification of left
ventricular function should only be attempted in
normal subjects or in patients with global dysfunction.

Record an M-mode (selected from a parasternal long
axis image) from just below the tips of the mitral
leaflets (*Figure 23a*). This is the standard position for
making measurements and the only one for which
normal values have been established. Measure the left
ventricular internal dimension at end-diastole (onset
of the QRS) and end-systole (conveniently taken as
the shortest point between the septum and posterior
wall). Percentage fractional shortening (normal
26−42%) is calculated from (end-diastolic − end-
systolic dimension) ÷ end-diastolic dimension. It
reflects the effect that circumferential shortening of
the myocardium has on cavity size at one particular
point in the ventricle. It is analogous to ejection
fraction but can only be regarded as representative of
overall left ventricular systolic function if the contraction pattern of the ventricle can be assumed to be
uniform.

10. Left ventricular hypertrophy?

The request may be to look for evidence of left
ventricular hypertrophy and/or to assess its severity.

Hypertrophy may be immediately obvious on cross-
sectional imaging and this may be all that is necessary
(*Figure 24*). If quantification is required use an M-
mode taken from the parasternal long axis view at the
level of the tips of the mitral leaflets. Measure the
thickness (leading edge to leading edge) of the interventricular septum and the posterior wall of the left
ventricle at end-diastole (onset QRS complex). A value
for either in excess of 12 mm is diagnostic of left
ventricular hypertrophy (*Figure 23*). Judge the severity
of left ventricular hypertrophy directly from these
measurements; estimates of left ventricular mass
derived from them are unhelpful in routine practice.
Although it is rarely necessary to calculate it for
ordinary clinical purposes, it is worth noting that the

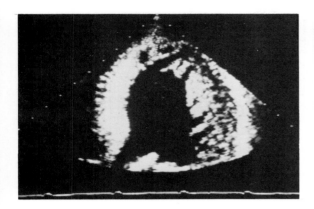

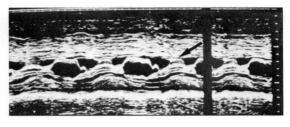

Figure 26. Hypertrophic cardiomyopathy; M-mode. Septal/posterior wall ratio 2.0. Arrow indicates systolic anterior motion of the mitral apparatus.

Figure 24. Apical long axis image (from an elderly patient with rheumatic heart disease) showing severe left ventricular hypertrophy.

a

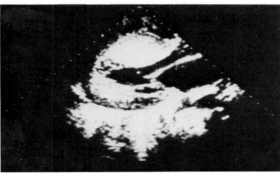

b

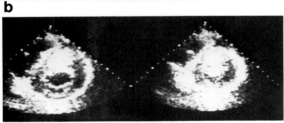

Figure 25. Hypertrophic cardiomyopathy. (a) Parasternal long axis image (diastole) showing bulky septum and narrow left ventricular cavity. (b) Parasternal short axis images in diastole (left) and systole showing hypertrophy particularly involving the septum, abnormal texture of septal echoes and cavity obliteration in systole.

ratio of wall thickness to cavity radius (normal 0.36) is a convenient expression of the difference between concentric and eccentric left ventricular hypertrophy: the ratio increases in the former and remains normal in the latter. It can be calculated from the M-mode

as 2 × posterior wall thickness ÷ the left ventricular internal dimension (all measurements at end-diastole).

10.1 *Differential diagnosis of the cause of hypertrophy*

In the case of concentric hypertrophy, systemic hypertension and aortic stenosis are the common causes. In their absence, consider hypertrophic cardiomyopathy. This is a disease of unknown cause which is characterized by massive left ventricular hypertrophy, impaired diastolic function and 'increased' contractility. Look on the cross-sectional images for a thick-walled left ventricle with hypertrophy particularly involving the septum (or occasionally being confined to the apex), and for a narrow left ventricular cavity which becomes obliterated in systole (*Figure 25*). On M-mode records (*Figure 26*) look for immobility of the interventricular septum, a septal to posterior wall ratio of 1:5 or more, systolic anterior motion of the mitral apparatus (reflecting papillary muscle displacement in the narrow cavity) and for systolic notching of the aortic valve (reflecting cavity obliteration). With the exception of the last, none of these features is specific. However, if most are present (and other causes of left ventricular hypertrophy can be discounted), the diagnosis can be regarded as established.

Eccentric hypertrophy occurs with volume overloading, for example aortic regurgitation. It also occurs in athletes. The athlete's heart is essentially a magnified version of the normal heart. On cross-sectional images myocardial texture looks normal and M-mode-derived indices of diastolic and systolic function are also normal. The increased wall thickness may particularly involve the septum and the septal/posterior wall ratio can be as high as 1.5.

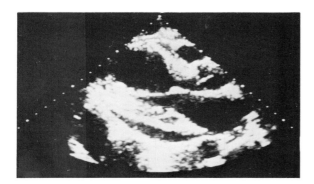

Figure 27. Parasternal long axis image (late diastole). Pericardial fluid has collected behind the left ventricle. There is also a small quantity of fluid anteriorly.

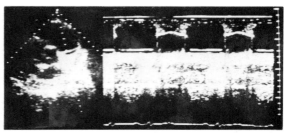

Figure 28. Parasternal short axis image and M-mode showing thickened stenotic pulmonary valve.

11. Pericardial effusion

Use cross-sectional echocardiography (*Figure 27*). Start by looking in the parasternal long axis view for an echo-free space behind the left ventricle (as a rule, pericardial fluid does not collect behind the left atrium). Then use other views to look for extension of the fluid around the lateral and anterior walls of the heart; estimate the size of the effusion. Small effusions have no effect on cardiac function but substantial effusions can, by compression, seriously compromise it (cardiac tamponade). If the effusion is substantial, look for early diastolic collapse of the right ventricular and right atrial free wall (usually best seen in the parasternal long axis and apical four-chamber views). The presence of either is reliable evidence of tamponade and is an indication for urgent aspiration of the effusion. Arrangements should be made for the aspiration to be performed under ultrasonic guidance.

Echocardiography can occasionally be helpful in determining the cause of an effusion. It may detect invasion of the pericardial space by metastatic tumour. Fibrous strands within the fluid suggest a subacute or chronic inflammatory process and if there is, in addition, thickening of the underlying pericardium, tuberculous pericarditis should be considered.

12. Congenital heart disease

The investigation of the complex lesions which present in infancy and early childhood is a specialized topic and beyond the scope of this chapter. Only congenital lesions likely to be encountered in adult practice are considered here. Investigation to exclude them should not be undertaken routinely.

12.1 *Co-arctation of the aorta*

Start with CW Doppler from the suprasternal probe position. Look for high velocity flow in the descending aorta; the peak velocity and the duration of the high velocity jet increase with increasing severity of the co-arctation. A co-arctation may be identified by cross-sectional imaging from the suprasternal notch. If feasible, simultaneous imaging and pulsed Doppler should be used to determine the precise site (usually just distal to the left subclavian artery) and extent of the lesion. A bicuspid aortic valve is a common association and should always be sought.

12.2 *Congenital aortic stenosis*

See Section 3.

12.3 *Pulmonary stenosis*

Start with CW Doppler (use a subcostal probe position if simultaneous imaging is not possible). If the peak velocity of pulmonary systolic flow indicates pulmonary stenosis, use pulsed Doppler to determine whether the stenosis is valvar or infundibular and look on parasternal short axis images (*Figure 28*) for systolic doming and/or calcification of the leaflets (indicating valvar stenosis) and for infundibular hypertrophy (it is usually secondary to valvar stenosis but is occasionally an isolated finding).

12.4 *Ventricular septal defect*

Most ventricular septal defects are perimembranous, that is they are situated close to the atrio−ventricular

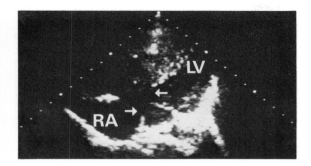

Figure 29. Apical four-chamber image showing both an ostium primum atrial septal defect and a perimembranous ventricular septal defect.

junction (*Figure 29*). Muscular defects, that is those confined to the muscular portion of the interventricular septum, are less common. With cross-sectional echocardiography, even quite large ventricular septal defects can be difficult to detect. A combination of imaging and Doppler is the best first approach. Use a subcostal and the apical four-chamber views; if colour flow Doppler is not available, place the sample volume to the right of the interventricular septum and, starting at the atrio−ventricular junction, look for a high velocity jet indicating a left to right shunt across the septum. Once the site of the defect has been pinpointed, use imaging (looking for echo drop-out) to define its anatomy (not always easy).

The presence of a ventricular septal defect, but not its precise site, can be detected by non-imaging Doppler; place the transducer at the left sternal edge (first try the point where the murmur is loudest) and look for high velocity turbulent flow directed towards it. Whatever Doppler method is used, the assessment should include an estimate of pulmonary artery pressure. This can be calculated from the tricuspid regurgitant flow (see Section 2) or by subtracting the gradient across the defect from the systolic blood pressure measured in the arm. Both of these calculations assume the absence of pulmonary stenosis which may be associated with a ventricular septal defect and should always be sought; the second further assumes the absence of obstruction to left ventricular outflow.

12.5 *Atrial septal defect*

Ostium secundum atrial septal defects are situated in the middle of the inter-atrial septum. The less common

ostium primum defects are situated close to the atrio−ventricular junction and are associated with a perimembranous ventricular septal defect (*Figure 29*). The diagnosis can be suspected when right ventricular dilatation and reversed motion of the interventricular septum are seen on parasternal long axis images and on M-modes derived from them; these features are not, however, specific. For definitive diagnosis, use the subcostal four-chamber or the parasternal short axis views. Look for echo drop-out at the site of the defect. If it is not obvious and simultaneous Doppler is feasible, place the sample volume (the parasternal short axis view may be best for this) to the right of the atrial septum and move it upwards, starting from the atrio−ventricular junction, looking for evidence of left to right shunting. Flow through an atrial septal defect occurs principally with atrial systole and early ventricular systole and has a maximum velocity of about 1 m s^{-1}. If these manoeuvres are unsuccessful, try contrast echocardiography. Use 5 ml of agitated saline (injected into a peripheral vein) to opacify the right atrium. Look for a negative contrast effect (failure of opacification at the site of a left to right jet) or, as there is usually an element of right to left shunting, look too for a few bubbles of contrast entering the left atrium. If an atrial septal defect is identified, complete the examination with a Doppler estimate of pulmonary artery pressure.

13. Acknowledgement

The technical assistance of Mary Kerr and Rhona Shanks is acknowledged.

14. Further reading

1. Feigenbaum,H. (1986) *Echocardiography.* 4th edition, Lee and Febiger, Philadelphia.
2. Hunter,S. and Hall,R. (1986) *Clinical Echocardiography.* Castle House Publications, Tunbridge Wells.
3. Hatle,L. and Angelson,B. (1985) *Doppler Ultrasound in Cardiology.* 2nd edition, Lee and Febiger, Philadelphia.
4. Williams,G.A. and Labovitz,J.A. (1985) Doppler hemodynamic evaluation of prosthetic (Starr Edwards and Bjork−Shiley) and bioprosthetic (Hancock and Carpentier Edwards) cardiac valves. *Am. J. Cardiol.,* **56**, 325−332.

Chapter 10

Gynaecology

Patricia Morley and Anne Hollman

1. Introduction

The early work of Donald and his colleagues dealt with the possibility of visualizing major lower abdominal structures by ultrasound and the application of ultrasound to differentiate between cystic and solid abdominal and pelvic masses (1,2). Modern ultrasound systems now display high resolution images of normal anatomy and the diagnosis of pathological changes depends on accurate demonstration of changes in morphology and alteration in the acoustic properties of soft tissues. Ultrasonic imaging is now one of the major diagnostic imaging procedures used in the assessment of pelvic pathology.

2. Examination techniques

Pelvic scans using conventional scanning systems are performed with a full bladder. This is achieved by the patient taking an oral water load about 1.5 h prior to the examination. Rarely retrograde filling may be necessary. With the bladder filled, gas distended loops of bowel are displaced out of the pelvis. The uterus, ovaries, pelvic blood vessels and other soft tissues may be visualized in detail by scanning through the bladder which acts as an acoustic window. It is preferable to have the distal bowel empty which may otherwise obscure pelvic detail or cause acoustic artefacts.

As well as acting as an acoustic window the distended bladder acts as an anatomic reference point. Additionally a mobile pelvic mass is displaced out of the pelvis as the bladder fills, whereas a fixed mass is retained in the pelvis. Assessment of pelvic pathology should not be made without definite identification of the bladder; failure to ensure this criterion can result in major diagnostic errors (3). If pelvic malignancy is suspected or proven the examination

should be extended to include the regional lymph nodes, the urinary tract, the liver, the gut and omentum and the peritoneal cavity.

Recently endoscopic transducers have been introduced and the pelvis can be examined from the vagina or rectum. These transducers are not as yet in general use but provide detail not obtained by conventional scanning through the abdominal wall. When using these transducers it is not always necessary to fill the bladder.

3. Ultrasonic anatomy

The uterus lies in mid position behind the full bladder. When anteverted the uterine fundus points towards the anterior abdominal wall, when retroverted it is directed towards the sacrum (*Figures 1a and b*). The outline of the uterine fundus and body is smooth and the myometrium gives rise to a fine uniform echo pattern. The central cavity comprises a thin highly reflective line surrounded by a halo from the two endometrial layers. The Fallopian tubes and round ligaments are incompletely defined due to their anatomic configuration.

The normal non-pregnant adult uterus is pear shaped and is approximately 8 cm long. Its overall size should not exceed $9 \times 5 \times 3$ cm. The pre-pubertal uterus has a smaller body but a cervix which is similar in size to that of the adult. After the menopause the uterus decreases in size by at least 50% by the age of 60.

Cyclic changes are evident in the endometrium (*Figure 1*). In the early proliferative phase the endometrium is seen as a thin line (*Figure 1a*). Hackelöer and other authors report on as many as six different patterns during the menstrual cycle which correlate well with maternal oestradiol and progesterone levels (4,5). In the late pre-ovulatory proliferative phase the

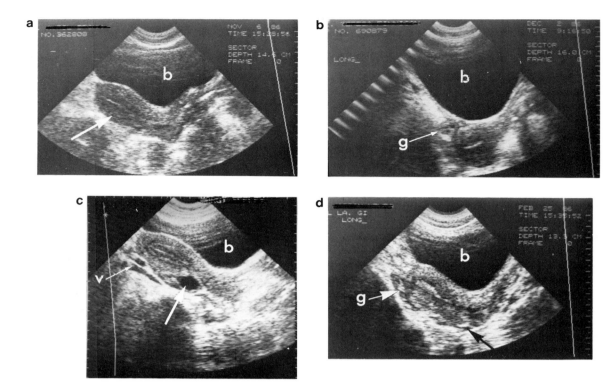

Figure 1. The normal uterus. (**a**) Normal mid position adult uterus. ↑ Sagittal section (ss) through the uterus, cervix and vagina. Note the thin cavity echo in the body of the uterus and the cervix. Early proliferative phase of the endometrium. The cavity of the vagina is also clearly defined. (**b**) Normal retroverted uterus. The endometrial cavity is only partly displayed. There is a loop of small bowel (g↑) lying above the fundus of the uterus. Sagittal section. (**c**) Normal uterus with a thick halo of endometrium surrounding the cavity. A mature ovarian follicle ↑ is seen behind the uterus and the internal iliac vessels behind the uterine fundus. Late proliferative phase of endometrium. Oblique sagittal section. (**d**) Normal uterus with thick echogenic endometrium. There is a very small volume of fluid in the cul de sac ↑ and a segment of small bowel g↑ posterior to the fundus. Secretory phase of endometrium. b = bladder; g = gut; v = vessels.

endometrium is represented by a hypoechoic ring around the normal cavity (*Figure 1c*). There may be associated dilatation of the uterine cavity which appears as a central ring-like structure. It is possible that this could represent the earliest sign of ovulation and should not be confused with an early gestation. In the secretory phase glandular elements produce large quantities of mucus which appear as dense central echoes (*Figure 1d*).

The cervix of the uterus is cylindrical in outline and is inserted into the vagina through the superoanterior wall (*Figure 1*). The cervix measures 2.5 cm approximately in the adult. The cervical canal is represented by a thin line.

The vaginal cavity is best demonstrated in the sagittal plane behind the base of the bladder (*Figure 1a*).

It measures 12 – 13 cm in length in the adult. It can also be visualized on transverse sections with the transducer angled caudad. An air-filled tampon in the vagina produces a highly reflective echo with strong distal shadowing (*Figure 2*); a fluid-filled tampon may simulate a cystic structure.

The ovaries are located within the triangle created by the lateral wall of the full bladder, the iliopsoas muscle and the lateral pelvic wall at the level of the uterine fundus (*Figure 3*). The ovaries are displaced superiorly by the pregnant uterus, and following delivery may return to a different position. Ultrasound consistently demonstrates blood vessels along the pelvic side walls (*Figure 3b*). The ovarian artery and vein in the infundibular pelvic ligament which show a constant relationship to the ovaries is a useful

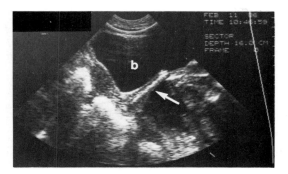

Figure 2. Vaginal tampon. Normal anteverted uterus with an echogenic air-containing tampon ↑ in the vagina. There is loss of soft tissue detail distal to the tampon due to its shadowing effect. Sagittal section. b = bladder.

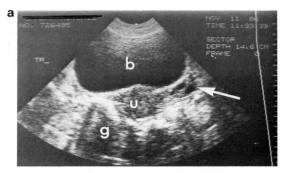

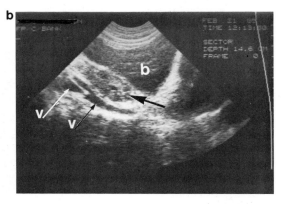

Figure 3. Normal ovaries. (**a**) The left ovary ↑ lies behind the bladder in the angle between the uterus and the lateral wall of the pelvis. Follicles are seen in the ovarian stroma. There are several gas-distended segments of bowel behind the uterus. Transverse section. (**b**) Oblique scan showing the normal ovary ↑ behind the bladder and the internal iliac vessels posterior to the ovary. The infundibulopelvic ligament lies above the ovary anterior to these vessels. b = bladder; g = gut; v = vessels; u = uterus.

reference plane for locating these structures (6).

The size of the normal ovary can easily be measured, and is approximately 20 × 30 × 20 mm. There is a great variation in size, but an ovary is considered enlarged if it exceeds 50 mm in one axis, or 30 mm in two axes. Other more accurate measurements have been calculated and include the ovarian volume [which approximates to (length × width × thickness)/2; normal range $1.8-5.7$ cm^3] or surface area (maximum 8.8 cm^2).

Within the cortex of the ovoid structure of the ovary several small follicles can be identified (*Figure 3a*). During the menstrual cycle usually one follicle progressively enlarges, becomes mature, and ruptures into the peritoneal cavity at mid cycle. Following ovulation the corpus luteum may persist and become cystic, or a follicle retention cyst may develop. These physiological changes can easily be monitored so that ultrasound now plays an important role in the assessment of developing ovarian follicles in the *in vitro* fertilization programmes $(7-10)$.

The development of improved resolution systems has resulted in the visualization of detailed pelvic soft tissue anatomy (*Figure 4a−c*). The rectus abdominis, iliopsoas, obturator internus and pelvic floor muscles are normally displayed and in detailed pelvic examinations it may be necessary to define these structures. Arteries and veins can be outlined and confirmed by the presence of pulsation and by Doppler studies. Normal ureters can sometimes be identified and dilated ureters may be clearly visualized. The ureteric orifice can be seen at the base of the bladder (*Figure 4c*) and the jet of urine into the bladder commonly identified. Small bowel loops are recognized by their peristaltic

activity (*Figure 1b* and *c*) but segments of colon may cause diagnostic difficulties and are not consistently identified except by use of a water enema. For a detailed discussion the reader is referred to more detailed texts on the subject.

4. Ultrasound assessment of gynaecological pathology

An ultrasound examination may be employed to confirm the clinical diagnosis of a pelvic mass to determine its size and to assess its consistency. Wherever possible an attempt is made at a pathological diagnosis but this is fraught with pitfalls and an ultrasound examination should be interpreted with a

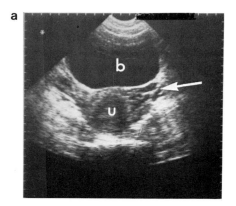

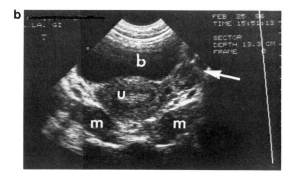

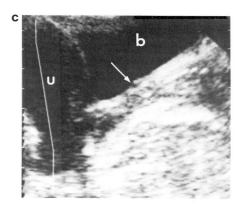

Figure 4. Normal pelvic soft tissues. (**a**) Transverse scan through the body of the uterus. Prominent uterine vessels ↑ are seen on the left. (**b**) Angled transverse scan through the body of the uterus with the piriformis muscles displayed posteriorly. The muscles are less echogenic than the pelvic connective tissue. Iliac and uterine vessels are outlined on either side of the uterus. The iliopsoas muscle ↑ lies anterolateral to the left side of the bladder. (**c**) Sagittal section through the bladder with the normal ureteric orifice ↑ anterior to the vagina. b = bladder; u = uterus; m = muscle.

Table 1. The pelvic mass.

Cystic mass	Solid
Functional cyst of the ovary	Uterus
Parovarian cysts	Uterine tumours
Endometriotic cyst	Ovarian tumours
Cystic ovarian tumour	Retroperitoneal tumours
Parasitic cysts	Enlarged lymph nodes
Lymphocyst	Intestinal tumours
Retroperitoneal cyst	Bladder tumours
Tubal abscess	Ectopic kidney
Hydrosalpinx	
Pelvic abscess	
Pelvic haematocele	
Ectopic pregnancy	
Bowel	
Bladder	

degree of caution and with a full knowledge of the clinical history. Hackelöer states that 60–70% of lower abdominal masses can be definitely assigned to a particular organ based on the sonographic criteria alone (5). Patients with pelvic masses who are not having early laparotomy should have the clinical diagnosis confirmed by ultrasound. Cystic pelvic masses are easily identified with ultrasound and are usually ovarian in origin. Similarly the majority of solid masses encountered are uterine fibroids. However there remains a significant differential diagnosis (*Table 1*).

Certain features are more commonly associated with malignancy (*Table 2*) but unless a tumour is advanced there is overlap in the appearance of benign and malignant lesions. Whenever a pelvic mass is thought to be malignant the ultrasound examination should be extended to examine the remainder of the abdomen to assess possible tumour dissemination.

When assessing the uterus for possible pathology, changes in uterine size, myometrial texture and endometrial thickness should be assessed (*Table 3*). Prior to the menarche myometrial texture is homogeneous but becomes more granular in adult life. Textural changes are seen with significant pathology (*Table 3*). However, a uterus may appear ultrasonically normal but be abnormal pathologically.

As well as identifying ovarian mass lesions the ovary may be assessed for changes in size and texture and for follicular size, number and distribution. These features are observed in the investigation of dysfunctional bleeding and infertility and in the *in vitro*

Table 2. Sonographic features associated with malignancy.

Cystic mass	Solid mass
Solid material in cyst	Soft vascular or necrotic tumours
Grossly thickened septae	Patchy variation in texture
Solid replacement of loculations	Mass filling pelvis
Complex internal structure	Loss of definition of uterine outline
Loss of definition of cyst wall	Fixation in the pelvis
Fixation in pelvis	Ascites
Ascites	Tumour dissemination in abdomen
Tumour dissemination in abdomen	

Table 3. Assessment of the uterus for possible pathology.

A. *Uterine enlargement*

Pregnancy and the puerperium
Leiomyomata
Adenomyosis
Myometrial hypertrophy
Endometrial hyperplasia
Haematometra, pyometra

Endometrial carcinoma
Leiomyosarcoma
Carcinoma of the cervix
Chorionic tumours
Secondary malignant involvement

B. *Endometrial thickening*

Menstrual and pre-menstrual phase
Early pregnancy and the puerperium
Ectopic pregnancy
Incomplete abortion

Endometrial hyperplasia
Endometrial polyp
Endometrial carcinoma
Chorionic tumours
Secondary malignant involvement

C. *Abnormal myometrial texture*

Adenomyosis
Leiomyoma
Acute myometritis

Endometrial carcinoma
Leiomyosarcoma
Trophoblastic disease
Secondary uterine involvement

fertilization programmes. Modern high resolution imaging systems are required for such detailed studies.

5. Pelvic inflammatory disease (PID)

Ascending infection from the vagina is the most common cause of PID though blood borne infection and direct spread from adjacent organs also occurs. Possible local causes include ruptured appendix abscess, diverticular abscess, Crohn's disease, pelvic surgery and puerperal and post-abortion complications. Intrauterine contraceptive devices (IUCD) predispose to PID, infection occurring most frequently at the time of insertion or as a result of perforation of the device into the pelvis.

5.1 Ultrasonic features

5.1.1 Acute PID

Uterine echogenicity is reported as being reduced in acute PID though this is a difficult feature to assess. There can be a *prominent decidual reaction* in association with endometritis or the endometrial echo may be prominent (11).

Small volumes of fluid may be seen behind the uterus (*Figure 5a*) and with severe infections a large collection can develop in the cul de sac (*Figure 5b*). Small volumes of pus in this region may be difficult to distinguish from the uterus as they are of similar echogenicity.

Pyosalpinx occurs at a later stage and a dilated fluid-filled tube is easily identified with ultrasound. Layered debris is often seen in these dilated tubes and there is commonly bilateral involvement (*Figure 5c*).

Tubovarian abscess may develop with progressive infection and at a later stage multiple abscesses may be seen in the pelvis. Abscesses on ultrasound have a shaggy outline and contain debris (*Figure 5d*) and sometimes gas. The uterine outline may become ill defined.

5.1.2 Chronic PID

With long-standing PID, widespread fibrosis with adhesions develops. These obliterate the outline of the pelvic organs and there is loss of definition of the pelvic soft tissue with ultrasound.

Hydrosalpinx may develop due to tubal adhesions and is often bilateral. Hydrosalpinx tends to adopt a characteristic configuration (11) with the ampullary

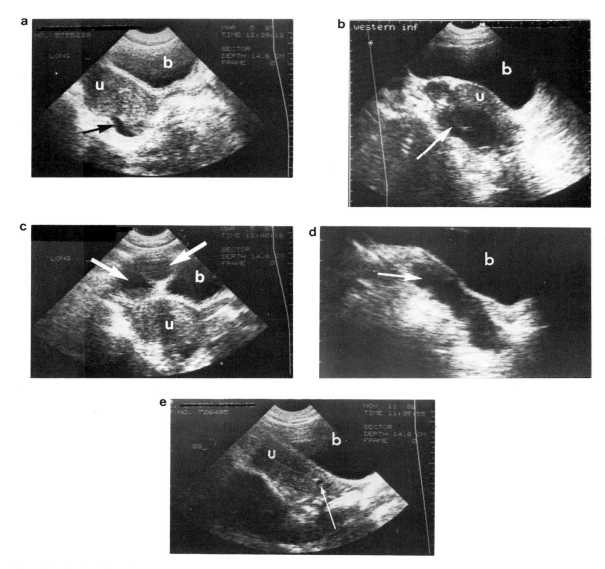

Figure 5. Sagittal sections of pelvic inflammatory disease. (**a**) Acute pelvic inflammatory disease. Bulky uterus and a small volume of fluid in ↑ the cul de sac. (**b**) Large pelvic abscess ↑ behind the uterus with a characteristic shaggy outline. This abscess developed post-acute appendicitis. (**c**) Pyosalpinx. Dilated tubes ↑ with a fluid:debris layer lying anterolateral to the uterus. Same patient as **a**. (**d**) Large tubal abscess ↑ with an elongated shaggy outline and intraluminal debris. (**e**) Nabothian cysts ↑ in the cervix. b = bladder; u = uterus.

end dilating more than the interstitial portion and appears fluid filled on ultrasound. With chronic salpingitis the tube is thick walled and appears solid.

Peritoneal adhesions are common in severe PID and inflammatory peritoneal retention cysts develop (12).

5.1.3 *IUCD and abscesses*

These abscesses tend to be unilateral and may result from perforation of the uterus but may also occur even when the IUCD is still in the uterus. IUCD-related abscesses are often due to actinomycosis and are multi-septate (11). Occasionally the IUCD can be seen in the abscess.

5.1.4 Nabothian cysts

Chronic non-specific cervicitis can be associated with inflammatory stenosis and cystic dilatation of cervical glands. These Nabothian cysts are sometimes seen with ultrasound (*Figure 5e*).

5.2 Other imaging techniques

Hysterosalpingography (HSG) is contraindicated in patients with active PID, in which cases ultrasound should be used. HSG is as accurate as laparoscopy in assessing tubal patency and remains an important technique to investigate patients infertile from previous PID. HSG may also demonstrate the complications of PID, namely hydrosalpinx, tubovarian abscess and pelvic adhesions.

Computerized tomography (CT) and magnetic resonance imaging (MRI) are not usually employed to assess PID, but may show its complications, particularly pelvic abscesses.

6. Endometriosis

In *endometriosis* stromal or glandular endometrial tissue is found in locations other than the lining of the endometrial cavity. It is thought to affect up to 15% of women during their active menstrual life and is associated with infertility in 40% of patients (13). In *internal endometriosis* (adenomyosis) endometrial cell rests are found in the myometrium.

External endometriosis refers to such deposits in the Fallopian tubes, ovaries, uterine ligaments, recto-vaginal septum, pelvic peritoneum or laparotomy scars.

6.1 Ultrasonic features

In adenomyosis the uterus may be slightly enlarged and a thickened Swiss cheese appearance has been reported (14) although there are usually no diagnostic features. External endometriosis may be demonstrated as a discrete pelvic mass (endometrioma) or in its more common diffuse form. Endometriomas are cystic or complex and may contain haemorrhagic debris (*Figure 6*). They are most commonly seen in the adnexae or cul de sac (15). Features of diffuse endometriosis are

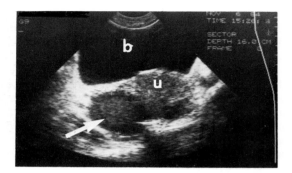

Figure 6. Endometriosis. Large endometrioma ↑ containing haemorrhagic debris lying to the right of the uterus. Smaller cysts are seen behind the uterus with some loss of definition of posterior uterine outline. Transverse section. b = bladder; u = uterus.

difficult to demonstrate. Focal areas of echogenicity with poor definition of pelvic structures have been described. There may be accentuation of pelvic arterial pulsation (13).

6.2 Other imaging techniques

There are no specific CT or MRI features of external endometriosis. An endometrioma is typically shown as an adnexal mass with CT and MRI. Both techniques, unlike ultrasound, will demonstrate acute haemorrhage within deposits, and indicate the chronicity of haemorrhagic lesions. Bowel wall involvement and pelvic side wall lesions have been shown with CT (13). MRI depicts the extent of disease in the pelvis as well as ultrasound (16) and is probably better than CT.

7. The uterus

7.1 Congenital anomalies

7.1.1 Hypoplasia of the uterus

This is a rare condition in which the degree of hypoplasia varies from the severe 'fetal' type associated with other genitourinary tract abnormalities to the 'small adult' type in which there is a small uterine body, often accompanied by hypoplasia of the cervix.

7.1.2 Bicornuate uterus

Failure of Müllerian duct fusion results in a wide spectrum of anomalies ranging from the slight defect in the fundus of the uterus to uterus didelphys, where there is complete failure of fusion of the Müllerian ducts resulting in duplication of the uterus, the cervix and the vagina.

Uterine bicornis unicollis or bicornuate uterus is the most common type, and clinically may become apparent when the patient presents with recurrent abortion, fetal malposition or retained placenta. Both uterine horns and the intervening septum can be identified with ultrasound and are most easily diagnosed in the post-partum period. Ultrasound is useful diagnostically in pregnancy when a second horn is identified and occasionally a second pregnancy. The kidneys should also be examined in patients with uterine anomalies for possible associated abnormalities of the urinary tract.

7.1.3 Imperforate hymen

This leads to retention of blood or mucus causing haematocolpos, haematometra, haematosalpinx and occasionally haematoperitoneum. Ultrasonically haematocolpos is seen as a large pear shaped fluid-filled mass displacing the uterus upwards. It may present in the neonatal period as a pelvic mass but the diagnosis is usually delayed until adolescence after the onset of puberty.

7.2 The intrauterine contraceptive device

The IUCD (or IUD) is in widespread use throughout the world. Ultrasound is usually employed in the management of complications. It is also used to check for correct placement following insertion (17).

7.2.1 Ultrasonic features

There is a wide range of devices in clinical use. Each device has a characteristic configuration which may be recognized ultrasonically (17). Most IUDs in current use produce a highly reflective shadowing, linear echo seen in the cavity of the uterus (*Figure 7a*). Without posterior shadowing it may be difficult to differentiate an IUD from prominent endometrium though manipulation of gain setting or use of post-processing facilities assist in identifying the device.

7.2.2 Complications associated with IUCD

Possible complications include pregnancy, unrecognized expulsion, uterine perforation, migration of the device, and irregular bleeding.

Pregnancy may occur in the presence of an IUCD.

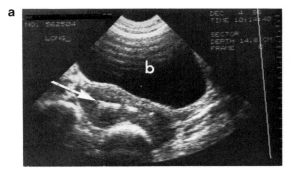

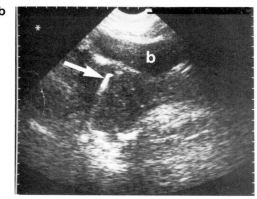

Figure 7. The intrauterine contraceptive device. (**a**) The IUCD ↑ is lying centrally in the cavity of the uterus. The device is strongly reflective and this echo pattern is typical of a 'copper 7'. (**b**) The IUCD ↑ has migrated into the myometrium and is in an eccentric position. Sagittal section. b = bladder.

As the uterus enlarges the device becomes drawn up into the uterine cavity and the patient may present with a 'lost' device. The IUCD is nearly always separate from the gestational sac, and can be demonstrated as such. Later it may prove impossible to distinguish the device from fetal echoes.

(i) *Lost intrauterine device.* If the threads of the IUCD can no longer be seen or palpated clinically as they protrude through the cervical os, the device is considered to be lost. This may be because the device has been expelled spontaneously, has perforated the myometrium, or the threads have become detached from the device or have migrated upwards. An ultrasound examination remains the initial investigation of choice to determine whether the IUCD is still in its normal position within the endometrial cavity. Eccentric positioning in the uterus is likely to be due to partial uterine perforation (*Figure 7b*) or misplacement by fibroids. Ultrasound is not good for locating an IUCD which has completely penetrated the uterus and lies in the pelvic or abdominal cavity. A plain abdominal film will confirm whether the device has been expelled, or lies in the peritoneal cavity.

7.2.3 Other imaging techniques

More recently CT has been advocated either for equivocal cases when ultrasound has been difficult or as a means of localizing an extrauterine device (18). MRI has also been able to demonstrate these devices and may become increasingly useful to demonstrate lost devices in the future.

7.3 Endometrial hyperplasia

Abnormal hyperplasia or atrophy of the endometrium accounts for 15−20% of gynaecological problems and is a common cause of uterine bleeding. Marked endometrial thickening may be seen on ultrasound and occasionally an associated cyst or functional tumour (*Table 3B*).

7.4 Myometrial hypertrophy

This is characterized by diffuse symmetric enlargement of the myometrium which may measure up to 3−4 cm. The ultrasound appearance is non-specific (*Table 3A*).

7.5 Uterine leiomyoma

The uterine leiomyoma (fibroids, myoma, fibroleiomyoma) is the most common benign tumour of the uterus, and is found in 20% of women over 30 years. They are responsible for one third of gynaecological admissions to hospital. Fibroids are usually multiple and may cause menstrual disorders, pain or pressure symptoms related to the bladder or bowel, but usually they are asymptomatic. Virtually always found in the myometrium of the body of the uterus (rarely in the cervix or uterine ligaments), they are described according to their situation as subserous, interstitial or submucous. Fibroids are usually multiple and large tumours may develop areas of red degeneration or necrobiosis. Occasionally these necrotic areas become liquified to produce cystic degeneration. Postmenopausally fibroids become firmer and often partially or completely calcified.

7.5.1 Ultrasound features

There is a wide variation in the appearance of

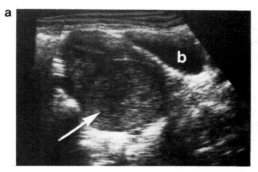
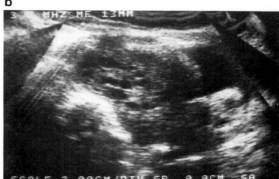
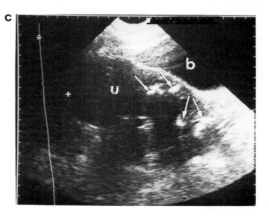

Figure 8. Sagittal sections of uterine fibroids. (**a**) The uterus is bulky and there is a large posterior fibromyoma ↑. (**b**) Enlarged uterus with cystic spaces due to cystic degeneration in a large fibromyoma. (**c**) Enlarged uterus with calcification ↑ and distal shadowing due to multiple calcified fibroids. b = bladder; u = uterus.

leiomyomata and only a small proportion of those present are seen ultrasonically. The 'typical' appearance is that of a moderately echogenic mass causing nodular uterine enlargement (*Figure 8a*). There is, however, a wide range of texture and echogenicity which is probably related to the presence of degeneration, the vascular supply and the organization of muscle and fibroid elements in the mass (19). During pregnancy fibroids tend to be less echogenic. When degeneration occurs fibroids are more transonic and may be less echogenic. Cystic areas are seen in cystic degeneration (*Figure 8b*) which can produce a very complex picture. Calcification is relatively common particularly in the elderly patient and may cause extensive shadowing (*Figure 8c*). In patients with a large pelvic mass, clinically diagnosed as fibroids, ultrasound should be used to confirm the diagnosis. In a series of patients with a clinical diagnosis of fibroids a small proportion were shown to have solid or cystic ovarian tumours (unpublished data).

7.5.2 Alternative imaging techniques

The most common CT finding of leiomyomata is distortion of the uterine outline, since some tumours (as with ultrasound), are not visualized as discrete masses distinct from the normal myometrium. CT, however, is reported as being a more sensitive technique in imaging the secondary changes that occur, such as calcification, degeneration or haemorrhage (20).

It seems likely that MRI will prove to be superior to ultrasound, both in demonstrating leiomyomata, particularly with small or multiple tumours, and in detecting complications with greater sensitivity and specificity. It may differentiate fibroids from endometrial carcinoma (21), which is not possible with either ultrasound or CT.

7.6 Trophoblastic disease

The hydatidiform mole (benign or invasive) and choriocarcinoma are the commonest placental tumours affecting one in 2000 pregnancies in Great Britain. There are at present two different classification schemes for gestational trophoblastic disease. The older histopathological classification divides trophoblastic disease into hydatidiform mole (chorioadenoma disease) and choriocarcinoma. This has been replaced with a more practical clinical classification of benign trophoblastic disease, malignant non-metastatic disease and malignant metastatic trophoblastic disease.

7.6.1 Hydatidiform mole (benign trophoblastic disease)

Swelling of the chorionic villi, and hyperplastic and

a

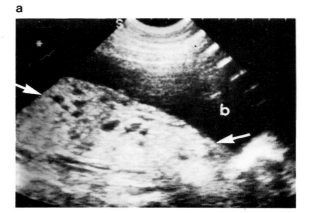

b

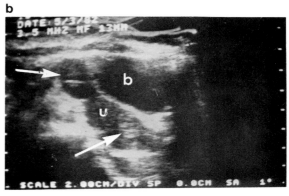

Figure 9. Trophoblastic disease. (**a**) Hydatidiform mole. Sagittal section through an enlarged uterus ↑ containing echogenic tissue with multiple cysts and cystic spaces. (Courtesy of Dr Margaret McNay, Queen Mother's Hospital, Glasgow.) (**b**) Malignant trophoblastic disease. Echogenic area in myometrium ↑ and theca lutein cysts ↑ above the uterine fundus. b = bladder; u = uterus. (Reproduced from The Uterus. In *Clinical Diagnostic Ultrasound,* Blackwell Scientific Publications by courtesy of the publishers.)

anaplastic proliferation of the chorionic epithelium characterize this tumour. The tumour may be benign, or it may be invasive if the uterine wall and vessels are infiltrated. Metastases are rare. In most cases no fetus develops, and ovarian lutein cysts occur in up to 50% (22). Bleeding during the third to fifth months of pregnancy, associated with a uterus which is large for dates herald the diagnosis. Serum chorionic gonadotropin (hCG) is characteristically raised, and provides an important indicator of recurrent disease after surgical removal of the lesion.

7.6.2 Ultrasonic features

In most cases the pattern consists of echogenic intra-uterine tissue interspersed with punctate cystic areas (*Figure 9a*). Irregular cystic areas may occur secondary to internal haemorrhage. The appearance varies with gestational duration and the size of the hydropic villi (23,24).

Partial molar pregnancy (i.e. hydatidiform mole with a co-existent fetus) and hydropic degeneration of the placenta are both associated with a fetus or a fetal pole. They may be differentiated ultrasonically with careful technique (25).

7.6.3 Choriocarcinoma and invasive mole (malignant trophoblastic disease)

Choriocarcinoma is a highly malignant tumour of the chorionic epithelium. Fifty per cent have an antecedent hydatidiform mole; 25% follow spontaneous or therapeutic abortion; and 25% follow a normal pregnancy. Direct pelvic spread notably affects the vaginal vault and parametria; and invasion of uterine vessels and lymphatics results in early metastatic disease, most commonly affecting lungs, liver, ovaries and brain. Chemotherapy is curative in over 70% of high risk patients. Invasive mole is suspected clinically if there are persistently elevated hCG levels and uterine bleeding. Patients may first present with haematogenous metastases.

7.6.4 Ultrasonic features

Malignant trophoblastic disease is usually seen as echogenic focal areas in the myometrium (*Figure 9b*) and persistent theca lutein cysts (25).

7.6.5 Other imaging techniques

CT now has a definitive place, over and above ultrasound, in staging patients with invasive trophoblastic disease, by excluding metastases and extrauterine spread. The examination, therefore, must include the pelvis, abdomen, chest and brain. CT may also have a place in distinguishing invasive mole from choriocarcinoma (26) which, until recently, could only be achieved by angiography. This technique also demonstrates associated bilateral ovarian cysts (theca lutein cysts).

Angiography has been widely and successfully employed to distinguish hydatidiform mole from choriocarcinoma; however, the more important distinction of invasive mole from choriocarcinoma has proved more difficult. CT now replaces angiography in staging the disease.

7.7 Carcinoma of the endometrium

Endometrial carcinoma occurs most frequently between the ages of 50 and 60. It is associated with a late menopause, fibroids, nulliparity and endometrial hyperplasia. The tumour is an adenocarcinoma which initially fills the uterine cavity, and later spreads through the myometrium to invade any of the pelvic structures. Lymphatic spread to the lungs, liver and bones may take place. The presenting symptom is nearly always post-menopausal bleeding, and the diagnosis is confirmed by curettage and histology. The prognosis depends on the histology and clinical staging.

7.7.1 Clinical staging

Stage I: carcinoma confined to the body of the uterus.
Stage IA: length of uterine cavity is 8 cm or less.
Stage IB: length of uterine cavity is greater than 8 cm.
Stage II: carcinoma has involved the body of the uterus *and* the cervix.
Stage III: carcinoma extends outside the uterus but not outside the true pelvis.
Stage IV: carcinoma extends outside the true pelvis or has obviously involved the mucosa of the bladder or the rectum.

Seventy-five per cent of patients are Stage I or II at presentation (29).

7.7.2 Ultrasound features

Ultrasound features of endometrial carcinoma are non-specific with a low sensitivity. The tumour may cause uterine enlargement, often symmetrical with a thickened endometrial cavity. Haematometra can produce dense echoes within the endometrial canal sometimes with layering. Sonography has not been shown to accurately depict myometrial invasion though an abnormal myometrial texture may be seen. A lobular contour of the uterus with alteration in uterine texture suggests a Stage III or Stage IV lesion (27). Ultrasound has been used to differentiate Stage IA and IB tumours by measuring the length of the uterine canal and when intracavitary radiotherapy is to be employed ultrasound can be used to determine the correct size and check the position of the applicator (28). Ultrasound has also been used in pre-treatment clinical staging but the sonographic appearances are not

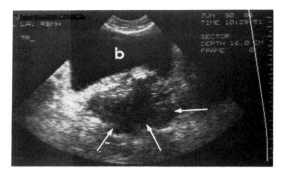

Figure 10. Recurrent endometrial carcinoma. Transverse section post-hysterectomy with a large soft tissue mass ↑ seen behind the bladder. Compare with *Figure 12*. b = bladder.

consistent except with advanced pelvic invasion. It is, however, a useful method of assessing recurrent or metastatic disease following surgery or radiotherapy (*Figure 10*). The examination should include pelvic examination for local recurrence and assessment of the regional lymph nodes, peritoneal cavity, renal tract and liver for evidence of metastatic spread.

7.7.3 Alternative imaging techniques

CT, like ultrasound, cannot be used to accurately delineate the primary tumour of endometrial carcinoma or the extent of myometrial invasion (27). However, CT has been shown to be valuable for staging with an accuracy of 83% for tumours confined to the uterus, and 86% for demonstrating extra-uterine spread (29). This compares with a 56% accuracy for ultrasound (30). CT is also more accurate than ultrasound in planning radiation portals, for evaluating the response to radiation or surgical treatment and in detecting recurrent tumour. However, difficulty occurs in distinguishing recurrence from radiation fibrosis.

It seems probable that with MRI, the extent of endo-metrial carcinoma will be accurately delineated from the normal uterus (31), particularly with the development and use of monoclonal antibodies attached to paramagnetic contrast agents (27). Adenopathy is as well demonstrated with MRI as with CT. Recurrent disease is more easily visualized with MRI as the image can be viewed in several planes (32).

7.8. Sarcoma of the uterus

Sarcoma of the uterus accounts for only 0.56% of cases of uterine malignancy (33). The commonest type is the leiomyosarcoma; the mixed mesodermal tumour

and the endometrial stromal sarcoma are rare. All types are associated with early blood borne metastases and a poor prognosis.

Leiomyosarcoma may arise from a leiomyoma or may develop diffusely in the uterine muscle. The patient is usually middle aged and presents with bleeding and abdominal pain. Treatment consists of total hysterectomy and bilateral oophorectomy. Local recurrence is common and the five year survival rate of about 30% reflects early metastatic spread.

7.8.1 Ultrasonic features

There are no specific ultrasonic signs to differentiate a leiomyoma from a leiomyosarcoma. The uterus tends to be significantly enlarged and the tumour echogenic with areas of cystic necrosis (34). An increase in size of a uterine leiomyoma post-menopausally is sugges-tive of malignant change. Ultrasound is used to monitor patients after diagnosis for local recurrence.

7.8.2 Other imaging techniques

Both CT and MRI are of value in staging the tumour for extrauterine spread and metastatic disease; and in following the response of metastases to chemotherapy.

7.9 Carcinoma of the cervix

Carcinoma of the cervix is the second most common malignancy in women after breast cancer. The incidence, which is increasing in younger women, reaches a maximum between 45 and 55 years of age. The tumour is most commonly squamous (90%) and usually arises from or near the squamo-columnar junction. When invasive, cervical ulceration, a fungating growth or tissue infiltration takes place. Direct extension affects contiguous pelvic structures, and lymphatic spread involves pelvic and para-aortic nodes. Haematogenous metastases affect lungs, liver and bone in particular and occur late. Cervical cytology followed by cone biopsy are essential to assess invasive disease. Early disease is treated by surgery; advanced pathology requires pelvic irradiation and abdominal irradiation with combination chemo-therapy.

7.9.1 Clinical staging

Staging of cervical carcinoma follows the International Federation of Gynaecology and Obstetrics (FIGO) classification.

Stage 0: carcinoma *in situ*.

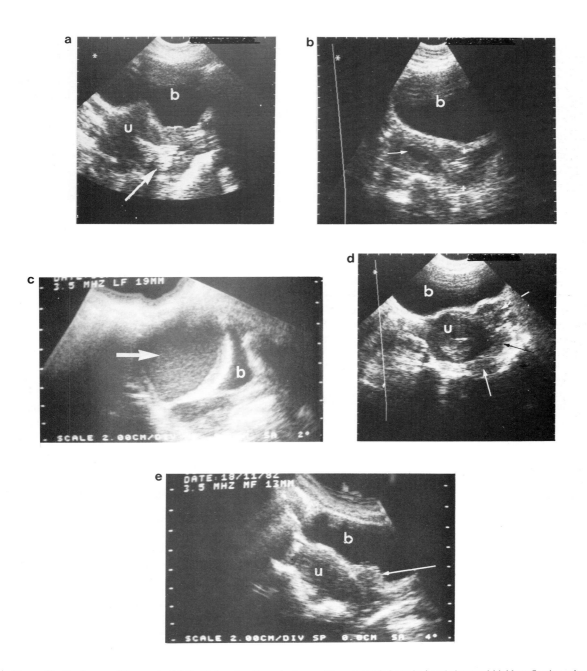

Figure 11. Carcinoma of the cervix. (**a**) Sagittal section through uterus with an expanded cervix ↑ and abnormal highly reflective echoes in the cervix and uterus. The upper vagina is also expanded from tumour invasion. (**b**) Carcinoma of the cervix with fluid in the endometrial cavity. Small pyometra ↑. Note the expanded cervix and abnormal central echoes. Sagittal section. (**c**) Massive expansion of the uterine cavity ↑ due to the haematometra secondary to carcinoma of the cervix. Sagittal section. (**d**) Extensive tumour infiltration ↑ into the left parametria extending up the pelvic side wall. Transverse section. (**e**) Bladder invasion ↑ from carcinoma of the cervix. Sagittal section. b = bladder; u = uterus.

Stage I: carcinoma confined to the cervix.

Stage II: spread to the parametrium (not extending to side wall) and/or involvement of upper vagina.

Stage III: extension to the pelvic side wall and/or involvement of lower third of vagina. All cases with hydronephrosis or a non-functioning kidney.

Stage IV: involvement of bladder or rectum or distant metastases.

Clinical staging consists of a full clinical examination, chest radiography, intravenous urography, cystoscopy, sigmoidoscopy and lymphography. The accuracy of clinical staging is approximately 55%. Ultrasound and CT have now been shown to provide a more accurate assessment.

7.9.2 *Ultrasonic features*

Ultrasound is employed to assist in staging the extent of cervical cancer at clinical presentation. However, like CT, it has not been accepted as an essential imaging technique to assist in the FIGO classification for clinical staging.

Sonography has also been widely used to follow up patients after surgery or chemotherapy, to detect recurrence of disease, to plan radiation portals, to assist with intracavitary radiation planning and to follow patients after radiation therapy.

The examination should include the cervical region, assessment of the regional lymph nodes of the pelvis and retroperitoneal area, the renal tracts, the liver and the peritoneal cavity.

At clinical presentation the primary tumour can usually be demonstrated and measured (*Figure 11a* and *b*). The endometrial cavity may be distorted by the tumour or, as a result of obstruction of the endometrial canal, there may be haematometra or pyometra (*Figure 11b* and *c*). Extension of tumour into the parametria can be shown as can tumour extending out of the true pelvis (*Figure 11d*). Bladder invasion is usually well seen (*Figure 11e*), but rectal involvement has proved more difficult to assess.

Recurrent tumour in the pelvis is usually seen as a residual posterior pelvic mass or as lymphadenopathy. However, it has proved difficult to distinguish recurrent tumour from radiation fibrosis (*Figures 10* and *12*) without serial observation or cytological confirmation from percutaneous biopsy.

The accuracy of ultrasound in staging cervical cancer is reported as approximately 56% (30). Ultrasound is consistently less accurate than CT for clinical

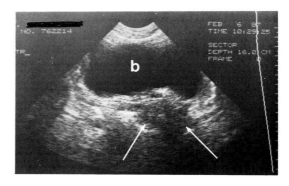

Figure 12. Post-irradiation fibrosis. Transverse section of a large mass behind the bladder ↑ due to extensive post-irradiation fibrosis following treatment for ovarian carcinoma. Compare with *Figure 10*. b = bladder.

staging, but seems to be as accurate as CT for detecting recurrent tumour.

7.9.3 *Other imaging techniques*

The overall staging accuracy of CT is probably over 80% (32). It therefore compares favourably with ultrasound (56% accuracy) and clinical staging (55% accuracy) and its accuracy improves with higher clinical staging. The problems of CT in staging this disease include difficulty differentiating a large cervical mass from vaginal involvement; determining early parametrial disease; establishing if the tumour involves only the parametria or the side wall as well; distinguishing bladder/rectal compression from involvement; and determining whether enlarged nodes are only secondary to reactive change, or whether normal size nodes contain disease. Ultrasound has similar problems. Following treatment, radiation fibrosis may be difficult to distinguish from recurrent tumour; but usually with fibrosis the only abnormality involves the parametria (35).

Lymphography remains the most accurate imaging modality (85% accuracy) (36) for evaluating adenopathy despite the fact that this examination cannot show internal iliac or obturator nodes. It remains a useful technique to detect metastases in normal sized nodes, and to differentiate metastases from reactive hyperplastic enlarged nodes found at CT (37).

MRI, at present, can be used to demonstrate the primary cervical tumour and stage the disease, at least as well as CT (32).

8. The vulva and vagina

8.1 *Tumours of the vulva and vagina*

These are uncommon tumours of the female genital tract. They comprise squamous cell carcinoma, adenocarcinoma, sarcoma or melanoma and they vary in behaviour and have different anatomic routes of spread. Assessment of local invasion and lymphatic involvement is of importance to treatment and prognosis.

Conventional and ultrasonic techniques (*Figure 13*) are used to assess local invasion and distant metastases. Pelvic lymph node involvement is common but the incidence of extra-pelvic nodal involvement is not known.

Ultrasound can be used after diagnosis to assess spread or after treatment to monitor the response of local disease and metastases. CT is also advocated for assessment.

8.2 *Vaginal cysts*

These originate from remnants of the Wolffian duct and are occasionally seen with ultrasound. Vaginal cysts are usually small (1−2 cm) and are located adjacent to the lateral wall of the vagina. Rarely these cysts may enlarge up to 5 or 6 cm.

9. The ovary

9.1 *Ovarian vein thrombophlebitis*

Puerpueral ovarian vein thrombophlebitis is an uncommon syndrome presenting within the first week following delivery. Symptoms include lower abdominal pain radiating to the ipsilateral flank and costovertebral angle. A palpable mass may be present at the lower pole of the kidney. The ultrasound features described are of a mass surrounding the lower pole of the kidney extending medially to the inferior vena cava (IVC) (38). The diagnosis has also been reported using CT (39).

9.2 *Benign ovarian disease*

9.2.1 *Functional cysts of the follicle and corpus luteum*

(i) *Follicular cysts.* Ovarian follicles may enlarge to 15 mm in diameter and then involve by resorption

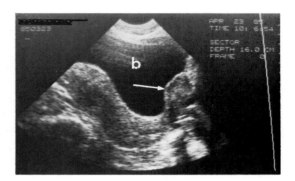

Figure 13. Vaginal carcinoma. Sagittal section through uterus. The vagina is expanded by the tumour. There is a large bulge ↑ into the bladder but no extension through the wall. Compare with *Figure 1a*. b = bladder.

of the fluid when the ovum dies. Not uncommonly, a follicular cyst may persist (*Figure 14a*) and can enlarge up to 100 mm in diameter. Unless this becomes large it tends to be asymptomatic and often regresses by spontaneous resolution. Multiple follicular cysts are found in certain conditions: metropathia haemorrhagica; cystic hyperplasia of the endometrium and the Stein−Leventhal syndrome. Ultrasonically these are seen as small, thin-walled unilocular cysts which protrude from the ovary.

(ii) *Corpus luteum cysts.* Corpus luteum cysts can develop from either the corpus luteum of pregnancy or the corpus luteum of menstruation. If a cyst results from the corpus luteum of menstruation, amenorrhoea can result, and the uterus may be slightly enlarged. Haemorrhage into the corpus luteum may cause pain, and the symptoms and signs resemble those of an ectopic pregnancy. At sonography, the cysts are small, anechoic, unilateral and thin-walled (*Figure 14b*). If haemorrhage has occurred then solid elements may be seen within the cyst.

(iii) *Theca lutein cysts.* These cysts are usually found in association with hydatidiform mole and choriocarcinoma (*Figure 9b*). They result from excess gonadotropin stimulation. Both ovaries are enlarged and contain multiple bilateral cysts. The ovaries return to normal after successful removal of the primary tumour.

9.2.2 *Ovarian hyperstimulation*

This is usually associated with therapy for infertility though it has been described in hydatidiform mole and

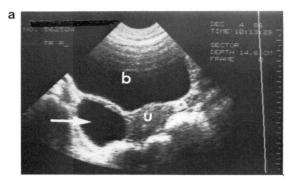

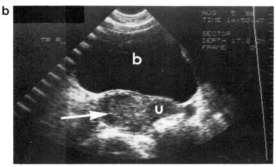

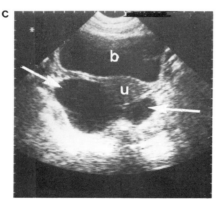

Figure 14. Transverse sections of functional cysts of the ovary. (a) Follicular cyst. Transverse section through unilocular follicular cyst ↑ of the right ovary. (b) Corpus luteum cyst. Large haemorrhagic cyst of the right ovary ↑ displacing the uterus to the left. These cysts are very similar to the endometriomas. (c) Bilateral ovarian cystic enlargement ↑ following therapy for infertility. b = bladder; u = uterus.

chorioepithelioma. Clinically hyperstimulation may produce abdominal heaviness, swelling and pain. Ovarian cystic enlargement develops which can be identified by ultrasound (*Figure 14c*). Severe

symptoms which are rare include ascites, hydrothorax, shock and thromboembolic phenomena.

9.2.3 *Parovarian cyst or broad ligament cyst*

Parovarian cysts are relatively common benign cystic tumours found either between the layers of the broad ligament or near the ovary. They are thought to arise from the vestigial remnant of the Wolffian body. They may reach a large size and present as a pelvic mass, or may cause pelvic pain if the cyst becomes haemorrhagic, undergoes torsion or ruptures.

9.2.4 *Ultrasonic features*

Most cysts seen on ultrasound are thin-walled and relatively large measuring up to 15−18 cm. The majority are unilocular though occasionally there is a thin septum. They are similar in appearance to large follicular cysts and serous cystadenomas.

9.2.5 *Polycystic ovarian disease — POD*

The classic syndrome of secondary amenorrhoea, obesity, hirsutism, infertility and bilaterally enlarged ovaries was described in 1935 by Stein and Leventhal. Variants on this syndrome have since been described and polycystic ovaries may be seen with idiopathic hirsutism (40), anovulation, oligomenorrhoea, obesity and menstrual irregularity. Ovaries of pre-pubertal girls can appear similar and polycystic ovaries may be seen without any of the classic signs or symptoms. Clinical and laboratory spectra of the syndrome may be diverse (40,41).

(i) *Ultrasonic features.* The ovaries are usually, but not invariably, enlarged, contain multiple cysts (microcysts) and have a thick tunica. Ovaries are defined as being polycystic on ultrasound if there are multiple cysts (10 or more) 1−8 mm and nearly always less than 1 mm (42) in diameter arranged either peripherally around a dense core of stroma or scattered throughout an increased amount of stroma (43) (*Figure 15a* and *b*).

(ii) *Other imaging techniques.* In POD, the ovaries themselves are shown as bilateral solid adnexal masses with CT. However, CT should be avoided as a means of assessing this benign condition since the cysts themselves are not imaged, and the radiation dose is high. MRI, however, may well prove useful in the future.

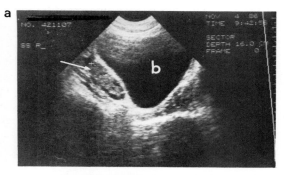

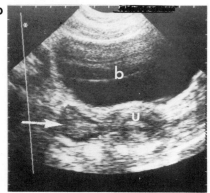

Figure 15. Polycystic ovaries. (**a**) Sagittal section of the right ovary ↑. There are multiple small peripheral cysts. This ovary was not enlarged. (**b**) Transverse section. The ovary ↑ is enlarged and cysts are arranged peripherally. b = bladder; u = uterus.

Table 4. Classification of ovarian tumours.	

Epithelial tumours

75% of ovarian tumours: 95% of malignant ovarian tumours.

Benign	*Malignant*
Serous cystadenoma	Serous cystadenocarcinoma
Adenofibroma	Mucinous cystadenocarcinoma
Cystadenofibroma	Endometrioid carcinoma
Mucinous cystadenoma	Clear cell carcinoma
Brenner tumour	Undifferentiated adenocarcinoma

Germ cell tumours

15% of all ovarian tumours: 1% of malignant ovarian tumours.

Benign	*Malignant*
Cystic teratoma	Malignant teratoma
Solid teratoma	Dysgerminoma
	Endodermal sinus tumour
Specialized: carcinoid	Choriocarcinoma
struma ovarii	

Sex-cord stromal tumours

10% of all ovarian tumours: 2% of malignant ovarian tumours.

Benign	*Malignant*
Granulosa−theca cell tumour	Sertoli−Leydig cell tumours
Thecoma	Gonadoblastoma
Fibroma	
Sertoli−Leydig cell tumours	

Unclassified tumours

Secondary tumours of the ovary

9.3 *Benign tumours of the ovary*

Only the more commonly encountered benign tumours of the ovary are discussed below. For a classification of ovarian tumours see *Table 4*.

9.3.1 *Serous cystadenoma*

These are benign cystic tumours, which have a tendency to become malignant. The cysts can vary in size from small to enormous in which case they may fill much of the pelvis and abdomen. On ultrasound, they appear as well-defined cystic adnexal masses, which are usually unilocular (*Figure 16a*). Occasionally small amounts of echogenic material are noted in the cysts. Both ovaries are involved in up to 50% of cases, but involvement of the second ovary may be overlooked on sonography.

9.3.2 *Mucinous cystadenoma*

This is another common benign epithelial tumour of the ovary, which is cystic. It uncommonly affects the other ovary or becomes malignant. The tumour size varies from 5 cm to large, and on ultrasound is seen typically as a multilocular cystic mass (*Figure 16b*). The septae tend to be thin-walled and well defined. Rupture of this tumour into the peritoneal cavity can cause pseudomyxoma peritonei (*Figure 16c*).

9.3.3 *Ovarian fibroma*

This relatively uncommon benign tumour of the ovary is composed of fibrous connective tissue. It presents either as a large pelvic mass in post-menopausal women, or as Meig's syndrome (ascites, hydrothorax and ovarian tumour). The usual appearance at sono-

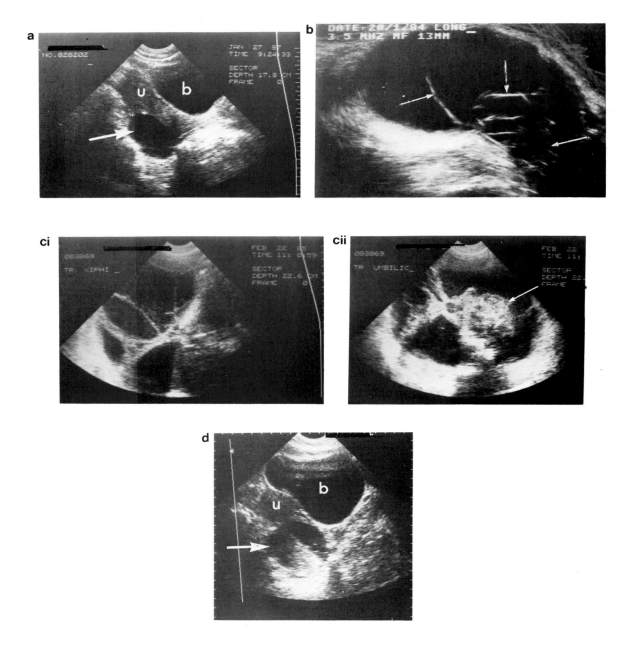

Figure 16. Benign tumours of the ovary. (**a**) Small serous cystadenoma ↑ behind the uterus. Sagittal section. (**b**) Large mucinous cystadenoma filling the lower abdomen and pelvis. There are multiple thin septae ↑. Sagittal section. (**c**) Pseudomyxoma peritonei. (**i**) High transverse scan in epigastrium. (**ii**) Transverse scan at umbilical level. Numerous septations and thick mucinous strands ↑ are seen with loculations of a thinner mucinous jelly. No true wall is defined. (**d**) Dermoid cyst. There is a well-defined cyst ↑ with two distinct compartments lying behind the uterus. This appearance is typical of a dermoid cyst. Sagittal section. b = bladder; u = uterus.

graphy is of a well-defined solid tumour, with striking attenuation of the ultrasonic beam, due to the fibrous component of the tumour. Occasionally, the tumour may be hypoechoic or even cystic. Ascites is found in up to 50% of patients whose tumours exceed 5 cm in diameter.

9.3.4 *Pseudomyxoma peritonei*

Pseudomyxoma peritonei results from rupture of a pseudomucinous cystadenoma of the ovary or more rarely the appendix. A ruptured mucocele of the appendix can also produce a similar appearance. Mucus is a chemical irritant and causes peritonitis with implantation of mucus cells in the peritoneum. The cells secrete a gelatinous material causing the 'jelly belly'. On ultrasonic examination the jelly simulates other fluids in the peritoneal cavity with punctate echoes and adhesive bands seen in the jelly (*Figure 16c*).

9.3.5 *Cystic teratoma (ovarian dermoid cyst)*

The cystic teratoma or ovarian dermoid cyst is a benign tumour of the ovary, which is thought to arise from the germ cells. The dermoid cyst represents approximately 15% of all ovarian tumours and occurs most commonly between the ages of 20 and 40. About 12% of cases are bilateral. Only 1−2% of all dermoids become malignant, squamous cell carcinoma being most common. The dermoid is a cystic tumour; it consists of a thick walled cyst filled with yellow sebaceous fluid. Intra-cystic solid components arise from the cyst wall. Ectodermal structures are most common, and include hair, skin and teeth.

(i) *Ultrasonic features.* Due to the variability in the contents of the cyst, teratomas give rise to a broad range of ultrasound appearances. The classic appearance of a well-defined cystic adnexal mass, containing solid (fat, calcium) areas which are highly reflective and produce distal shadowing posteriorly, is only found in one third of cases (44,45). Fluid levels which are virtually diagnostic of a dermoid cyst (*Figure 16d*) are found only in a minority of cases. The majority of tumours do not produce such pathognomonic features. About 20% of cases are almost entirely cystic in appearance, 25% are predominantly solid, whilst 25% are not detected by ultrasound (this is because small solid tumours merge acoustically with the pelvic soft tissues) (3). Malignant change can be suspected

if there is loss of definition of the cyst wall, fixation of the mass in the pelvis or secondary features such as ascites, omental deposits or liver metastases.

9.3.6 *Other imaging techniques in benign ovarian tumours*

CT has proven to be a useful imaging modality for ovarian cancer, but for benign ovarian disease (apart from a few exceptions), it has not been shown to provide additional diagnostic information. However, the CT appearances of these benign conditions have been documented and these lesions may first be imaged with this technique (46,20). Characteristic CT features are found in about 70% of ovarian dermoids, including the fat−fluid level; calcification in the dermoid plug or cyst wall or a predominantly fat density mass (47). CT may prove useful when the ultrasound findings are equivocal. Malignant change produces a mass with loss of normal tissue planes and involvement of nearby structures.

Ovarian mass lesions are more easily imaged by MRI than CT due to the greater potential of the former for multiplanar imaging. Apart from a few exceptions (e.g. ovarian fibroma) the appearances are non-specific. MRI, as yet, has proved less specific than CT in the diagnosis of dermoids as calcification is not identified.

9.4 *Carcinoma of the ovary*

Carcinoma of the ovary is the fourth most common cause of cancer death in women. Approximately 90% of tumours are epithelial in origin and 10% are derived from germ or stromal cells. The peak incidence occurs between 40 and 65 years of age.

Between 60 and 70% of patients present with Stage III or IV (widespread) disease (27), because Stage I and II disease is often asymptomatic. Direct peritoneal spread (to the pouch of Douglas and greater omentum in particular) is common and causes ascites. The pattern of lymphatic spread is unpredictable and primarily involves the retroperitoneal nodes and, less commonly, the iliac and inguinal nodes.

Most patients undergo total abdominal hysterectomy, bilateral salpingo-oophorectomy and removal or debulking of residual tumour. Radiotherapy or combination chemotherapy then follow, often with a 'second look' laparotomy to assess the success of treatment and to remove residual tumour.

The 5 year survival rate is 20−25% (27) and depends on disease severity at presentation and the response to treatment.

9.4.1 *Clinical staging of ovarian carcinoma*

Federation of Gynaecology and Obstetrics (FIGO) classification.

Stage I: growth limited to the ovary:
(a) IA limited to one ovary
(b) IB limited to both ovaries
(c) Stage IA or IB with ascites
Stage II: growth involving one or both ovaries with pelvic extension.
Stage III: growth involving one or both ovaries, with intraperitoneal metastases, limited to the abdomen.
Stage IV: growth involving one or both ovaries with distant metastases.

9.4.2 *Screening for ovarian cancer*

With the high resolution systems now in current use it is possible to delineate small cystic masses of ovarian origin. As 94% of all carcinomas contain cystic components and the majority of epithelial tumours are cystic, early detection of cystic carcinomas could be possible. Solid carcinomas that invade the capsule are more difficult to assess. Asymmetry of the ovaries by measurement of volumes can be detected and in a screening study of post-menopausal women useful information was obtained in a high proportion of patients scanned. Out of 1077 subjects scanned, 1041 were considered normal and 15 abnormal. Of these 15 cases one proved to be a Stage I carcinoma and nine others were classified as potentially malignant (48,49). Though considered cost-effective no screening programme has as yet been set up in the UK.

9.4.3 *Ultrasound features of ovarian cancer*

Ultrasound is usually the first imaging investigation to be carried out when the patient presents clinically with a pelvic mass. Sonography can confirm the mass is of ovarian origin, delineate its size and position for the surgeon, and may be able to indicate whether the lesion is malignant or not (*Table 2*). Features which suggest malignancy include: bilateral tumours; loss of definition of the cyst wall; fixation of the mass; a complex mass (particularly with large amounts of solid tissue within it) (*Figure 17*); ascites; omental plaques; and hepatic deposits (3). It has however proved impossible by this technique to predict accurately the

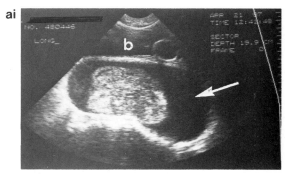

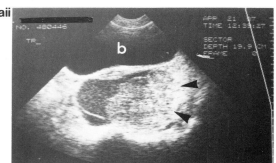

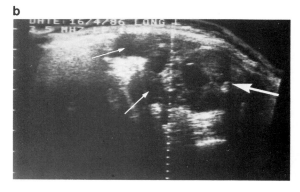

Figure 17. Ovarian carcinoma. Patients presenting with a pelvic mass. (**a**) Large ovarian cyst ↑ fixed in the pelvis. The bladder is seen anteriorly. There is a large volume of solid material in the cyst which on transverse scan appears to be invading the cyst wall▲ . (**i**) Sagittal section. (**ii**) Transverse section. (**b**) Complex cyst ↑ with loss of definition of the cyst wall superiorly and with massive direct invasion of tumour through the wall into the peritoneal cavity ↑
b = bladder.

histology of the different malignant ovarian tumours (51) (*Table 5*).

Patients with ovarian carcinoma may also present clinically with ascites and ultrasound will be the

Table 5. Ovarian masses.

Single cyst

Follicular cyst
Corpus luteum cyst
Parovarian cyst
Serous cystadenoma

Multiple cysts

Endometriosis
Theca lutein cysts
Polycystic ovaries
Pelvic inflammatory disease

Complex cysts

Cystic teratoma
Mucinous cystadenoma
Clear cell tumour
Cystadenofibroma
Malignant cysts
Infected cysts
Secondary tumours

Solid mass

Cystic teratoma
Fibroma
Brenner tumour
Solid teratoma
Granulosa—theca—luteal cell tumour
Arrhenoblastoma
Primary malignant tumours
Secondary tumours

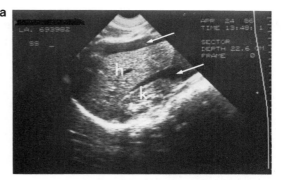

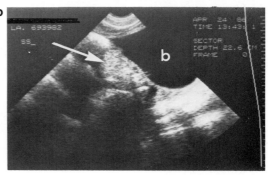

Figure 18. Ovarian carcinoma. Patient presenting with ascites and an enlarged ovary. (**a**) Sagittal section in the right upper quadrant showing ascitic fluid ↑ surrounding the liver. (**b**) Enlarged ovary ↑. No significant volume of fluid in the cul de sac which is obliterated by metastatic tumour. b = bladder; h = liver; k = kidney.

primary imaging investigation. Careful evaluation of the peritoneal cavity is essential and in the absence of an obvious large pelvic mass both ovaries should be carefully assessed for enlargement or textural changes (*Figure 18*).

Peritoneal involvement can be inferred if the patient develops ascites (*Figure 19*). Small amounts of ascitic fluid can be most easily found above the liver, or in the right paracolic gutter or in the pelvis above the bladder. The cul de sac is commonly obliterated with advanced ovarian cancer. Discrete peritoneal tumour deposits are much more difficult to detect, and probably need to be at least 2 cm in diameter before they can be imaged. Extensive involvement of the peritoneum is recognized as solid omental deposits, localized bowel involvement or as extensive matting of the bowel to the posterior abdominal wall (*Figure 19*). This is more easily recognized in the presence

of ascites. Intrahepatic metastases may occasionally be demonstrated, which indicates haematogenous spread. However, the more common metastases on the liver surface from direct peritoneal involvement are more easily identified at laparotomy. Enlarged para-aortic and para-caval nodes may be seen as lobulated echo-poor masses, but involved pelvic nodes are infrequently demonstrated. The examination is only complete with an assessment of the renal tracts to detect ureteric obstruction.

Imaging techniques, including ultrasound and CT, have not proved as useful to the clinician in the initial staging work-up of patients with ovarian cancer as they have for other pelvic malignancies. The accuracy of ultrasound and CT for staging ovarian cancer is about 70% (30). A meticulous staging laparotomy remains necessary in virtually every patient (51), as the accuracy of this procedure is still higher than any imaging modality. However, ultrasound still plays an

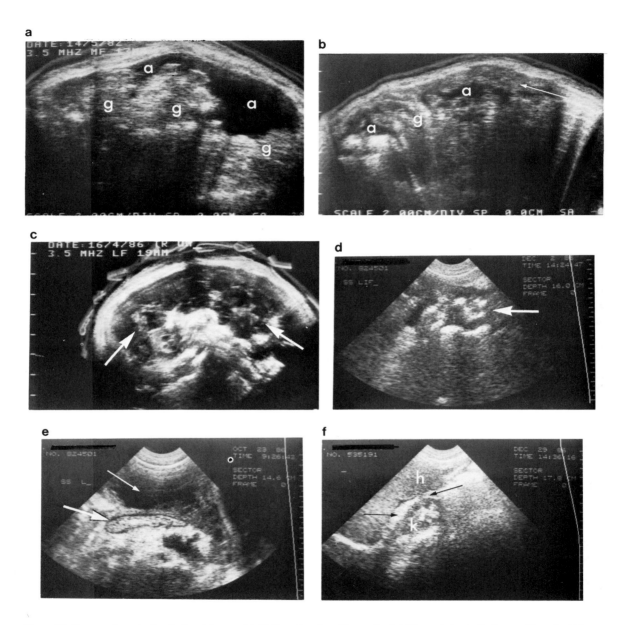

Figure 19. Tumour dissemination in the abdomen. (**a**) Malignant ascites. The ascitic fluid lies anterior to the loops of bowel which are becoming matted together and bound down posteriorly. Transverse section. (**b**) Omental infiltration. Mass of tumour involving the omentum ↑. Ascites is also present. Transverse section. (**c**) Massive tumour infiltration. There is a huge mass of tumour ↑ extending out of the pelvis involving bowel and omentum. Transverse section. (**d**) Localized fixation of bowel. Adherent segments of bowel ↑ with an associated small pocket of ascites. Sagittal section. (**e**) Bowel wall involvement. Fixed segment of bowel ↑ with adherent cystic tumour ↑. The patient had a faecal fistula. Transverse section. (**f**) Peritoneal involvement. Thickened peritoneum ↑ anterior to the right kidney. The peritoneum is extensively involved with tumour seedlings. There is a small volume of ascites in the right subhepatic space. Sagittal section. a = ascites; g = gut; h = liver; k = kidney.

important role in evaluating this disease.

Following therapy, whether by surgery or chemo-therapy, ultrasound has proved useful in the follow-up of patients to monitor response of established disease to treatment, and to detect recurrent disease. It is of value to the surgeon prior to a 'second look' laparotomy when it may be used to direct his attention to areas of suspicion in the abdomen, or to areas difficult to assess pre-operatively such as the liver and the retroperitoneal nodes. Sonography may also provide accurate information for the planning of radio-therapy.

9.4.4 Other imaging techniques

(i) *CT features.* After ultrasound assessment a patient suspected of having ovarian cancer usually undergoes a diagnostic and staging laparotomy. CT is rarely used for staging, although it can provide a fairly accurate assessment of the pelvic mass and the extent of tumour spread.

Like ultrasound, CT has a role in patient follow-up after treatment. However, as a negative CT does not exclude minor residual or recurrent disease (52) it has as yet not been found accurate enough to avoid a 'second look' laparotomy.

Ascites is well demonstrated by CT (53) and peritoneal and omental metastases are more easily detected than with ultrasound. Tumours as small as 1 cm in diameter have been demonstrated (54), however smaller lesions may be found at laparotomy.

(ii) *Lymphography.* Forty-six per cent of lymphograms performed for ovarian cancer are abnormal (55), whilst CT only demonstrated lymphadenopathy in 9% of patients in a different series. However lymphography is an invasive technique and is not used routinely in this disease, despite its greater accuracy than CT.

9.5 Secondary tumours of the ovary

The ovary is a frequent site for secondary malignancy. Carcinoma is most common and the primary site is usually the stomach, colon, uterus or breast. Metastatic leukaemia or lymphoma also occur. Most secondary tumours of the ovary probably arise from haema-togenous or lymphatic spread; less commonly peri-toneal implantation is implicated. In most cases the tumour affects both ovaries; however, if unilateral, the right side is most often affected.

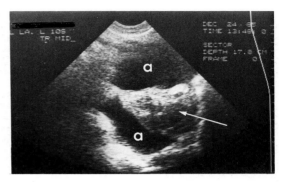

Figure 20. Secondary tumour of the ovary (Krukenberg). Patient presenting with ascites. There is tumour replacement of the left ovary ↑ and involvement of the uterus. The cul de sac is not obliterated and contains ascitic fluid (a).

9.5.1 Ultrasonic features

At ultrasound the affected ovary is enlarged and the ovarian mass is often indistinguishable from a primary ovarian tumour. The texture of the mass is variable and may be cystic, mixed or solid. If bilateral solid ovarian masses are found, the examination should be extended to search for a primary cause, for ascites and peritoneal involvement. An ovarian mass associated with ascites and with a clear pouch of Douglas is more likely to represent a secondary tumour than a primary of the ovary (personal observation) (*Figure 20*).

9.5.2 CT features

The CT appearance of secondary tumour of the ovary, like ultrasound, is of an ovarian mass lesion or lesions of variable internal structure. Some tumours are cystic, some are solid and others have a mixed appearance, again resembling a primary ovarian tumour. Ascites, and peritoneal deposits may be demonstrated by CT. On occasion, the primary neoplasm may also be identified (56).

10. References

1. Donald,I., MacVicar,J. and Brown,T.G. (1958) Investigation of abdominal masses by pulsed ultrasound. *Lancet*, **1**, 1188−1195.
2. Donald,I. (1965) Ultrasonic echo sounding in obstetrical and gynaecological diagnosis. *Am. J. Obstet. Gynecol.*, **93**, 935−941.

3. Morley,P. and Barnett,E. (1985) The ovarian mass. In *The Principles and Practice of Ultrasonography in Obstetrics and Gynaecology.* Sanders,R.C. and James,A.E. (eds), 3rd edition, Appleton-Century-Crofts, Norwalk, Connecticut, pp. 473−515.

4. Bald,R. (1983) Studien über die sonographische Endometrium-darstellung. *Med. Diss Marburg.*

5. Bald,R. and Hackelöer,B.J. (1983) Ultraschalldarstellung verschiedener Endometriumformen. In *Ultraschalldiagnostik 82.* Otto,R. and Jan,F.X. (eds), Thieme, Stuttgart, New York.

6. Hackelöer,B.J. and Nitschke-Dabelstein,S. (1980) Ovarian imaging by ultrasound: an attempt to define a reference plane. *J. Clin. Ultrasound,* **8**, 487−500.

7. Hackelöer,B.J., Fleming,R. and Robinson,H.P. (1979) Correlation of ultrasonic and endocrinologic assessment of human follicular development. *Am. J. Obstet. Gynecol.,* **135**, 122−128.

8. Fleischer,A.C., Daniell,J., Radier,J., Lindsay,H. and James, A.E. (1981) Sonographic monitoring of ovarian follicular development. *J. Clin. Ultrasound,* **9**, 275−280.

9. Hoult,I.J., Crespigny,L.Ch.de and O'Herlihy,C. (1981) Ultrasound control of clomiphene human chorionic gonadotrophin stimulated cycles for oocyte recovery and in-vitro fertilisation. *Fertil. Steril.,* **36**, 316−322.

10. Buttery,B., Trounson,A., MacMaster,R. and Wood,C. (1983) Evaluation of diagnostic ultrasound as a parameter of follicular development in an in-vitro fertilisation program. *Fertil. Steril.,* **39**, 458−463.

11. Sanders,R.C. (1985) Pelvic inflammatory disease and endometriosis. In *The Principles and Practice of Ultrasonography in Obstetrics and Gynaecology.* Sanders,R.C. and James,A.E. (eds), 3rd edition, Appleton-Century-Crofts, Norwalk, Connecticut, pp. 583−596.

12. Lees,R.F., Feldman,P.S. and Brenbridge,N.A. (1978) Inflammatory cysts of the pelvic peritoneum. *Am. J. Roentg.,* **131**, 633.

13. Friedman,H., Vogelzang,R.L., Mendelson,E.B., Neiman, H.L. and Cohen,M. (1985) Endometriosis detection by US with laparoscopic correlation. *Radiology,* **157**, 217−220.

14. Fleischer,A.C., Entman,S.S., Porrath,S.A. and James,A.E. (1985) Sonographic evaluation of uterine malformations and disorders. In *The Principles and Practice of Ultrasonography in Obstetrics and Gynaecology.* Sanders,R.C. and James,A.E. (eds), 3rd edition, Appleton-Century-Crofts, Norwalk, Connecticut, pp. 531−568.

15. Walsh,J.W., Taylor,K.J.W. and Rosenfield,A.T. (1979) Grey scale ultrasonography in the diagnosis of endometriosis and adenomyosis. *Am. J. Roentg.,* **132**, 87−90.

16. Dooms,C.G., Hricak,H. and Tscholakoff,D. (1986) Adnexal structures: MR imaging. *Radiology,* **158**, 639−646.

17. Cochrane,W.J. (1985) Ultrasound and the uterine device. In *The Principles and Practice of Ultrasonography in Obstetrics and Gynaecology.* Sanders,R.C. and James,A.E. (eds), 3rd edition, Appleton-Century-Crofts, Norwalk, Connecticut, pp. 597−602.

18. Richardson,M.L., Kinard,R.E. and Watters,D.H. (1982) Location of intrauterine devices: evaluation by computed tomography. *Radiology,* **142**, 690.

19. Fleischer,A.C., Porrath,S.A., Julian,C. and James,A.E. (1980) Sonographic evaluation of benign uterine disorders. In *The Principles and Practice of Ultrasonography in Obstetrics and Gynaecology.* Sanders, R.C. and James,A.E. (eds), 3rd edition, Appleton-Century-Crofts, Norwalk, Connecticut, pp. 387−406.

20. Gross,B.H., Moss,A.A., Mihara,K., Goldberg,H.I. and Glazer, G.M. (1983) Computed tomography of gynaecologic disease. *Am. J. Roentg.,* **141**, 765−773.

21. Hamlin,D.J., Pettersson,H., Fitzsimmons,J. and Morgan,L.S. (1985) MR imaging of uterine leiomyomas and their complications. *J. Comput. Assist. Tomogr.,* **9**, 902−907.

22. Kent Davis,W., McCarthy,S., Moss,A.A. and Braga,C. (1984) Computed tomography of gestational trophoblastic disease. *J. Comput. Assist. Tomogr.,* **8**, 1136−1139.

23. Reuter,K., Michelwitz,H. and Kahn,P.C. (1980) Early appearance of hydatidiform mole by ultrasound. A case report. *Am. J. Roentg.,* **134**, 588−589.

24. Wittmann,B.K., Fulton,L., Cooperberg,P.L., Lyons,E.A., Miller,C. and Shaw,D. (1981) Molar pregnancy: early diagnosis by ultrasound. *J. Clin. Ultrasound,* **9**, 153−156.

25. Reid,M.H., McGahan,J.P. and Oi,R. (1983) Sonographic evaluation of hydatidiform mole and its look-alikes. *Am. J. Roentg.,* **140**, 307−311.

26. Miyasaka,Y., Hachiya,J., Furuya,Y., Seki,T. and Watanabe, H. (1985) CT evaluation of invasive trophoblastic disease. *J. Comput. Assist. Tomogr.,* **9**, 459−462.

27. Choyke,P.L., Thickman,D., Kressel,H.Y., Lynch,J.H., Jaffe, M.H., Clark,L.R. and Zeman,R.K. (1985) Controversies in the radiologic diagnosis of pelvic malignancies. *Radiol. Clin. North Am.,* **23**, 531−549.

28. Brascho,D.J., Kim,R.Y. and Wilson,E.E. (1978) Use of ultrasonography in planning intracavitary radiotherapy of endometrial carcinoma. *Radiology,* **129**, 163−167.

29. Walsh,J.W. and Goplerud,D.R. (1982) Computed tomography of primary, persistent and recurrent endometrial malignancy. *Am. J. Roentg.,* **139**, 1149−1154.

30. Sanders,R.C., McNeil,B.J., Finberg,H.J., Hessel,S.J., Siegelman,S.S., Adams,D.F., Alderson,P.O. and Abrams, H.L. (1983) A prospective study of computed tomography and ultrasound in the detection of and staging of pelvic masses. *Radiology,* **146**, 439−442.

31. Butler,H., Bryan,P.J., Lipuma,J.P., Cohen,A.M., Yousef, S.E., Andriole,J.G. and Lieberman,J. (1984) Magnetic resonance imaging of the abnormal female pelvis. *Am. J. Roentg.,* **143**, 1259−1266.

32. Bies,J.R., Ellis,J.H., Kopecky,K.K., Sutton,G.P., Klatte, E.C., Stehman,F.B. and Ehrlich,C.E. (1984) Assessment of primary gynaecological malignancies. Comparison of 0.15T resistive MRI with CT. *Am. J. Roentg.,* **143**, 1249−1257.

33. Whitehouse,G.H. and Wright,C.H. (1986) Gynaecology. In *Diagnostic Radiology.* Grainger,R.G. and Allison,D.J. (eds), Churchill Livingstone, Edinburgh, UK, Vol. 3, pp. 1593−1629.

34. Husband,J.E. (1986) Computed tomography in gynaecology. In *Diagnostic Radiology.* Grainger,R.G. and Allison,D.J. (eds), Churchill Livingstone, Edinburgh, UK, Vol. 3, pp. 1624−1626.

35. Walsh,J.W., Amendola,M.A., Hall,D.J., Tisnado,J. and Goplerud,D.R. (1981) Recurrent carcinoma of the cervix: CT diagnosis. *Am. J. Roentg.,* **136**, 117−122.

36. Fuchs,W.A. and Seiler-Rosenberg,G. (1975) Lymphography in carcinoma of the uterine cervix. *Acta Radiol. (Diagn.),* **16**, 353−361.

37. Ginaldi,S., Wallace,S., Jing,B. and Bernardino,M.E. (1981) Carcinoma of the cervix: lymphangiography and computed tomography. *Am. J. Roentg.*, **136**, 1087−1091.

38. Warhit,J.M., Fagelman,D., Goldman,M.A., Weiss,L.M. and Sachs,L. (1984) Ovarian vein thrombophlebitis: diagnosis by ultrasound and CT. *J. Clin. Ultrasound*, **12**, 301−303.

39. Shaffer,P.B., Johnson,J.C., Bryan,D. and Fabri,P.J. (1981) Diagnosis of ovarian vein thrombophlebitis by computed tomography. *J. Comput. Assist. Tomogr.*, **5**, 436−439.

40. Adams,J., Polson,D.W. and Franks,S. (1986) Prevalence of polycystic ovaries in women with anovulation and idiopathic hirsutism. *Br. Med. J.*, **293**, 355−359.

41. Hann,L.E., Hall,D.A., McArdle,D.R. and Seibel,M.M. (1984) Polycystic ovarian disease: sonographic spectrum. *Radiology*, **150**, 531−534.

42. Parisi,L., Tramonti,M., Casciano,S., Zurli,A. and Gazzarrini, O. (1982) The role of ultrasound in the study of polycystic ovarian disease. *J. Clin. Ultrasound*, **10**, 167−172.

43. Swanson,M., Sauerbrei,E.E. and Cooperberg,P.L. (1981) Medical implications of ultrasonically detected polycystic ovaries. *J. Clin. Ultrasound*, **9**, 219−222.

44. Guttman,P.H. (1977) In search of the elusive benign cystic ovarian teratoma: application of the ultrasound 'Tip of the Iceberg sign'. *J. Clin. Ultrasound*, **5**, 403−406.

45. Laing,F.C., Van Dalsem,V.F., Marks,W.M., Barton,J.L. and Martinez,D.A. (1981) Dermoid cysts of the ovary: their ultrasonographic appearance. *Obstet. Gynecol.*, **57**, 99−104.

46. Sawyer,R.W., Vick,C.W., Walsh,J.W. and McClure,P.H. (1985) Computed tomography of benign ovarian masses. *J. Comput. Assist. Tomogr.*, **9**, 784−789.

47. Feldberg,M.A.M., Van Waes,P.F.G.M. and Hendriks,M.J. (1984) Direct multiplanar CT findings in cystic teratoma of the ovary. *J. Comput. Assist. Tomogr.*, **8**, 1131−1135.

48. Campbell,S., Goesseus,L., Goswamy,R. and Whitehead,M.I. (1982) Real time ultrasonography for determination of ovarian morphology and volume. A possible early screening test for ovarian cancer. *Lancet*, **1**, 425−426.

49. Goswamy,R.K., Campbell,S. and Whitehead,M.I. (1983) Screening for ovarian cancer. In *Clinics in Obstetrics and Gynaecology*. Campbell,S. (ed.), W.B.Saunders, London, Vol. 10, p. 621.

50. Moyle,J.W., Rochester,D., Sider,L., Shrock,K. and Krause,P. (1983) Sonography of ovarian tumors: predictability of tumor type. *Am. J. Roentg.*, **141**, 985−991.

51. Mamtora,H. and Isherwood,I. (1982) Computed tomography in ovarian carcinoma: patterns of disease and limitations. *Clin. Radiol.*, **33**, 165−171.

52. Levitt,R.G., Sagel,S.S. and Stanley,R.J. (1981) Detection of neoplastic involvement of the mesentery and omentum by computed tomography. *Am. J. Roentg.*, **131**, 835−838.

53. Kalovidoris,A., Gouliamos,G.P., Gennatas,K., Dardoufas,K. and Papavasiliou,C. (1984) Computed tomography of ovarian carcinoma. *Acta Radiol. (Diagn.)*, **25**, 203−208.

54. Johnson,R.J., Blackledge,G., Eddleston,B. and Crowther,D. (1983) Abdomino-pelvic computed tomography in the management of ovarian carcinoma. *Radiology*, **146**, 447−452.

55. Athey,P.A., Wallace,S., Jing,B., Gallager,H.S. and Smith,J.P. (1975) Lymphangiography in ovarian cancer. *Am. J. Roentg.*, **123**, 106−113.

56. Megibow,A.J., Hulrick,D.H., Bosniak,M.A. and Balthazar, E.J. (1985) Ovarian metastases: computed tomographic appearances. *Radiology*, **156**, 161−164.

11. Further reading

1. Sanders,R.C. and James,A.E., eds (1985) *Principles and Practice of Ultrasonography in Obstetrics and Gynaecology*. 3rd edition, Appleton-Century-Crofts, Norwalk, Connecticut.

2. Hansmann,M., Hackelöer,B.J. and Staudach,A. (1985) *Ultrasound Diagnosis in Obstetrics and Gynaecology*. Springer Verlag, Berlin, Heidelberg, New York, Tokyo.

3. Barnett,E. and Morley,P., eds (1985) *Clinical Diagnostic Ultrasound*. Blackwell Scientific Publications, Oxford, London, Edinburgh, Boston, Palo Alto, Melbourne.

4. Moss,A.A., Gamsin,G. and Genant,H.K. (1983) *Computed Tomography of the Body*. W.B.Saunders, Philadelphia, London.

5. Magnetic Resonance Imaging. *Radiol. Clin. North Am.* (1984) Volume 22, No. 4, W.B.Saunders, Philadelphia, London.

Chapter 11

Vascular system

C.D.Sheldon

1. Introduction

In recent years increasing use has been made of Doppler ultrasound techniques in the investigation of vascular disease. Techniques employed vary considerably from centre to centre, and utilize a wide spectrum of equipment, ranging from a simple Doppler stethoscope costing around £100 to a colour-coded duplex scanner costing over £100 000.

There have been conflicting reports on the relative merits of the various modalities, and in most cases Doppler techniques are not used as the definitive investigation. However, they have been shown to provide a cost-effective initial screening, which results in an overall reduction in other techniques such as angiography and provide a non-invasive method to follow-up the progress of treatment.

This chapter reviews the different techniques used to assess a variety of vascular diseases and includes comments based on the author's own experience where appropriate.

2. Lower limb arterial disease

Chronic occlusive arterial disease of the lower limb is a debilitating disease which in its early stages results in symptoms of intermittent claudication (cramp-like pain in the leg muscles after walking a certain distance). The disease is often progressive, with a rapid decrease in the claudication or walking distance until pain at rest supervenes. Eventually this rest pain may become severe and ulceration and gangrene occur. Reconstructive vascular surgery may be performed for disabling claudication, but is usually reserved as a limb salvage procedure, when the only other course of treatment would be amputation.

Arteriography remains the definitive investigation for lower limb arterial disease—it gives an anatomical representation of the disease considered essential by the vascular surgeon. However, it is unable to estimate the haemodynamic significance of a particular lesion and tends to underestimate the extent of the disease in some situations. Arteriography is invasive and, despite recent advances, still has a small morbidity and mortality associated with the technique. Arteriography should only be performed when surgical intervention is planned.

2.1 Measurement of blood pressure

A simple Doppler stethoscope can detect flow in the peripheral arteries of an ischaemic leg, such as the posterior tibial, dorsalis pedis and peroneal arteries, in situations in which pulses cannot be detected by conventional methods. The systolic blood pressure at the ankle level can be determined by inflating a standard 12 mm wide sphygmomanometer pressure cuff placed just above the ankle until the flow signal ceases. Systolic pressure is measured as the pressure at which flow is just detected as the cuff is slowly deflated. In the normal resting subject in a supine position this ankle systolic pressure is equal to or greater than the brachial systolic pressure.

Yao (1) defined pressure index as the ratio of the ankle systolic pressure to the brachial systolic pressure. Normally, the pressure index is greater or equal to one; in the patient with significant occlusive arterial disease it is less than one. When this index was used to assess the severity of disease, it was found to have a significant correlation with the patient's symptoms.

Problems are encountered in a small number of patients where it is not possible to compress the arteries of the leg due to extensive calcification of the vessel wall and hence a measurement of systolic blood pressure cannot be obtained. Furthermore, fluctuations in the blood pressure between readings may produce

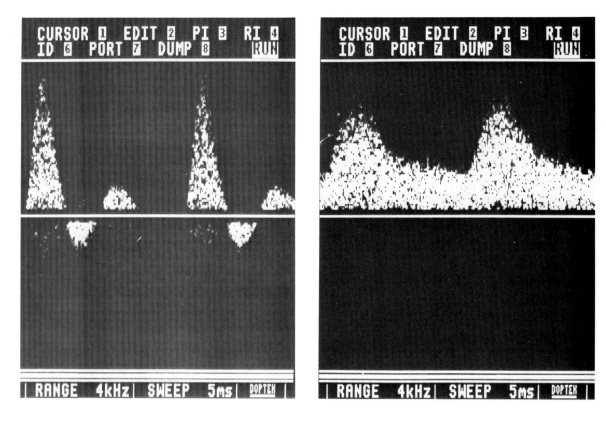

Figure 1. Doppler sonograms from the posterior tibial artery of (**a**) a normal subject and (**b**) a patient with a superficial femoral artery occlusion.

errors, but these can be reduced by taking simultaneous measurements of brachial and ankle pressure with two operators.

A measurement of resting pressure index is invaluable in excluding sources of leg pain which are not vascular in origin and preventing unnecessary arteriography. It provides the clinician with a method of quantifying the severity and progression of the disease and is useful in assessing the success of treatment and as such should be used in every vascular clinic and ward. However, such a simple measurement does have its limitations which need to be recognized.

A simple resting measurement of ankle pressure index gives no information about the anatomical position of the disease. Some further information can be obtained by performing segmental pressures using a cuff of suitable dimensions around the thigh. Many centres monitor the ankle pressure at regular intervals following exercise on a treadmill. Since the pressure

difference across a stenotic segment of artery is proportional to blood flow, a standard exercise test will improve the sensitivity of the test to minor disease, particularly in the detection of stenoses in the aorto-iliac segment. It is also claimed that some information about the position of the disease may be inferred from the exercise pressure response.

2.2 *Waveform analysis*

Audible interpretation of the Doppler ultrasound signal is adequate for simple applications, such as blood pressure measurement, but when attempts are made to characterize changes in the audio signal which may occur in disease it becomes subjective and dependent upon the skill of the operator.

Spectral analysis is the preferred method of analysing the Doppler signal and may be performed in real time (see Chapter 4). The output from such a system is normally displayed as a frequency versus time sonogram and an example obtained from the posterior tibial vessel of a normal resting subject is shown in *Figure 1a*. In such a display the x-axis represents time, the y-axis is the Doppler shift frequency, which is proportional to velocity, and the darkening or grey scale of the display represents the amplitude of the signal, which is approximately proportional to the number of scattering red blood cells moving at each velocity. This particular display is a directional sonogram, with velocities towards the probe shown above the line and velocities away from the probe shown below the line.

The arteries of the lower limb supply the relatively high impedance tissues of the leg and therefore in the normal, resting subject there is a portion of reverse flow following systole. However, in the presence of significant arterial narrowing, the characteristic shape of the sonogram is altered. *Figure 1b* shows the posterior tibial sonogram in a patient with a superficial femoral artery occlusion and demonstrates a damping of the waveform with loss of the reverse flow component and continuous flow in diastole.

Doppler sonograms may be obtained from many sites in the leg. Visual inspection by an experienced operator will be able to identify the location of the more severe lesions by noting the degree of damping at the various sites, as reported by Walton *et al.* (2) for the aorto-iliac segment. However, to provide a more objective approach to the analysis, some method of quantifying the sonogram is required.

The Doppler sonogram contains information about the velocities of all the moving scatterers within the sensitive area or sample volume of the transducer. The amount of data contained within a heart cycle is considerable, and therefore it is usual to simplify the sonogram for analysis. A zero crossing meter (see Chapter 4) provides a low cost method of processing the audio signal and gives a voltage output which is proportional to the root-mean-squared frequency of the input signal. Such systems are amplitude dependent and are subject to artefactual disturbance.

In the absence of significant disturbed flow, sonograms from the lower limb arteries usually exhibit waveforms which have a well-defined maximum frequency envelope and it is this single-value waveform which is almost exclusively used to quantify the sonogram. Although this maximum frequency envelope may be quite adequately estimated from eye-balling the sonogram, many of the commercial Doppler spectrum analysers will estimate the instantaneous maximum frequency and compute the Doppler

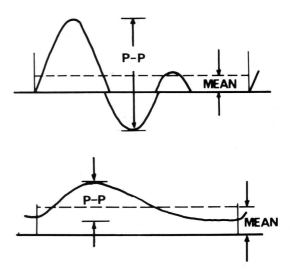

Figure 2. Calculation of Pulsatility Index (PI) in a normal and damped arterial waveform, where PI = peak-to-peak amplitude (P−P)/time averaged mean.

indices. It is essential to superimpose this waveform onto the original sonogram to ensure that spurious noise or low amplitude signals are not producing an incorrect envelope.

There are several reported methods of quantifying the maximum frequency envelope, but there appears to be no general agreement over which method gives the most clinical information. Pulsatility index (PI), as described by Gosling and King (3), has gained widespread popularity because of its simplicity. It is defined as the peak to peak frequency (allowing for any reverse flow) divided by the mean frequency over one cardiac cycle (*Figure 2*). It is a dimensionless index and independent of the angle of insonation. Pulsatility index is normally calculated from signals obtained from the femoral, popliteal and pedal pulse sites in the supine patient. Prior to taking the measurement the patients should be allowed to rest for at least 10 min, since pulsatility index depends on peripheral resistance (4). Several cardiac cycles should be analysed to reduce variations caused by heart rate changes and respiration. Analysis of disease below the femoral artery is usually accomplished by measuring the ratio of the proximal to the distal pulsatility index, known as the pulsatility index damping factor (3).

Published results on the accuracy of pulsatility index and damping factor vary. The author's own studies show that in the detection of aorto-iliac disease, a

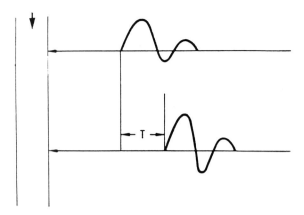

Figure 3. Measurement of the pulse wave transit time (T).

measurement of common femoral pulsatility index can detect occlusions with a sensitivity of 94% and specificity of 90%, but is less successful in detecting significant arteriographically demonstrated stenotic disease (sensitivity 71%, specificity 69% for stenoses $\geq 20\%$ diameter reduction) (5). When compared with another haemodynamic assessment, the papaverine test (6), agreement was improved (sensitivity 81%, specificity 82%) (7). Some advocate the routine use of common femoral pulsatility index to detect unrecognized proximal disease prior to femoro-popliteal grafting (8).

Results for the detection of superficial femoral disease were less satisfactory. From a measurement of femoral to popliteal damping factor, the sensitivity for the detection of occlusions was 77% and specificity 76% (5).

Several reasons have been put forward to explain the limitations of pulsatility index measurements in the detection of peripheral arterial disease. There is agreement in the literature that it is not only disease proximal to the site of measurement that affects pulsatility index; increasing distal stenosis has been shown to reduce pulsatility index (9), thereby affecting the accuracy of detecting both aorto-iliac and femoro-popliteal disease (10).

Pulse wave transit time is another parameter which has been advocated for assessing segmental arterial disease. It is measured as the foot-to-foot time between the onset of the velocity waveform at the proximal and distal site under investigation and is normalized by the mean blood pressure (*Figure 3*) (3). Transit time is dependent on the compliance of the vessel and will be decreased by age but increased by severe occlusive disease. Transit time gave similar results to damping factor for the detection of femoro-popliteal disease, and may be useful in detecting more distal disease (5).

There are two methods of analysing the maximum frequency envelope which have not received the same popularity because of their more mathematical nature. The first of these is the transfer function analysis method described by Skidmore and Woodcock (11). In this technique the velocity/time waveform is digitized and the discrete Fourier transform is calculated. The equivalent Laplace transform may be obtained by a curve-fitting procedure and described by a third order polynomial. The roots of the polynomial are calculated to obtain three coefficients δ, γ and ω_o. This procedure is relatively straightforward and may be carried out on a small digital computer.

δ, later known as the Laplace transform damping, has been compared with pulsatility index for the detection of aorto-iliac disease and was found to be more accurate at detecting stenoses greater than 50% (sensitivity 85%, specificity 84%) (12). The ω_o gradient (proximal ω_o/distal ω_o) has been compared with pulsatility index damping factor for the detection of femoro-popliteal occlusion and showed an improved accuracy in one study (13), but not in another (14).

The other method, principal component analysis, was first described for the analysis of carotid artery waveforms (15) and has the advantage that the waveform can be accurately described by only two or three terms once the principal components for the population being studied has been obtained. It has been shown to be more sensitive than pulsatility index or Laplace transform damping in the assessment of aorto-iliac disease (16).

3. Cerebrovascular disease

Cerebrovascular disease is the third most common cause of death in the UK after heart disease and malignant neoplasm. There is no treatment for cerebral infarction once it has occurred and some patients will have no warning of an impending stroke. Others may have a short lived weakness in an arm or leg or suffer from a transient visual disturbance, known as transient ischaemic attacks. The management of these patients is controversial, with some clinicians advocating anti-platelet therapy, without identifying the reason for the attack. Others propose surgical intervention, since Hass *et al.* (17) found that 75% of patients with

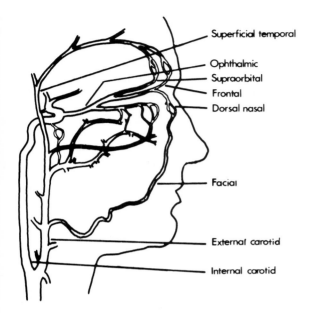

Figure 4. The periorbital arterial circulation.

Labels in figure: Superficial temporal, Ophthalmic, Supraorbital, Frontal, Dorsal nasal, Facial, External carotid, Internal carotid

symptoms of cerebrovascular ischaemia had arterial disease of the extracranial vessels in the neck at sites accessible to surgery. When surgical treatment is being considered, it is necessary to identify the presence and extent of any arterial disease. Doppler techniques are used in many centres as a routine screening test, before other investigations such as arteriography are performed.

3.1 *Periorbital tests*

The direction of flow in the terminal branches of the ophthalmic artery (*Figure 4*) may be monitored by a simple directional Doppler instrument placed on the eyelid. Normally the blood flow in these periorbital vessels is out of the orbit, but severe internal carotid artery disease may cause reversal of the flow direction, with blood flowing into the ophthalmic artery from the anastomotic connections with the various branches of the external carotid artery. Although flow reversal usually indicates severe pathology, a normal flow direction does not preclude disease since collateral pathways can maintain the periorbital flow direction.

The existence of collateral pathways may be determined by sequential compression of the possible collateral pathways. For example the well known temporal artery occlusion test, where the temporal artery is momentarily occluded by external compression, will often result in an augmentation of the supraorbital or frontal artery flow in the normal. However, when the internal carotid artery is diseased and the superficial temporal artery is the dominant collateral vessel, compression will reduce or reverse an abnormal orbital flow. Compression of the other branches of the external carotid artery or contralateral common carotid may also be performed and the functional collateral pathways established.

Published data on the accuracy of the periorbital Doppler test in detecting carotid artery disease show wide variations, but one report by Barnes *et al.* (18) gave a sensitivity of only 59% for the detection of disease of more than a 50% diameter stenosis. It is generally agreed that less significant disease which could be considered suitable for operative treatment cannot be detected. Therefore, periorbital Doppler tests alone do not provide a sufficiently accurate method for cerebrovascular screening, however, they are useful in confirming occlusions of the internal carotid artery detected by imaging techniques.

3.2 *Waveform analysis*

Doppler signals may be obtained from many of the vessels in the neck using either hand-held non-imaging systems or Doppler or duplex imaging scanners. *Figure 5* shows the Doppler sonograms obtained with a pulsed Doppler instrument from the common, internal and external carotid arteries of a normal person. Each artery has its own characteristic waveform shape, with the internal carotid having a higher mean frequency but less pulsatile waveform than the external carotid artery, due to its lower cerebrovascular resistance. The shape of the maximum frequency envelope changes with disease and a number of parameters have been described to detect these changes.

Planiol and Pourcelot (19) described their resistance index as $(S-D)/S$ where S is the peak systolic velocity and D is the end diastolic velocity (*Figure 6*). Values of resistance index above 0.75 in the common carotid artery indicated increased peripheral resistance, but the accuracy of this measurement in determining internal carotid artery disease was not established.

Baskett *et al.* (20) measured the ratio of the heights of two systolic peaks A/B (*Figure 7*) to assess carotid disease. In normal vessels it was found that A/B was greater than 1.05 in the common carotid artery, but only 31% of diseased vessels had a value less than

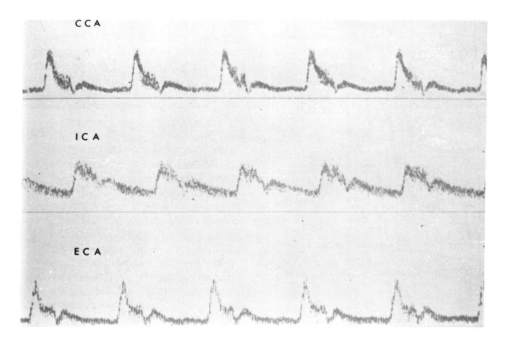

Figure 5. Doppler sonograms obtained from the common carotid artery (CCA), internal carotid artery (ICA) and external carotid artery (ECA) of a normal person.

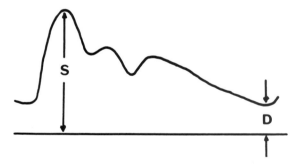

Figure 6. Measurement of Resistance Index (RI) where RI = (S−D)/S.

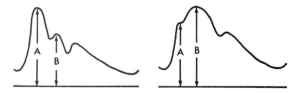

Figure 7. Location of the A and B systolic peaks in a normal and abnormal common carotid waveform.

1.05. However, when combined with a measurement from the ipsilateral supraorbital artery a sensitivity of 75% and specificity of 90% was obtained for the detection of all grades of disease.

Principal component analysis was originally applied to the maximum frequency envelope of the common carotid waveform (15), and gave slightly better results than the A/B ratio.

3.3 *Vessel imaging*

Ultrasonic imaging of the carotid arteries was first attempted by Olinger (21) using a pulse−echo B-scan system. The technique improved with the development of real-time systems, but difficulty was experienced in visualizing lesions which appeared in the acoustic shadow produced by wall calcification. Furthermore, it was not always possible to distinguish a non-calcified plaque or stationary thrombus from blood (22) due to their similar acoustic impedance, with the result that

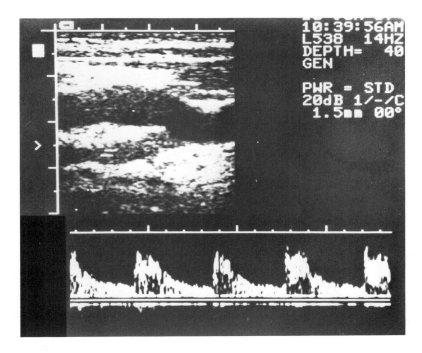

Figure 8. Duplex scan of a normal carotid bifurcation. The position of the Doppler sample volume in the external carotid artery is shown by the two parallel lines.

even totally occluded vessels sometimes appeared patent on the pulse—echo image.

The technique was improved by combining a real-time pulse—echo system with a pulsed Doppler unit mounted in single probe assembly (see Chapter 4). These so-called duplex systems produce a conventional pulse—echo image with the position of the Doppler beam and its sample volume superimposed (*Figure 8*). Duplex systems allow an occlusion to be diagnosed from the absence of a Doppler signal when the sample volume is positioned within the lumen of the imaged vessel.

Ultrasonic imaging of the carotid vessels may also be obtained by Doppler techniques using either continuous (24) or pulsed (25) ultrasound (see Chapter 4); an example of a normal carotid bifurcation obtained with a pulsed system is shown in *Figure 9*. Continuous wave systems have the advantage that there is no frequency or range ambiguity, therefore high velocities obtained from a stenosis can be accurately recorded, however, they provide no depth resolution which can be a significant problem when two vessels lie on top of each other.

Like the pulse—echo systems, reliance on the image alone is misleading, particularly when non-visualization of segments of vessels occurred as a result of an unfavourable vessel to probe angle or wall calcification (26). The resolution of the Doppler imaging system is poor, especially with some of the continuous wave systems, and the quality of the image is dependent on the skill of the operator and the time spent on each view. The pulse—echo imaging system has the potential to detect minor, haemodynamically insignificant disease but can be more difficult than a Doppler imaging system to produce a diagnostically relevant image in the presence of disease. Some of the duplex scanners have large probes which makes it difficult to obtain views of the proximal, common carotid, vertebral or subclavian vessels.

Whichever imaging modality is used, confirmation of a haemodynamically significant stenosis is usually based on the audio or spectral interpretation of the Doppler signal obtained from the vessel under investigation.

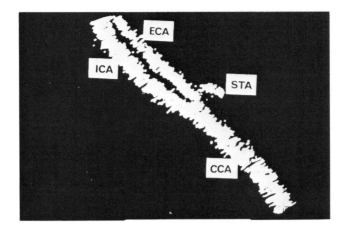

Figure 9. A normal scan of the carotid bifurcation obtained with a pulsed Doppler imaging system and showing the common carotid artery (CCA), internal carotid artery (ICA), external carotid artery (ECA) and superior thyroid artery (STA).

3.4 *Spectral interpretation*

As well as changes in the shape of the maximum frequency envelope discussed earlier, measurements taken in the vicinity of a stenosis will show other spectral changes. *Figure 5* showed the sonogram from a normal internal carotid artery sonogram obtained with a pulsed Doppler system where the sample volume was positioned in the centre of the vessel. There is a well-defined maximum frequency envelope with a narrow band of frequencies and a clear area or window below. This is due to the fact that within the sample volume of the instrument, most of the red blood cells are moving with the same velocity. By comparison, *Figure 10* shows the sonograms from a patient with a significant internal carotid artery stenosis. Within the stenosis, the Doppler frequency increases due to the reduction in cross-sectional area. Downstream from the stenosis, disturbed or turbulent flow produces broadening of the spectrum which results in both loss of the well-defined maximum frequency and loss of the clear window. Subjective assessment of these spectral changes has been used to detect stenotic lesions (27) and overall accuracies of around 90% have been obtained. More objective measurements have been made of both maximum frequency and spectral broadening.

Measurement of the maximum Doppler frequency produced by the stenosis has enabled residual lumen diameters of 3.2 mm or less to be diagnosed with a sensitivity of 92% and specificity of 94% (28). However, less significant disease cannot be detected from a single measurement of peak frequency due to variations in the insonation angle or variations in the carotid blood flow, caused for example by severe contralateral disease. These errors may be reduced by measuring the vessel/probe angle on a real-time image to calculate the velocity and using the ratio of the internal carotid velocity to the common carotid velocity. Using these corrections, the sensitivity of the technique for detecting less significant disease improves (29).

A number of methods of quantification of the degree of spectral broadening have been described. Some of the best results have been obtained by taking the ratio of the maximum to mean frequency at peak systole (30)—this was able to detect stenoses which reduced the lumen diameter by 25% or more with a sensitivity of 90% and specificity of 98%.

Ultrasonic imaging combined with Doppler spectral analysis provides a non-invasive and accurate method of screening for carotid artery disease.

3.5 *Intracranial Doppler assessment*

Although severe intracranial disease can be inferred from an abnormal internal carotid waveform, most intracranial disease will not be detected by an examination of the extracranial circulation. Visualization of the intracranial vessels is possible in the neonate,

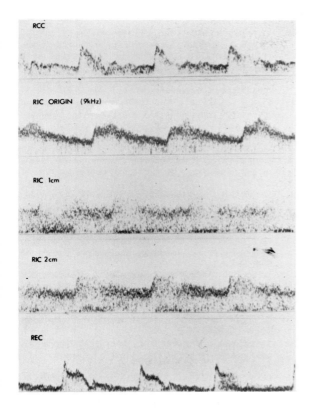

Figure 10. Doppler sonograms obtained from the right common (RCC), right internal carotid (RIC) and right external carotid (REC) arteries of a patient with an internal carotid artery stenosis. The frequency scale is 0−4.5 kHz on all recordings except for RIC origin, which is 0−9 kHz.

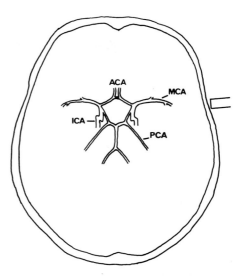

Figure 11. Transcranial approach showing the middle cerebral artery (MCA), anterior cerebral artery (ACA), posterior cerebral artery (PCA) and internal carotid artery (ICA).

4. Conclusions

Doppler ultrasound techniques have achieved wide acceptance as a diagnostic tool in the assessment of vascular disorders. The presence of arterial disease of the lower limb can be determined with a simple measurement of pressure index and Doppler imaging and spectral analysis provides an accurate method of screening for extracranial disease.

Conflicting reports on the accuracy of lower limb waveform analysis, and the reluctance of clinicians to accept new techniques involving mathematical analysis, has meant that its impact on the management of the ischaemic limb has been slow. However, the increasing use of duplex scanners and the development of new techniques such as transcranial Doppler will ensure that Doppler will play an increasing role in the diagnosis of vascular disease in the years to come.

before closure of the anterior fontanelle, but in the adult this is difficult because of the high attenuation of bone. Recently, Aaslid *et al.* (31) have used a 2 MHz pulsed Doppler to insonate the intracranial vessels through thin areas of temporal bone above the zygomatic arch (*Figure 11*). By changing both the angulation of the transducer and the depth of the sample volume it is possible to obtain Doppler recordings from the middle cerebral, anterior cerebral and posterior cerebral arteries of most patients.

The applications of a Doppler assessment of intracranial haemodynamics are slowly being realized. It has been used to assess the intracranial vessels and circle of Willis in patients with atherosclerosis and cerebral spasm and to evaluate various neonatal disorders such as hydrocephalus and birth asphyxia.

5. References

1. Yao,S.T. (1970) Haemodynamic studies in peripheral arterial disease. *Br. J. Surg.*, **57**, 761−766.
2. Walton,L., Martin,T.R.P. and Collins,M. (1984) Prospective assessment of the aorto-iliac segment by visual interpretation

of frequency analysed Doppler waveforms—a comparison with arteriography. *Ultrasound Med. Biol.*, **10**, 27 – 32.

3. Gosling,R.G. and King,D.H. (1975) Ultrasonic angiology. In *Arteries and Veins*. Harcus,A.W. and Adamson,L. (eds), Churchill Livingstone, Edinburgh, pp. 61 – 98.

4. Evans,D.H., Barrie,W.W., Asher,M.J., Bentley,S. and Bell, P.R.F. (1980) The relationship between ultrasonic pulsatility index and proximal arterial stenosis in a canine model. *Circ. Res.*, **46**, 470 – 475.

5. Booth,D.B., Forrest,H., Morrice,J.J., Quin,R.O., Sheldon, C.D. and Vallance,R. (1981) The role of non-invasive Doppler sonography in the assessment of peripheral vascular disease. In *Haemodynamics of the Limbs-2*. Puel,P., Boccalon,H. and Enjalbert,A. (eds), G.E.P.E.S.C., Toulouse, pp. 73 – 79.

6. Quin,R.O., Evans,D.H. and Bell,P.R.F. (1975) Haemo-dynamic assessment of the aorto-iliac segment. *J. Cardiovasc. Surg.*, **16**, 586 – 589.

7. Sheldon,C.D. (1980) *Ultrasonic Doppler Arteriography. Evaluation Report.* Scottish Home and Health Department.

8. Charlesworth,D., Harris,P.L., Cave,F.D. and Taylor,L. (1975) Undetected aorto-iliac insufficiency: a reason for early failure of saphenous vein bypass grafts for obstruction of the super-ficial femoral artery. *Br. J. Surg.*, **62**, 567 – 570.

9. Aukland,A. and Hurlow,R.A. (1982) Spectral analysis of Doppler ultrasound: its clinical application in lower limb ischaemia. *Br. J. Surg.*, **69**, 539 – 542.

10. Baker,A.R., Evans,D.H., Prytherch,D.R., Morton,D.B., Bentley,S., Asher,M.J. and Bell,P.R.F. (1986) Some failings of pulsatility index and damping factor. *Ultrasound Med. Biol.*, **12**, 875 – 881.

11. Skidmore,R. and Woodcock,J.P. (1980) Physiological interpre-tation of Doppler-shift waveforms—I Theoretical considera-tions. *Ultrasound Med. Biol.*, **6**, 7 – 10.

12. Baird,R.N., Bird,D.R., Clifford,P.C., Lusby,R.J., Skidmore, R. and Woodcock,J.P. (1980) Upstream stenosis. Its diagnosis by Doppler signals from the femoral artery. *Arch. Surg.*, **115**, 1316 – 1322.

13. Campbell,W.B., Baird,R.N., Cole,S.E.A., Evans,J.M., Skid-more,R. and Woodcock,J.P. (1983) Physiological interpre-tation of Doppler shift waveforms: the femorodistal segment in combined disease. *Ultrasound Med. Biol.*, **9**, 265 – 269.

14. Baker,A.R., Prytherch,D.R., Evans,D.H. and Bell,P.R.F. (1986) Doppler ultrasound assessment of the femoro-popliteal segment: comparison of different methods using ROC curve analysis. *Ultrasound Med. Biol.*, **12**, 473 – 482.

15. Martin,T.R.P., Barber,D.C., Sherriff,S.B. and Prichard,D.R. (1980) Objective feature extraction applied to the diagnosis of carotid artery disease uing a Doppler ultrasound technique. *Clin. Phys. Physiol. Meas.*, **1**, 71 – 81.

16. Macpherson,D.S., Evans,D.H. and Bell,P.R.F. (1984) Common femoral artery Doppler wave-forms; a comparison

of three methods of objective analysis with direct pressure measurements. *Br. J. Surg.*, **71**, 46 – 49.

17. Hass,W.K., Fields,W.S., North,R.R., Kricheff,I.I., Chase, N.E. and Bauer,R.B. (1968) Joint study of extracranial arterial occlusion. *J. Am. Med. Assoc.*, **203**, 159 – 166.

18. Barnes,R.W., Rittgers,S.E. and Putney,W.W. (1982) Real-time Doppler spectrum analysis. *Arch. Surg.*, **117**, 52 – 57.

19. Planiol,T. and Pourcelot,L. (1974) Doppler effect study of the carotid circulation. In *Ultrasonics in Medicine*. deVlieger,M., White,D.N. and McCready,V.R. (eds). Excerpta Medica, Amsterdam, pp. 104 – 111.

20. Baskett,J.J., Beasley,M.G., Murphy,G.J., Hyams,D.E. and Gosling,R.G. (1977) Screening for carotid junction disease by spectral analysis of Doppler signals. *Cardiovasc. Res.*, **11**, 147 – 155.

21. Olinger,C.P. (1969) Ultrasonic carotid echoarteriography. *Am. J. Roentg. Rad. Ther. Nucl. Med.*, **106**, 282 – 295.

22. Blackshear,W.M., Phillips,D.J., Thiele,B.L., Hirsch,J.H., Chikos,P.M., Marinelli,M.R., Ward,K.J. and Strandness,D.E. (1979) Detection of carotid occlusive disease by ultrasonic imaging and pulsed Doppler spectrum analysis. *Surgery*, **86**, 698 – 706.

23. Barber,F.E., Baker,D.W., Nation,A.W.C., Strandness,D.E. and Reid,J.M. (1974) Ultrasonic duplex echo-Doppler scanner. *IEEE Trans. Biomed. Eng.*, **BME 21**, 109 – 113.

24. Reid,J.M. and Spencer,M.P. (1972) Ultrasonic Doppler technique for imaging blood vessels. *Science*, **176**, 1235 – 1236.

25. Mozersky,D.J., Hokanson,D.E., Baker,D.W., Sumner,D.S. and Strandness,D.E. (1971) Ultrasonic arteriography. *Arch. Surg.*, **103**, 663 – 667.

26. Berry,S.M., O'Donnell,J.A. and Hobson,R.W. (1980) Cap-abilities and limitations of pulsed Doppler sonography in carotid imaging. *J. Clin. Ultrasound*, **8**, 405 – 412.

27. Murie,J.A., Sheldon,C.D., Forrest,H. and Quinn,R.O. (1983) Pulsed Doppler imaging for carotid artery disease. *Scot. Med. J.*, **28**, 21 – 24.

28. Brown,P.M., Johnston,K.W., Kassam,M. and Cobbold,R.S.C. (1982) A critical study of ultrasound Doppler spectral analysis for detecting carotid disease. *Ultrasound Med. Biol.*, **8**, 515 – 523.

29. Blackshear,W.M., Phillips,D.J., Chikos,P.M., Harley,J.C., Theile,B.L. and Strandness,D.E. (1980) Carotid artery velocity patterns in normal and stenotic vessels. *Stroke*, **11**, 67 – 71.

30. Sheldon,C.D., Murie,J.A. and Quin,R.O. (1983) Ultrasonic Doppler spectral broadening in the diagnosis of internal carotid artery stenosis. *Ultrasound Med. Biol.*, **9**, 575 – 580.

31. Aaslid,R., Markwalder,T. and Nornes,H. (1982) Noninvasive transcranial Doppler ultrasound recording of flow velocity in basal cerebral arteries. *J. Neurosurg.*, **57**, 769 – 774.

Chapter 12

Superficial structure scanning

David Cooke and Henry C.Irving

1. Introduction

As the frequency of ultrasound increases, so the wavelength decreases and hence the axial resolution along the beam will be improved. High resolution is needed for adequate imaging of small viscera with detailed internal tissue architecture, but since the attenuation of high frequencies also increases, the depth of penetration of high frequency ultrasound beams is restricted. Fortunately, many small viscera are superficially located and so high resolution 'small parts' or 'superficial structure' scanning is not only possible but has also become an important application of ultrasonic imaging.

Different types of ultrasound equipment, whether static or real-time, linear array or sector, mechanical or electrical, have their high frequency transducers which may be used for direct contact or stand-off techniques.

1.1 Direct contact scanning

Many B-scanners were supplied with high frequency probes which were ideal for superficial structure scanning, with a well-focused beam of ultrasound from the single crystal transducer. The degree of both axial and lateral resolution available from such equipment produced images of exceptional detail and quality (*Figure 1*). Although B-scanners remain popular with some workers for superficial structure scanning (1), their cumbersome design and the time needed to thoroughly interrogate an organ, has led most centres to abandon static scanners in favour of the more convenient and easy-to-use real-time machines.

Whatever the format of the ultrasound apparatus, direct contact scanning is associated with several difficulties. By definition, much of the anatomy under investigation is so superficial that it lies very close to the ultrasound transducer, resulting in loss of detail

due to poor beam geometry and obscuring of the echo pattern by artefacts (e.g. transmitter break-through). In addition, the skin contours are often irregular and/or curved making good skin-to-transducer contact difficult, if not impossible, to achieve. However, the ability to palpate the structures at the same time as scanning does add an extra dimension to the examination, and many ultrasonographers find direct contact scanning invaluable in the assessment of small nodules in the scrotum, thyroid and breast.

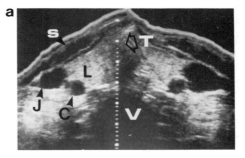

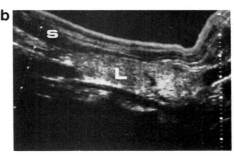

Figure 1. Normal thyroid (static B-scan with 7.5 MHz short focus transducer). (**a**) Transverse; (**b**) sagittal. (L = lateral lobe of thyroid, S = strap muscles, C = common carotid artery, J = internal jugular vein, T = trachea, V = vertebral body).

1.2 Stand-off scanning

The purpose of a stand-off is to displace the anatomy further into the near-field and into the focal zone of the beam, by interposing material through which ultrasound will travel without attenuation. There is also the added bonus of better skin contact since the distal face of the stand-off device will consist of a pliable rubber membrane which can mould to the skin contour.

Various forms of stand-off are available. The earliest methods used water baths balanced over the patient and the transducer was simply dipped in. Such contraptions were prone to leaking over both patient and ultrasonographer and it became apparent that purpose-built equipment was needed.

Many manufacturers designed probes with an in-built water bath, so that the transducer was positioned within the water bath which was enclosed by a rubber membrane. Dedicated small-parts scanners employing these principles thus became available and proved popular and useful (*Figures 4* and *7*), although problems did occur. These included leak of air bubbles into the water bath (necessitating frequent bleeding of the system), accumulation of debris and fungal contaminants in the water (due to permeation through the rubber membrane) giving rise to spurious echoes, and perishability of the rubber.

Recently blocks of a silicone material have become available ('Kitecko' −3M) and this is a substance that is ideal for superficial structure scanning since it not only moulds to the patient and transmits ultrasound (2), but is convenient to handle and apply (*Figure 2*). The major disadvantage is its rather high cost since it does deteriorate and has to be replaced every few months.

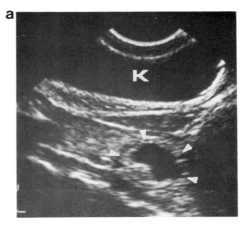

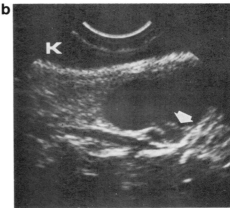

Figure 2. Sagittal scans through lateral lobes, using Kitecko (K) as a stand-off. (**a**) Degenerating adenoma with a rim of solid tissue (arrowheads) around a cystic centre. (**b**) Cystic mass with debris (arrow).

2. Thyroid

2.1 Anatomy

This bilobed gland is easily identified ultrasonically as an organ of homogeneous mid-level echoes straddling the trachea and lying antero-medial to the great vessels of the neck. The two lobes of the thyroid are joined across the mid-line by the isthmus. The strap muscles of the neck are seen antero-laterally as echo-poor structures. The normal parathyroid glands are embedded in the posterior surface of the thyroid lobes but are indistinguishable from the normal thyroid.

Enclosed in a sheath of pre-tracheal fascia the gland moves with the larynx during swallowing, and this movement can be observed with real-time scanning and is useful in delineating the margins of the gland. Similarly, masses lying along the thyroglossal tract, such as thyroglossal cysts, can be identified and shown to move appropriately on swallowing.

Ultrasound is able to distinguish thyroidal from extra-thyroidal masses, is excellent at distinguishing solid from cystic lesions, and is able to determine whether glandular enlargement is due to a diffuse process or to discrete masses, either single or multiple (3). However, such structural assessment often needs to be combined with the functional information that is available from scintigraphic scans, in order to provide the clinician with a meaningful pathological diagnosis.

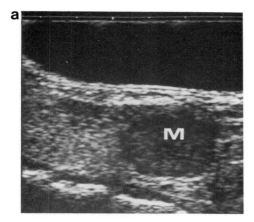

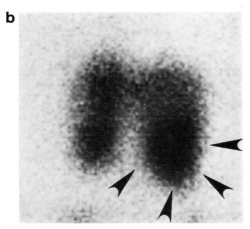

Figure 3. (a) Sagittal scan showing an echo-poor solid mass (M). (b) Isotope scan shows that the nodule is 'hot' (arrowheads), confirming that it is an adenoma.

2.2 Focal lesions

Focal abnormalities may be single or multiple, and clinical assessment is notoriously unreliable in making the distinction. In up to 40% of cases with glandular enlargement thought clinically to be due to a solitary mass, ultrasound will reveal either multiple nodules or diffuse abnormality of the gland. Clinical management is greatly influenced by this distinction as multiple lesions are much more likely to indicate a benign nature (4).

Solitary thyroid masses may be any of the following.

(i) *Simple cysts*. Truly simple cysts of the thyroid are unusual but are recognized by the classical ultrasound findings of completely echo-free contents,

smooth margins and increased through transmission of sound (enhancement). Cystic masses in the thyroid are usually, however, degenerating follicular adenomas, and do not fulfill these criteria.

(ii) *Follicular adenomas*. These may undergo cystic degeneration leaving a rim of solid material around a cystic centre containing debris (*Figure 2*). Haemorrhage into the nodule is not unusual, causing rapid and sometimes painful enlargement of the gland, and follow-up sonography will show progressive diminution in size. When solid, follicular adenomas are usually echo-poor in comparison with surrounding normal thyroid tissue, homogeneous in echo texture and well demarcated in outline (*Figure 3a*). The 'halo' sign observed around these adenomas probably represents a pericapsular inflammatory infiltrate causing an echo-free rim around the mass (*Figure 4a*). When the 'halo' is complete all the way around the lesion it is a strong indicator of benignity, although there are recorded instances where a 'halo' has been seen around carcinomas (5). Calcification may also occur in these adenomas producing areas of high echogenicity with distal acoustic shadowing.

(iii) *Malignant thyroid tumours*. These are usually solitary, solid and may be ill-defined in outline (*Figure 4b*). There may be metastasis to local lymph nodes which can be demonstrated on ultrasound. They are not reliably differentiated from benign adenomas by ultrasound alone, and current practice in the authors' department is to perform a radio-isotope scintiscan on all solitary solid nodules detected. If the nodule is 'hot' it is assumed to be benign (*Figure 3b*), while 'cold' or poorly functioning solid nodules are subjected to surgical excision biopsy in view of the high chance of malignancy.

2.3 Diffuse disease

Generalized enlargement of the gland occurs in both simple goitre, when the echo texture is smooth and homogeneous (*Figure 5a*), and in multinodular goitre. In the latter condition the echo texture is heterogeneous and nodular in pattern, the lateral lobes may be asymmetrically affected, and there may be varying numbers of larger, more discrete masses—which may be either cystic or solid, or a mixture of both (*Figure 5b*). Clinically there is often only one of these larger

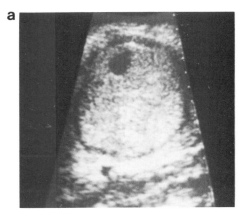

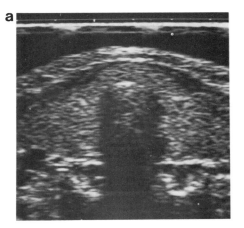

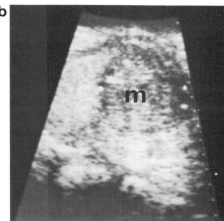

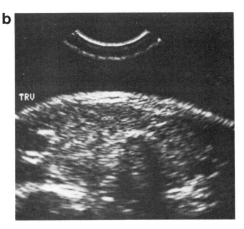

Figure 4. Sagittal scans using a 'small-parts' scanner with integral waterbath. (a) Adenoma containing an area of cystic degeneration and surrounded by a 'halo'. (b) Medullary carcinoma (m)—an echo-poor mass.

Figure 5. Transverse scans showing diffuse enlargement of the thyroid. (a) Simple goitre; (b) multinodular goitre.

masses palpable, but the true diagnosis is soon elucidated by the ultrasound scan.

The various forms of thyroiditis are indistinguishable from each other on ultrasound. The findings are of a diffusely enlarged gland with generalized reduction in echo amplitude (*Figure 6*), as is found with inflammatory conditions in other organs. Sometimes, however, the echo texture may be diffusely nodular, and the appearances may then be confused with multinodular goitre.

2.4 *Parathyroids*

As mentioned above (Section 2.1), the normal para-

thyroid glands are not identifiable on ultrasound, but parathyroid adenomas can be demonstrated on the posterior aspect of the thyroid gland as echo-poor solid masses (*Figure 7*), and ultrasound is now regarded as the best initial investigation in the search for the source of abnormal parathormone production (6).

2.5 *The role of thyroid ultrasound*

In summary, the importance of ultrasound lies in its ability to distinguish solitary from multiple lesions, and solid from cystic masses. Clinically impalpable thyroid and parathyroid tumours can be detected, and ultrasound can be used to screen patients at high risk of thyroid cancer or in the hunt for the unknown primary in patients with metastatic disease. Ultrasound is also

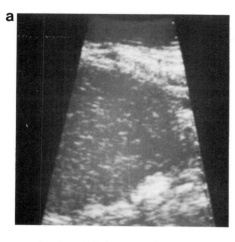

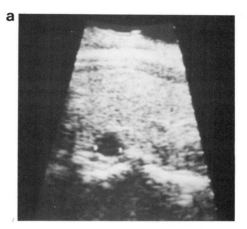

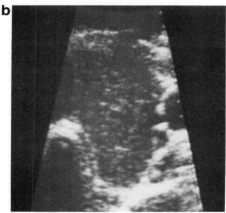

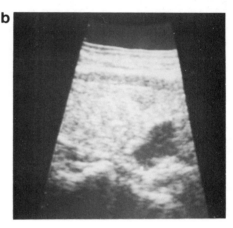

Figure 6. Thyroiditis with diffusely enlarged echo-poor gland. (**a**) Sagittal; (**b**) transverse.

Figure 7. (**a**) 6 mm parathyroid adenoma posterior to upper pole of thyroid. (**b**) 12 mm parathyroid adenoma posterior to lower pole.

sensitive at detecting recurrence of thyroid carcinoma following surgical resection and regular follow-up sonography in patients with locally invasive disease has recently been recommended (7).

3. Scrotum

3.1 *Anatomy and ultrasound appearances*

The testis is one of a pair of endocrine glands lying in the scrotum, their size being dependent on the age of the individual and his hormonal status. Ultrasonically the normal testis consists of a homogeneous and smooth echo texture which transmits sound well

so that only a fairly flat swept gain slope is required (*Figure 8*).

The gland is enclosed in a fibrous sheath—the tunica albuginea—which projects into the posterior part of the testis to form the mediastinum testis, and the whole organ is invaginated into the posterior part of a closed serous sac—the tunica vaginalis. Normally, this sac contains a thin layer of fluid and increased quantities of fluid cause formation of a hydrocoele (*Figure 8*).

The epididymis overlies the superior and posterolateral surface of the testis. It is a convoluted tube in which spermatozoa mature as they pass to the ductus deferens and is identified on ultrasound as a cap of higher reflectivity in comparison to the testis upon which it sits (*Figure 8*).

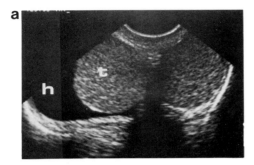

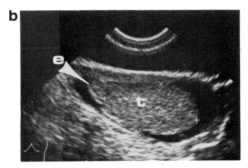

Figure 8. (a) Transverse scan of scrotum (direct contact) — hydrocoele (h) with normal testes (t). (b) Longitudinal scan through a normal testis (t) and epididymis (e).

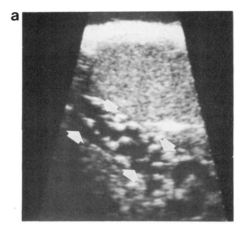

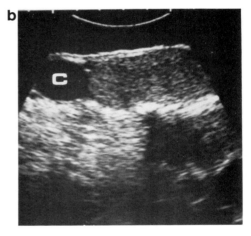

Figure 9. (a) Varicocoele (arrows). (b) Epididymal cyst (c).

Numerous veins leave the posterior border of the testis to form the pampiniform plexus. These veins may distend to produce a varicocoele of the spermatic cord if the valves of the internal spermatic vein are absent or incompetent. The vessels of the normal plexus have a diameter of 1−2 mm, but become dilated and easily visible on ultrasound if a varicocoele is present (8), (*Figure 9*). Embryological remnants, such as the appendix testis, may also be identified by ultrasound, especially in the presence of a hydrocoele.

3.2 *Epididymo-orchitis*

Inflammation of the epididymis causes diffuse enlargement (diameter normally <15 mm) and it becomes reduced in reflectivity (*Figure 10a*). Testicular involvement may result in enlargement of the organ which is also generally hypo-echoic (in comparison to the normal contra-lateral testis) (9), and hydrocoeles are a frequent association. Differentiation from testicular torsion requires reference to clinical and laboratory data, although the finding of a normal epididymis points to the diagnosis of torsion rather than epididymo-orchitis, and the use of Doppler and isotope studies for direct evaluation of blood flow in these conditions is encouraging (10). Epididymitis may go on to become a chronic infection, and the epididymis gradually increases in reflectivity and may shrink (*Figure 10b*).

3.3 *Trauma*

The sonographic appearance may range from normal to complete disruption of the testis with no recognizable normal tissue (11), (*Figure 11a*). Hydrocoeles are not uncommon and may contain blood (haematocoeles). Early recognition of a disrupted testis is important as the salvage rate of these testes is increased by early surgical intervention. Conversely, a normal

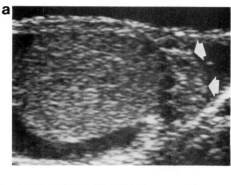

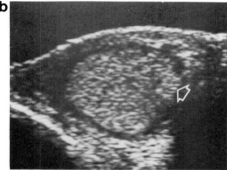

Figure 10. (a) Swollen epididymis (arrows) in acute epididymitis. (b) Increased reflectivity in a shrunken epididymis of chronic epididymitis (open arrow).

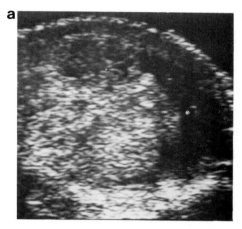

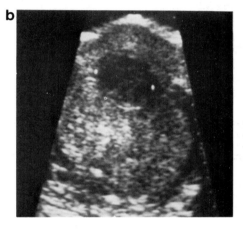

Figure 11. (a) Disrupted testis with diffuse abnormality and haematocoele due to trauma. (b) Focal echo-poor area due to haematoma in absence of history of trauma.

ultrasound would point to conservative management. Intra-testicular haematomas may present as an intra-testicular mass at a time remote from the episode of trauma, and excision biopsy becomes necessary for differentiation from a neoplasm (*Figure 11b*).

3.4 Scrotal masses

The key role of ultrasound is to differentiate intra-testicular from extra-testicular pathology.

Hydrocoeles may be simple or, following infection and haemorrhage, they may be complex with septa. The hydrocoele may be so large and/or tense that the underlying testis is impalpable and can only be examined with ultrasound.

Epididymal cysts and spermatocoeles show identical ultrasound appearances but may be distinguished by their location—the former in the head of the epididymis (*Figure 9*), the latter lying anywhere along its course.

All intra-testicular masses should be considered malignant unless the history and examination clearly points to a non-neoplastic process such as abscess or infarction. Malignant masses within the testis usually appear hypo-echoic when compared with normal tissue (*Figure 12a*), but calcification, haemorrhage and fibrosis may all produce increased echoes within the tumour. Indeed, 'burned out' primary tumours may appear solely as a focal area of increased echoes (12). Seminomas are usually more homogeneous in echo-texture than teratomas, and some forms of the latter may contain frankly cystic areas within the tumour (*Figure 12b*). Such ultrasonic distinctions are of purely academic value, and all testicular tumours must be examined histologically for accurate typing.

Not all testicular tumours are primary neoplasms;

a

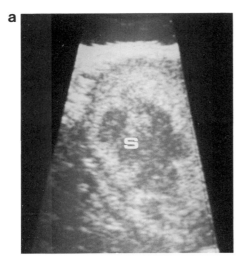

b

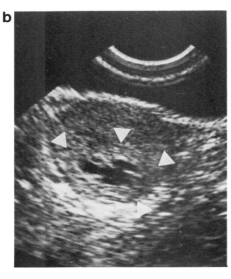

Figure 12. Testicular tumours: (**a**) seminoma (S); (**b**) teratoma (arrowheads).

metastases to the testes may be multiple and bilateral, while lymphoma and leukaemia may cause either focal or diffuse areas of hypo-echogenicity (*Figure 13*). Ten percent of tumours are associated wtih a hydrocoele, again emphasizing the need for accurate ultrasound evaluation of the underlying testis.

3.5 *Role of scrotal ultrasound*

Ultrasound of the scrotum is such a quick, cheap and

a

b

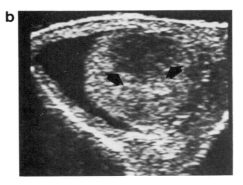

Figure 13. Lymphoma in testis (arrows): (**a**) longitudinal; (**b**) transverse.

accurate test that it is becoming a routine add-on to the clinical examination of all patients presenting to the urologist with any complaint relating to the scrotum and its contents. Furthermore, young and middle-aged males who are found to have lymph node enlargement in the abdomen or metastases in the lungs may have their testicular primary revealed by ultrasound—even when the testes are clinically normal (13).

Since the treatment of testicular tumours has improved dramatically over the last decade due to major advances in chemotherapy and radiotherapy, and since the most successful outcomes are in those patients in whom the tumours are detected early, discussion is now focusing upon the possibility of using ultrasound as a screening tool.

4. Other superficial structures

There are other viscera which are suitable for high resolution scanning and many workers are busy documenting the value of ultrasound in various clinical settings. Breast ultrasound might have lost favour as a screening method, but has gained support in the investigation of the breast lump as an adjunct to mammography (1). Ultrasound of the penis is being used in the assessment of Peyronie's disease (14), and ultrasound of the salivary glands may prove to be of value in the assessment of parotid and submandibular swellings (15). There is no doubt that further advances in the design and technology of ultrasound machines will ensure continuing expansion of this clinically valuable field of endeavour.

5. References

1. Guyer,P.B. and Dewbury,K.C. (1985) Ultrasound of the breast in the symptomatic X-ray dense breast. *Clin. Radiol.*, **36**, 69−76.
2. Lees,W.R. and McDicken,W.N. (1985) Kitecko Jelly blocks; end of the water bath. *Br. J. Radiol.*, **58**, 693.
3. Walker,J., Findlay,D., Amar,S.S., Small,P.G., Wastie,M.L. and Pegg,C.A.S. (1985) A prospective study of thyroid ultrasound scan in the clinically solitary thyroid nodule. *Br. J. Radiol.*, **58**, 617−619.
4. Scheible,W., Leopold,G.R., Woo,V.L. and Gosink,B. (1979) High-resolution real-time ultrasonography of thyroid nodules. *Radiology*, **133**, 413−417.
5. Propper,R.A., Skolnick,M.L., Weinstein,B.J. and Dekker,A. (1980) The nonspecificity of the thyroid halo sign. *J. Clin. Ultrasound*, **8**, 129−132.
6. Moreau,J.-F. (1984) Parathyroid ultrasonography. *Clin. Diagn. Ultrasound*, **12**, 93−100.
7. Simeone,J.F., Daniels,G.U., Hall,D.A., McCarthy,K., Kopans,D.B., Butch,R.J., Mueller,P.R., Stark,D.D., Ferucci,J.T. and Wang,C.A. (1987) Sonography in the follow-up of 100 patients with thyroid carcinoma. *Am. J. Radiol.*, **148**, 45−49.
8. Rifkin,M.D., Kurtz,A.B., Pasto,M.E. and Goldberg,B.B. (1985) Diagnostic capabilities of high-resolution scrotal ultrasonography: prospective evaluation. *J. Ultrasound Med.*, **4**, 13−19.
9. Hricak,H. and Hoddick,W.K. (1986) Scrotal ultrasound. *Clin. Diagn. Ultrasound*, **18**, 219−240.
10. Hricak,H., Taylor,K.J.W., Marich,K., Lue,T.F. and Burns,P. (1986) Doppler in urology. *Clin. Diagn. Ultrasound*, **18**, 241−270.
11. Jeffrey,R.B., Laing,F.C., Hricak,H. and McAnninch,J.W. (1983) Sonography of testicular trauma. *Am. J. Radiol.*, **141**, 993−999.
12. Shawker,T.H., Javadpour,N. and O'Leary,T. (1983) Ultrasonographic detection of burned-out primary testicular germ cell tumours in clinically normal testes. *J. Ultrasound Med.*, **2**, 477−479.
13. Fowler,R.C., Chennells,P.M. and Ewing,R. (1987) Scrotal ultrasonography: a clinical evaluation. *Br. J. Radiol.*, **60**, 649−654.
14. Isaacs,J.L., Ewing,R. and Irving,H.C. (1987) Ultrasound in Peyronie's Disease. *Br. J. Radiol.*, **60**, 626.
15. Whyte,A.M. and Byrne,J.V. (1986) A comparison of computed tomography and ultrasound in the assessment of parotid masses. *Clin. Radiol.*, **38**, 339−343.

Chapter 13

The use of pulsed ultrasound in ophthalmology

M.Restori

1. Introduction

Effective use of ultrasound in ophthalmic diagnosis requires a knowledge of both ophthalmic anatomy and pathology. Many ultrasonic techniques are used in diagnosis (1−5) although only A-scan and B-scan techniques are used routinely in most centres. Some authors (1) have placed great stress on measuring the amplitudes of echoes arising from abnormalities using A-mode. In Moorfields grey-scale 'real-time' B-mode imaging is the preferred method of diagnostic examination and A-mode is used for measurement of the axial length of the eye.

Ultrasound is now used routinely:-

(i) to measure the axial length of the eye,
(ii) to assess eyes in which opacity prevents direct ophthalmoscopic examination, and
(iii) to study both intra-ocular and orbital tumours.

1.1 B-mode coupling

Coupling may be 'direct'—to the closed eyelid via a liquid gel—or 'stand-off'—usually through a saline bath coupled to the open anaesthetized eye but, alternatively, through either a sealed bag/condom filled with water or, a solid gel pad. A 'stand-off' type of coupling is only required if the very anterior structures of the eye need to be imaged.

1.2 'Real-time' B-mode imaging

Mechanical or array scanners are suitable for ophthalmic scanning. Generally, 10 MHz probes are used, although lower frequencies may prove adequate for many conditions. Serial B-scan sectioning at regular intervals (approximately every millimetre) is performed with many directions of gaze. Dynamic studies are vital in the assessment of eyes with vitreo-retinal disease and are performed by asking the patient to deviate the eyes and observing the induced movements of abnormal echoes on the 'real-time' B-mode display.

2. Axial length measurement

Implantation of a plastic lens following cataractous lens extraction is nowadays a routine surgical procedure. Calculation of the optimum power of lens implant to give the patient the desired post-operative refraction requires measurement of the axial length of the eye using an A-scan technique. The patient is generally examined in a seated position with the chin placed on a rest and the forehead against a bar. The direction of gaze must be directly ahead; this is easily achieved if either eye is capable of fixing a target. The transducer, which is mounted on a spring-loaded assembly kept at a set tension so as not to indent the globe, is brought into direct contact with the anaesthetized cornea. The tear film provides a sufficient coupling medium. Many systems make allowances for the varying velocities of sound within the aqueous, lens and vitreous. An A-scan along the central axis of an eye is shown in *Figure 1a* and along an oblique meridian in *Figure 1b*. The sound beam should be arranged to strike the ocular interfaces perpendicularly so that the echoes arising from them will be both high in amplitude and steeply rising; this A-scan pattern should be retained for a range of settings of sensitivity.

3. Opacity of the ocular media

Opacification of the ocular media may prevent direct examination of an eye using optical instruments (6,7) and thus necessitate ultrasonic assessment of the eye.

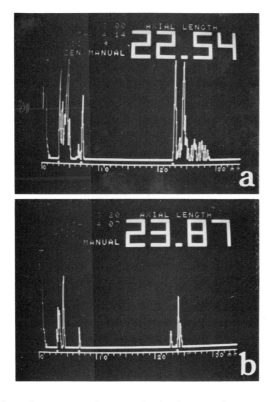

Figure 1. (a) A-scan along central axis of eye; small crosses on x-axis mark anterior lens, posterior lens and vitreo−retinal interfaces. (b) Oblique A-scan through eye; system rejects vitreo−retinal echo (see crosses on axis) and measurement is to a more posterior echo; reading is too long.

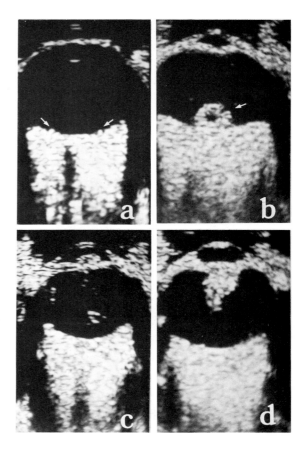

Figure 2. Transverse linear 'saline-bath' coupled B-scans. (a) Central section; normal eye. 'Baums bumps'—arrows. (b) Abnormal cornea; dislocated cataractous lens (arrow). (c) Abnormal cornea; ruptured anterior lens capsule, lens material in anterior chamber; vitreous opacity. (d) Abnormal cornea; ruptured posterior lens capsule, lens material in the vitreous cavity.

3.1 *The normal eye in B-scan section*

A central B-scan of the normal ophthalmic anatomy in transverse section taken through a saline bath using a linear mechanical scanner is shown in *Figure 2a*. Linear and small angle sector scanners produce a partial outline of the globe as the smooth ocular interfaces specularly reflect the interrogating sound pulses (8) such that only echoes from structures lying approximately perpendicularly to them arrive back at the transducer; rotation of the eye presents the curved structures of the eye more favourably to the sound beam. Internally the aqueous, lens substance and vitreous are echo free but are outlined by echoes at the interfaces between the ocular tissues (cornea, iris, lens capsule and retina). If the sound beam traverses the lens then the retina lying directly posterior to the

lens appears elevated due to the higher velocity of sound within the lens compared with surrounding tissues. The discontinuities of this artefactual elevation seen in B-scan section are known as 'Baums Bumps' (after Gilbert Baum). The retinal interface is otherwise smooth. The higher velocity of sound within the lens compared with the surrounding ocular tissues, its diverging action on the sound pulses and the fact that the sound pulses are more strongly absorbed in the lens, has led many authors to recommend by-passing it when examining the eye; this is necessary in strongly attenuating cataractous lenses, for example calcified cataracts. Orbital fat scatters the sound

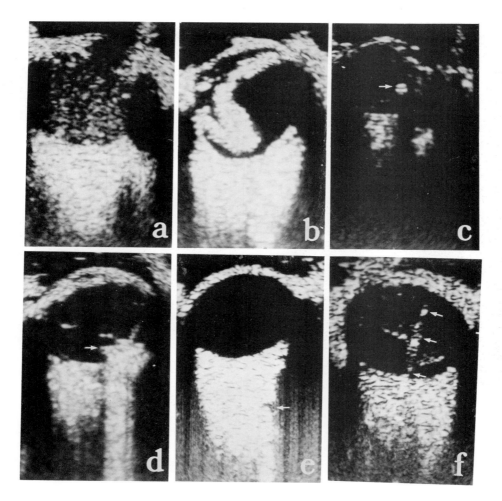

Figure 3. Transverse linear 'saline-bath' coupled B-scans. (**a**) Deviated gaze; dense vitreous haemorrhage. (**b**) Deviated gaze; intra-gel asteroid hyalitis. (**c**) Deviated gaze; short eye; foreign body (arrow) attenuating sound. (**d**) Large foreign body (arrow); 'ringing' artefact. (**e**) Orbital foreign body (arrow); 'ringing' artefact. (**f**) Foreign body track (arrows); intra-gel haemorrhage.

strongly; the optic nerve appears as a dark strip within the fat pad and the ocular muscles as dark bands bounding the major portion of the fat pad.

3.2 The pathological eye

3.2.1 The lens

Internal lens echoes suggest opacification of the lens. The absence of the anterior and posterior lens interface echoes together with the presence of an elliptical group of echoes within either the anterior chamber or the vitreous cavity (*Figure 2b*) suggests dislocation of a lens which has become cataractous; a posteriorly dis-

located lens can be seen to roll within the vitreous cavity during deviations of gaze. Rupture of either the anterior or posterior lens capsule results in echoes within the lens region which extend into either the anterior chamber (*Figure 2c*) or the vitreous cavity (*Figure 2d*), respectively.

3.2.2 The vitreous

Vitreous opacity generates echoes within the otherwise acoustically clear vitreous cavity. In general inflammatory cells and haemorrhage within the vitreous (*Figure 3a*) cannot be differentiated. Asteroid hyalitis

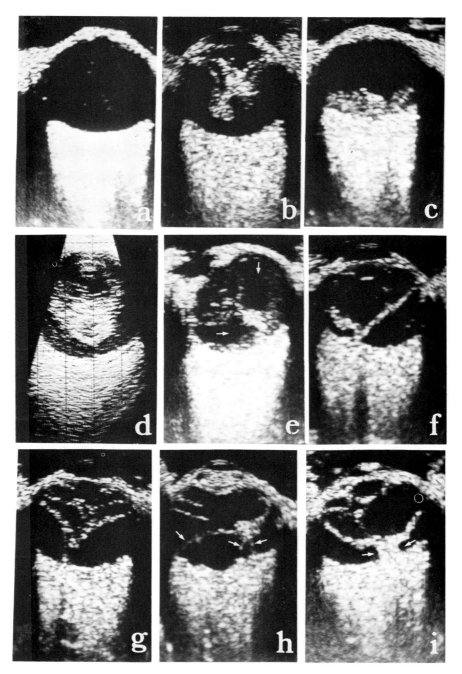

Figure 4. Transverse linear 'saline-bath' coupled and sector 'direct' coupled (**d**) B-scans. (**a**) PVD. (**b**) PVD; intra-gel haemorrhage. (**c**) PVD; retrohyaloid haemorrhage. (**d**) PVD; intra-gel and retrohyaloid haemorrhage. (**e**) Deviated gaze; retrohyaloid and central gel haemorrhage; clear cortical gel (arrows). (**f**) PVD; compacted haemorrhage along gel boundary. (**g**) PVD tethering to optic nerve head; intra-gel opacity. (**h**) PVD with multiple adhesions (arrows) to posterior retina; cataractous lens. (**i**) PVD with fibrotic membrane along gel boundary; wide residual vitreo−retinal adhesion (arrows); cataractous lens.

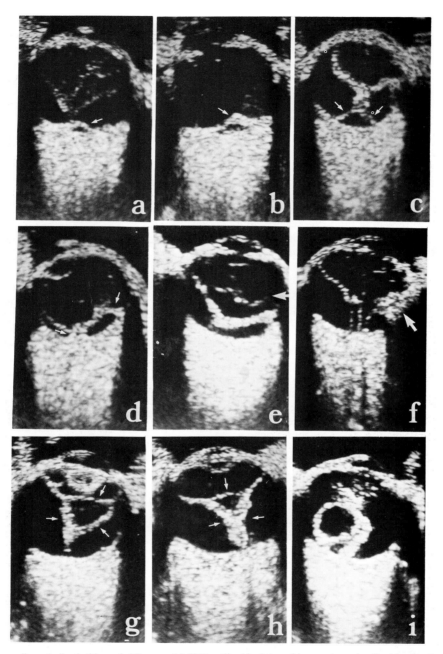

Figure 5. Transverse linear 'saline-bath' coupled B-scans. (**a**) PVD outlined by intra-gel haemorrhage; localized tractional retinal detachment (arrow). (**b**) PVD outlined by retrohyaloid haemorrhage; localized tractional retinal detachment (arrow). (**c**) Deviated gaze; fibrotic membrane along PVD interface; fibrotic vitreo−retinal adhesion into tractional retinal detachment (arrows). (**d**) Deviated gaze; PVD inserting by wide adhesion into a tractional retinal detachment (arrows). (**e**) Total retinal detachment; PVD (arrow); intra-gel opacity. (**f**) Sub-total rhegmatogenous retinal detachment; surgical explant (arrow)—unsuccessful surgical repair. (**g**) Total rigid retinal detachment; detached retracted vitreous gel; 'triangle sign'—arrows; cataractous lens. (**h**) 'Triangle sign'—arrows; retinal leaves in apposition at level of optic nerve head; cataractous lens. (**i**) PVD; intra-gel haemorrhage; giant retinal tear rolled up.

(*Figure 3b*), however, gives rise to characteristically high amplitude echoes. Discrete high amplitude echoes may arise from foreign bodies which may produce attenuation (*Figure 3c*) or 'ringing' artefacts (*Figure 3d* and *e*). A track of haemorrhage (*Figure 3f*) may indicate the pathway of a foreign body through the eye.

3.2.3 *Posterior vitreous detachment (PVD)*

Posterior vitreous detachment (the vitreous gel always remains adherent to its annular anterior insertion at the vitreous base) is an important ultrasonic diagnosis prior to attempted surgical removal of the gel (vitrectomy). In the simple case of PVD with no vitreous opacity (*Figure 4a*) the difference in impedance between gel and retrohyaloid fluid (the watery fluid which fills the space behind a detached vitreous gel) is extremely small; the echo from the interface between the detached gel and retrohyaloid fluid is therefore very weak and may only be detected during dynamic B-mode studies as pulses of sound strike portions of the interface perpendicularly. If vitreous opacity is confined within the detached vitreous gel (*Figure 4b*) or to the space behind the detached vitreous gel (*Figure 4c*) then the corporate mobility of the compartmentalized opacity suggests PVD. In the presence of both intragel and retrohyaloid opacity, differences in echogenicity of opacity within compartments (*Figure 4d*) may outline the gel boundary; commonly the opacities fill the retrohyaloid space but only partially fill the gel (*Figure 4e*). Sometimes echoes from compacted haemorrhage or fibrous tissue lie along the gel boundary (*Figure 4f*) and PVD may mimic a detached retina in both echo amplitude and topographically if the gel remains attached at the optic nerve head. In such cases dynamic studies are of paramount importance; an interface between a mass of gel and watery retrohyaloid fluid moving in a different manner compared with a membrane (detached retina) undulating within watery fluids (retrohyaloid and sub-retinal). PVD may be complete or incomplete. Often a residual adhesion (*Figure 4g*) of the detached vitreous gel to the optic nerve head occurs. The gel may also remain adherent to other parts of the retina (*Figure 4h* and *i*). The site and extent of such vitreo−retinal adhesions are best demonstrated during dynamic 'real-time' B-mode studies and may indicate not only the aetiology of the disease process but may also suggest the possible prognosis following surgery. The surgeon also must be aware of such adhesions in order to avoid exerting traction on the retina via such adhesions during vitrectomy. The violence of motion of the incompletely detached vitreous gel gives a fascinating insight into the tractional forces on the retina exerted via vitreo−retinal adhesions during ocular movements.

3.2.4 *Tractional retinal detachment*

Tractional retinal detachments (*Figure 5a−d*) usually occur in association with incomplete PVD (7) with residual vitreo−retinal adhesions. Of particular importance to the surgeon is whether or not the macular region of the retina is involved in the detachment. The macula lies directly temporal to and very slightly (~1 mm) inferior to the optic nerve head.

3.2.5 *Rhegmatogenous retinal detachment*

In total rhegmatogenous retinal detachment (*Figure 5e*) the retina remains adherent anteriorly in an annulus at its origin (the ora serrata) and posteriorly to the optic nerve head; it is detached elsewhere. Rhegmatogenous retinal detachment (*Figure 5e* and *f*) usually occurs secondary to PVD and retinal tear formation so, prior to an ultrasonic diagnosis of rhegmatogenous retina detachment, an ultrasonic diagnosis of PVD should be sought. A fresh mobile rhegmatogenous retinal detachment undulates following deviations of gaze. Mobility of the retinal detachment determines the operability and the type of surgical repair procedure deemed most appropriate. Restriction of mobility of the detached retina suggests that the detachment has been complicated by fibrous tissue proliferation on the retinal surface (9), often producing the 'triangle sign' (10) (*Figure 5g* and *h*). In giant retinal tears the echoes arising from the detached retina become discontinuous (*Figure 5i*), often folding over on themselves, and demonstrating an excessive mobility during dynamic studies.

3.2.6 *Choroid/sclera*

The choroid may detach shallowly (*Figure 6a*) in an annulus anterior to the ora serrata. In complete bullous choroidal detachment convex interfaces are seen which extend anterior to the ora serrata but seldom to the optic nerve head posteriorly. The pathognomonic sign of choroidal detachment is the persistent attachment of the choroid to the vortex ampullae of the vortex veins at the equator of the eye (*Figure 6b*). Haemorrhage (*Figure 6c*) may fill the supra-choroidal space. Signs of thickening of the coats of the eye either

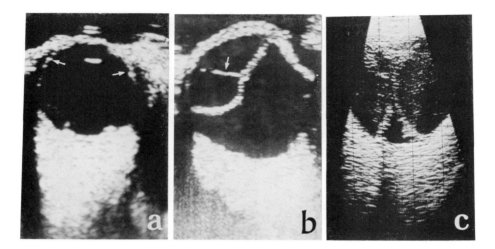

Figure 6. Transverse 'saline-bath' coupled and sector 'direct' coupled (**c**) B-scans. (**a**) Shallow anterior choroidal detachment (arrows). (**b**) Deviated gaze; bullous temporal choroidal detachment; vortex vein attachment (arrow). (**c**) Bullous choroidal detachment with supra-choroidal haemorrhage.

choroidal or scleral, are important in the diagnosis of phthisis bulbi and scleritis, respectively.

4. Intra-ocular and orbital tumours

4.1 Intra-ocular tumours

Malignant melanoma is the most common intra-ocular tumour. Ultrasonically a melanoma is detected as a rounded or collar-stud' shaped mass (*Figure 7a and b*) containing echoes of various amplitudes. An internal area of lower amplitude echoes (*Figure 7c*) is sometimes present. Replacement of the scattered high amplitude echoes arising from normal choroidal tissue by echoes of lower amplitudes arising from tumour cells [sometimes described as 'choroidal excavation' (11)] may be detected even in very small masses (*Figure 7d*). Secondary deposits of malignant cells (*Figure 7e*) tend to give rise to higher amplitude echoes compared with malignant melanomas; they may have irregular borders and generally do not show choroidal excavation. Sometimes a secondary deposit may be difficult to differentiate from a gross disciform lesion (that is mass of haemorrhage and fibrous tissue under the macular retina). Retinoblastoma (*Figure 7f*), is a tumour of babies and children and often contains calcium which attenuates sound very strongly and also reflects sound strongly. Extra-scleral extension of a malignant tumour is seen as either a difference

in echogenicity within the strongly scattering orbital fat (*Figure 7g*) or as scattered echoes within the normally acoustically-empty optic nerve (*Figure 7h*). In either case the observed ultrasonic feature is continuous with the intra-ocular portion of the tumour. Commonly, tumour dimensions (for example, maximum tumour base and maximum elevation) are measured to assess progression or the effects of treatment; Doppler studies (*Figure 7i*) are performed to assess blood flow characteristics.

4.2 Orbital tumours

Orbital tumours are relatively rare and therefore are only briefly mentioned here. Computerized tomography and magnetic resonance imaging are sometimes preferred for orbital imaging. Computerized tomography, for example, has the advantage over ultrasonic techniques in the orbit that it displays not only soft tissue but also bone (this is of particular value to the orbital surgeon). Ultrasound is, however, useful in determining the exact nature of the lesions (12). Generally, orbital tumours do not have characteristic appearances but more commonly display a range of possible ultrasonic features. The location, the distribution in both number and amplitude of internal echoes together with the attenuating and blood flow features of a lesion may indicate its likely nature. B-scans of selected orbital lesions are shown in *Figure 8*.

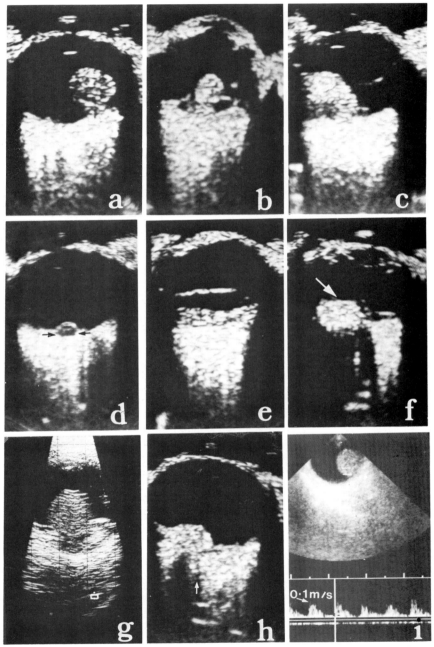

Figure 7. Transverse linear 'saline-bath' coupled and sector 'direct' coupled (**g** and **i**) B-scans. (**a**) Malignant melanoma. (**b**) 'Collar stud' malignant melanoma; overlying retinal detachment. (**c**) Large malignant melanoma; overlying retinal detachment. (**d**) Small malignant melanoma; choroidal excavation (arrows). (**e**) Posterior secondary deposit from breast; overlying retinal detachment. (**f**) Calcium-containing retinoblastoma (arrow) producing attenuation. (**g**) Extra-scleral extension of malignant melanoma into orbit. (**h**) Extra-scleral spread of malignant melanoma into optic nerve sheath (arrow). (**i**) Doppler spectrogram (velocity of blood flow versus time) on blood vessel within marked region in a B-scan of a malignant melanoma.

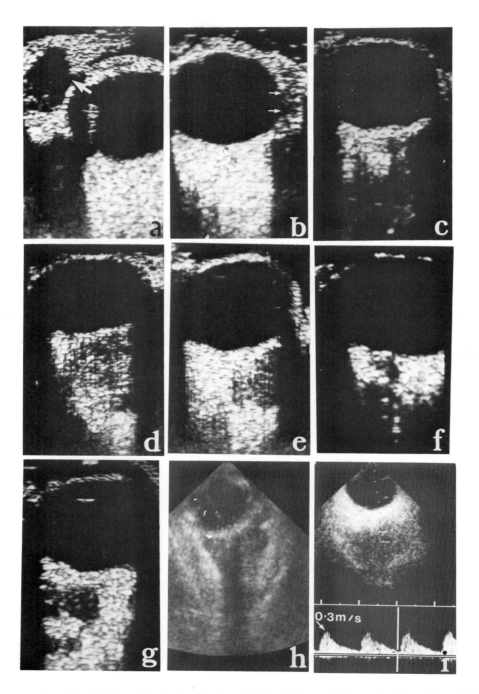

Figure 8. Transverse linear 'saline-bath' coupled and sector 'direct' coupled (**h** and **i**) B-scans. (**a**) Lacrimal gland cyst (arrow). (**b**) Adenocystic carcinoma of lacrimal gland (arrow). (**c**) Secondary deposit—orbital. (**d**) Orbital cavernous haemangioma. (**e**) Orbital neurilemmoma. (**f**) Optic nerve meningioma. (**g**) Diffuse orbital pseudotumour. (**h**) Orbital lymphoma around optic nerve. (**i**) Doppler spectrogram on blood vessel within marked region on B-scan of an orbital lymphoma.

5. Conclusions

Ultrasonic measurement of the axial length of an eye is valuable in predicting the optimum power of lens implant to use following cataractous lens extraction.

Ultrasound is an invaluable tool in the study of eyes in which opacity prevents direct ophthalmoscopic examination. It may aid in the decision as to whether or not to intervene surgically and in anticipating the likely surgical problems.

In eyes with intraocular tumours ultrasound is useful in diagnosing the type of tumour and in measurement of tumour dimensions, for example, to assess growth or regression following treatment. In the orbit, ultrasound is useful in determining the likely nature of tumours.

6. References

1. Ossoinig,K. (1977) Echography of the eye, orbit and periorbital region. In *Orbit Roentgenology*. Arger,P.H. (ed.), John Wiley and Sons, New York, p. 223−269.

2. Coleman,D.J. and Weininger,R. (1969) Ultrasonic M-mode technique in ophthalmology. *Arch. Ophthalmol.*, **82**, 475−479.

3. Restori,M. (1985) 'Real-time' immersion B-scan and C-scan techniques in ophthalmic diagnosis. *Ultrasound Med. Biol.*, **11**, 185−192.

4. Restori,M. and Wright,J.E. (1977) C-scan ultrasonography in orbital diagnosis. *Br. J. Ophthalmol.*, **61**, 735−740.

5. Restori,M. (1978) Ultrasonic holography in ophthalmic diagnosis. *Br. J. Clin. Equipment* (March), 71−75.

6. Fuller,D.G. and Hutton,W.C. (1982) *Pre-surgical Evaluation of Eyes with Opaque Media*. Grune and Stratton, New York.

7. McLeod,D. and Restori,M. (1979) Ultrasonic examination in severe diabetic eye disease. *Br. J. Ophthalmol.*, **63**, 533−538.

8. Restori,M., Leeman,S. and Weight,J.P. (1983) Ultrasound interaction in the eye. In *Ophthalmic Ultrasonography, Proceedings of the 9th SIDUO Congress*. Hillman,J.S. and Le May,M.M. (eds), Dr W.Junk, The Hague, pp. 423−432.

9. Gregor,Z., Restori,M. and McLeod,D. (1979) B-scan ultrasound in massive pre-retinal retraction. *Trans. O.S.U.K.*, **99**, 38−42.

10. Fuller,D.G., Laqua,H. and Machemer,R. (1979) Ultrasonic diagnosis of massive periretinal proliferation in eyes with opaque media (triangular retinal detachment). *Am. J. Ophthalmol.*, **83**, 460−464.

11. Coleman,D.J. (1974) Ultrasonic diagnosis of tumours of the choroid. *Arch. Ophthalmol.*, **91**, 344−354.

12. Chavis,R.M., Garner,A. and Wright,J.E. (1978) Inflammatory orbital pseudotumour: a clinicopathological study. *Arch. Ophthalmol.*, **96**, 1817−1822.

Chapter 14

Interventional ultrasound

Henry C. Irving

1. Introduction

The term 'Interventional Ultrasound' describes the use of ultrasound to guide needles into patients for various diagnostic and therapeutic purposes. Diagnostic uses include biopsy of solid masses and aspiration of fluid-containing lesions, whilst therapeutic procedures encompass drainage of both fluid-containing masses (cysts or abscesses) and organ systems such as the urinary and biliary tracts.

The potential for ultrasonic needle-guidance was recognized in the early days of ultrasound imaging, and specialized biopsy transducers were developed contemporaneously in centres on both sides of the Atlantic (1,2). These workers focused their attention upon the adaptation of static B-scan apparatus so that needles could be passed through a central channel in the transducer head to a predetermined depth whilst the transducer was aimed at the target on the storage oscilloscope (*Figure 1*). In expert hands this technique was capable of considerable accuracy, but suffered from the inability to visualize the needle tip during its pathway through the patient and the necessity for great manual dexterity on the part of the operator, and was superseded by the introduction of real-time needle-guidance systems as soon as these became available in the late 1970s.

It is the ability of ultrasound to give cross-sectional images of the abdominal and other viscera that makes possible its application to puncture-guidance. The only other imaging method that competes with ultrasound for this role is computed tomography. However, the advantages of ultrasound, with its greater versatility of image plane, use of non-ionizing radiation, cheapness and ready availability, ensure that ultrasound is the first choice method for needle-guidance in most centres, and computed tomography is usually reserved for those cases in which the ultrasound visualization of the target is hampered by intervening bowel gas or bone.

2. Needle-guidance techniques

2.1 Linear array puncture transducers

With the development of real-time imaging systems, it soon became apparent that they too could be adapted for needle-guidance techniques. Linear array puncture transducers gained early acclaim; there are two basic designs in which there is either a central canal through a gap within the transducer array (*Figure 2*), or a needle-guide attachment which is clipped or screwed onto the end of the array. In the former system, the needle pathway can be identified by a linear defect in the image where a crystal has been omitted from the array (*Figure 3a*), and the needle tip will be seen as a bright echo passing vertically downwards along this

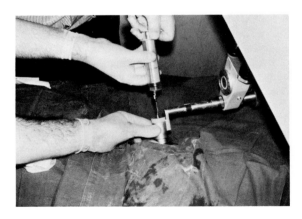

Figure 1. Aspiration of a liver abscess using a static B-scanner puncture transducer with a central canal. (The author's first case, 1978).

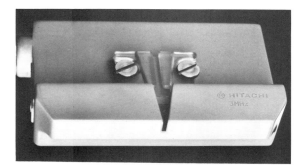

Figure 2. Linear array puncture transducer (central slot type).

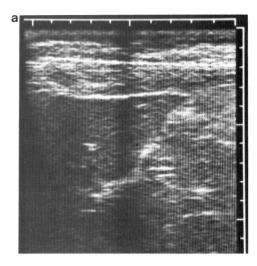

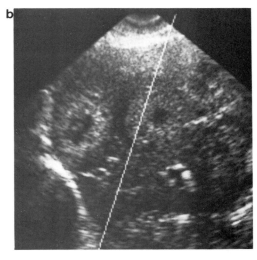

line, whereas in the latter system the needle pathway will angle across the image enabling a greater length of the needle to be visualized. However, the physical size of linear array transducers severely restricts their applicability to abdominal imaging, and this disadvantage becomes even more significant when needle-guidance systems are considered. Many targets within the abdomen require an angled approach, and the optimum needle entry point may be close to ribs or other protruberances, so that adequate transducer—skin contact becomes impossible for linear array probes. For this reason alone, most interventional ultrasonologists have abandoned such equipment in favour of sector scanners, although linear array puncture transducers have remained popular in some centres on the continent of Europe (3).

2.2 Sector scan puncture attachments

Sector scan transducers, whether mechanical or electrical (phased array), offer the major advantage of small head size, thus facilitating increased accessibility to many sites which are targets for interventional procedures (e.g. intercostal approaches to subphrenic spaces, liver and retroperitoneal structures), and also offer the bonus of proximity of needle entry site to the entry portal of the ultrasound beam. Purpose-built needle-guide attachments are available from most manufacturers, either as clip-on plastic or screw-on metal assemblies (*Figure 4*) and, although designs vary, the essential features will include rigidity of the attachment to ensure reproducibility of the direction of the needle pathway, and a choice of size of needle-guide inserts to permit the use of needles of differing calibres. There should be no need to purchase a special transducer, and the needle-guide assembly must be

Figure 3. Needle pathways indicated in: (**a**) linear array puncture transducer—linear defect in image due to gap in array; (**b**) sector scanner with needle-guide attachment—line electronically superimposed upon image.

capable of easy detachment and sterilization. The needle pathway is usually illustrated by electronically superimposing a line over the image on the TV monitor (*Figure 3b*). This process presupposes that the needle will travel along a fixed direction and is dependent upon the construction of the puncture attachment being sufficiently solid and rigid so that the angle or entry of the needle is constant, and all operators should regularly perform water bath tests

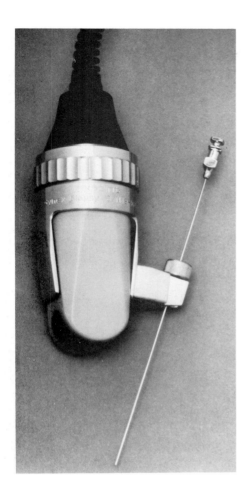

Figure 4. Sector scanner transducer with needle-guide attachment.

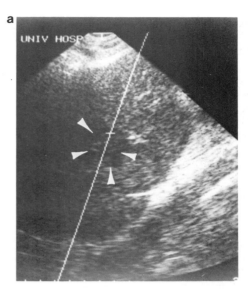

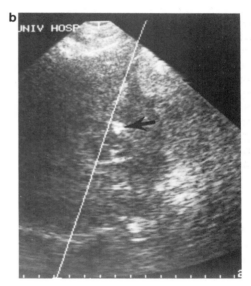

Figure 5. Biopsy of small mass, deep in the liver, via an intercostal approach: (**a**) mass in liver (arrowheads); (**b**) needle tip within mass (arrow).

on their equipment to check that this arrangement does hold true. In this way, needles can be guided with precision into targets which are both small and deeply situated, and the puncture can be performed despite such limited portals of access as intercostal spaces (*Figure 5*).

2.3 *Needle visualization*

With these sector scan systems, needle visualization will again be restricted to the bright echo emanating from the tip of the needle and this is adequate for most clinical circumstances. The amplitude of the needle tip echo does vary with the construction of the needle, and there have been attempts to increase the reflectivity of the needle by using a roughened surface coating to increase the number of interfaces available for reflection, and by the incorporation of an air-containing notch near to the needle tip, but neither of these methods has gained any widespread support. Recent attempts to convert the needle tip into a minute ultrasound transducer, emitting a signal on the one

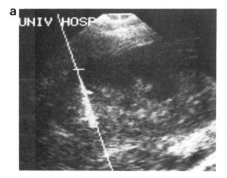

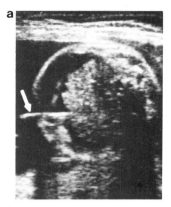

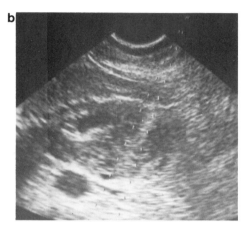

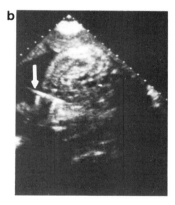

Figure 6. Echo-cluster produced by 'Biopty-cut' needle: (a) renal transplant biopsy; (b) pancreatic mass biopsy.

Figure 7. Intra-uterine transfusions using free-hand needle guidance: (a) linear array guidance; (b) sector scanner guidance (arrows indicate the needles).

hand (4) or acting as a receiver on the other (5), are ingenious and effective, but also suffer through their dependence upon the use of specially adapted needles. Some types of biopsy needle with a cutting action generate a cluster of echoes as the core of tissue is ensnared within the device, and this is a useful aid to the operator who can then be confident as to the exact origin of the tissue sample (*Figure 6*).

2.4 Free-hand techniques

In routine clinical practice, there will be many occasions when the accuracy provided by the above-mentioned needle-guidance systems is unnecessary. The target is either sufficiently large or superficial to permit a free-hand approach to the puncture procedure. This can be performed by simply marking the skin at the proposed entry site after careful ultrasound scanning has been performed to localize and

characterize the target, to assess the proposed needle pathway, to measure the correct depth of needle insertion and to gauge the effect of respiration (so that the procedure can be performed with respiration suspended in the correct phase). Free-hand punctures should be performed with the minimum possible needle angulation, since accuracy rapidly diminishes as angulation away from the truly vertical or horizontal approach increases. With the proviso that such preliminary scanning is meticulously performed, free-hand needle guidance is a quick and effective method, although small and deep targets are not amenable to this method.

2.5 Indirect needle guidance

A compromise technique that has many advocates is the use of real-time ultrasound apparatus to monitor a free-hand puncture without a needle-guide attachment. The needle is introduced into the patient whilst the ultrasound probe is held nearby and, with practice, the needle tip can be visualized passing towards the target. The advantage of this method is the freedom from the constraint of a rigid needle-guide conferring greater versatility of approach and opportunity for modification of angulation along the needle pathway. The problems concern the difficulty of locating the needle tip (greater in solid tissue than within fluid), and the ease with which the tip is lost from sight during needle manipulations. As intimated, this indirect method of needle guidance is particularly effective when the needle is passing through fluid and is therefore often employed for the obstetric applications of interventional ultrasound (*Figure 7*).

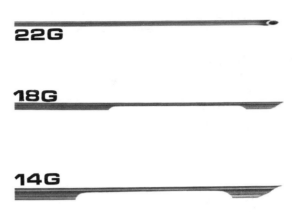

Figure 8. Needle tips—fine-needle (22-gauge), 18-gauge and 14-gauge 'Biopty-Cut'.

3. Biopsies

Pathological examination of a sample of tissue from a solid organ or tumour is often required for diagnosis, and the gathering of such a sample is rapidly and atraumatically performed by ultrasound-guided needle biopsy. The pathological information required from the biopsy will determine the size of tissue sample needed and hence the type of biopsy needle to be used.

3.1 *Fine-needle aspiration biopsy*

Certain tumours and inflammatory conditions may be diagnosed from a cytological examination in which a small cluster of cells has been aspirated into a fine needle and then smeared onto a microscope slide, fixed and stained. Such a sample can be obtained using a 22-gauge (or finer) needle which can be used without fear of damaging either the organ concerned or intervening structures such as bowel or other adjacent viscera, and is free from risk of causing haemorrhage. Accurate placement of the needle tip within the target is achieved by using one of the guidance methods outlined above, and cells are encouraged to enter into the needle by applying a suction pressure and at the same time jiggling the needle back and forth until the suction pressure is released, the needle withdrawn, the material expelled and the specimen prepared according to the instructions of the local cytologist. The interpretation of fine-needle aspiration biopsies is dependent upon the expertise of the cytologist, who is usually able to decide if the cells are malignant or inflammatory, and may also be able to type the malignancy,

although certain types of tumour (most notably lymphomas, cholangiocarcinomas and retroperitoneal sarcomas) and inflammatory conditions (e.g. of liver and kidney) require larger tissue samples for histological examination.

3.2 *Core (or large needle) biopsy*

Larger cores of tissue are required if a histological diagnosis is needed, when the arrangement of the cells and the tissue architecture can be studied. The needles for such core biopsies are of necessity larger and more traumatic than the aforementioned fine needles, the more effective types incorporating a cutting edge to facilitate detachment of the tissue core from its surroundings. A popular needle is the 'Tru-Cut' (Travenol) which is 14-gauge in diameter and therefore can only be used in limited situations and has a mechanism which requires much practice for smooth operation. A recent arrival on the market provides an alternative in which there is automatic triggering of the biopsy mechanism by a spring-loaded 'gun'. This 'Biopty' system (Radiplast) employs needles with a cutting notch similar to that of the 'Tru-Cut', but is available in both 14-gauge (2.0 mm) and 18-gauge (1.2 mm) diameters (*Figure 8*). The needle tip is clearly seen as it approaches the target (*Figure 9*), and the gun is 'fired' so that the inner notched obturator and outer cutting cannula are advanced in rapid succession, one over the other, into the lesion. The 'Biopty' gun thus increases the rapidity and reliability

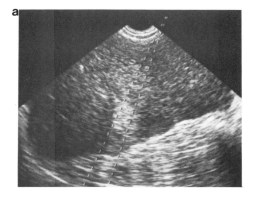

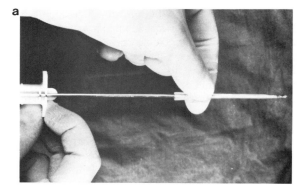

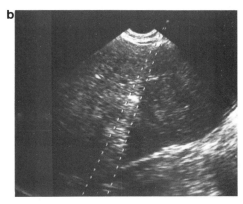

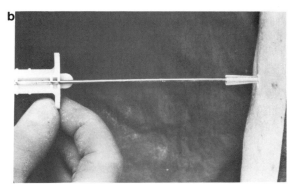

Figure 9. (a) Mass in liver. (b) 'Biopty-cut' needle tip (arrow) approaching mass.

of the biopsy, and the smaller 18-gauge needles considerably widen the range of applications (6,7).

3.3 *Plugged biopsy*

The risk of intra-peritoneal haemorrhage following core biopsy of the liver may be unacceptably high in patients with liver disease, in whom a prolongation of the prothrombin time or lowering of the platelet count may accompany the disease process, and yet these are often the very patients in whom an accurate histological diagnosis is crucial to further management and treatment. A method for safe biopsy of this group of patients has been described (8), involving the introduction of the biopsy needle through an outer sheath with subsequent plugging of the track with embolic material (e.g. absorbable gelatin sponge— 'Sterispon') (*Figure 10*). In this way, disordered blood coagulation has ceased to be a contra-indication to large needle liver biopsy.

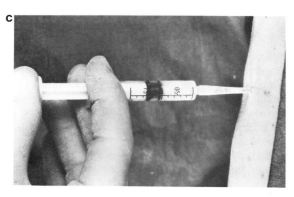

Figure 10. Plugged biopsy technique: (a) sheath over 'Tru-cut' needle; (b) sheath and needle within patient; (c) Sterispon plugs (in syringe) injected through sheath.

3.4 *Complications of biopsy*

Fine-needle aspiration biopsy is virtually free from complications and there are thus hardly any contra-indications. With such small needle diameters, haemorrhage and perforation are insignificant complications, and even vascular tumours, such as capillary haemangiomas of the liver, can be safely biopsied—as long as the needle pathway traverses normal liver tissue prior to entering the tumour (9). However, it is prudent to avoid puncturing distended gall bladders which may leak (if not decompressed with a drainage catheter), and phaeochromocytomas which may produce a hypertensive crisis if traumatized—even with a fine needle. Tumour seeding along the needle track has been reported following fine-needle biopsy (10,11) but cumulative experience shows that it is an extremely rare phenomenon and is probably avoidable by restricting the number of needle passes through the tumour.

The larger size of the core biopsy needles mitigates for more cautious usage. These needles cannot be passed through bowel or other viscera without appreciable risk of perforation or haemorrhage. Coagulation factors should be within normal limits (unless a plugged method is used), and a safe needle pathway must be carefully determined during the preliminary ultrasound scan.

4. Aspiration and drainage procedures

Ultrasonically-guided puncture of fluid-containing lesions or structures may be for either diagnostic or therapeutic purposes.

4.1 Diagnostic tap

The ultrasound diagnosis of a simple cyst, whether in kidney, liver or other viscus, requires an echo-free content, increased through-transmission of ultrasound (enhancement) and smooth, clearly defined walls. When all these features have been convincingly demonstrated, further action is rarely needed. However, if there is any doubt about these diagnostic criteria, or if the clinical features are sufficiently worrying, then the diagnosis needs to be confirmed by aspirating fluid from the cyst for pathological examination. Using a 22-gauge needle, the suspected cyst can be aspirated without risk, and this is usually performed on an out-patient basis. Care must be taken not to introduce any infection into the cyst, and unnecessary trauma to the cyst wall is prevented by attaching flexible tubing to the needle so that aspiration may proceed while the patient continues to breath quietly. Enough fluid is removed for the appropriate pathological tests to be performed, and it is standard

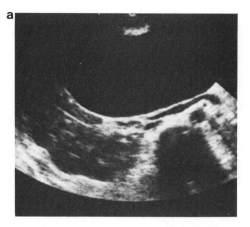

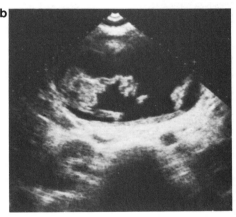

Figure 11. (a) Transverse scan of liver showing a large symptomatic cyst which was aspirated. (b) The cyst has re-accumulated and blood clots are seen within the cyst which was eventually removed surgically.

practice to go on to withdraw as much of the cyst content as is possible. Cyst fluid should be sent for cytological and microbiological examination and, in particular circumstances, fluid may also be sent for biochemical analysis (e.g. amylase estimation if pancreatic origin is suspected, or protein content if a lymphocele is a possibility). The presence of blood in the aspirate should raise the index of suspicion for malignancy, and other diagnostic studies such as computed tomography or arteriography may then be indicated.

Abscesses often present with the clinical features of fever and tenderness over the mass, or are suspected on ultrasound evidence due to the presence of debris

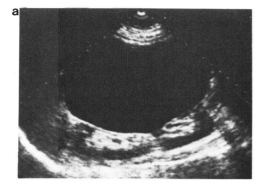

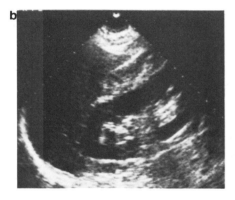

Figure 12. (a) Longitudinal scan through a large simple renal cyst which was causing polycythaemia. (b) Following cyst sclerosis with ethanol, the residual cyst has not re-expanded over a 4-year period of observation.

within the fluid or ill-defined margins of the mass. Occasionally none of these clues are apparent, and the diagnosis is only made after aspirating pus from the mass. Experience has shown that enough pus for diagnosis can always be aspirated through a 22-gauge needle, but a larger needle or catheter will be required if any therapeutic benefit is sought.

4.2 Therapeutic aspiration and drainage of cysts

Simple cysts invariably reaccumulate after aspiration (*Figure 11*), although it may take weeks or months for their former volume to be regained (12). Aspiration of cysts which are symptomatic, by virtue of their size or pressure effects on adjacent organs, will afford temporary relief, and repeated aspirations at intervals can be performed and may be an acceptable form of treatment—especially in the elderly. If a more permanent ablation of the cyst is required, instillation of a sclerosant is easily performed (13,14). An illustrative case is shown (*Figure 12*) in which a large renal cyst was causing polycythaemia. Repeated aspirations resulted in temporary remissions of the haematological disorder, but sclerosis of the cyst with absolute alcohol has prevented recurrence during a period of observation of 4 years to date.

4.3 Therapeutic aspiration and drainage of abscesses

The management of intra-abdominal abscesses has been revolutionized by the use of ultrasound for early diagnosis and for guiding drainage procedures (15). Gone are the days of submitting a debilitated patient to diagnostic laparotomy in a last ditch attempt to find the source of sepsis—non-invasive imaging will provide the answer. Nor is there any longer the necessity to operate to achieve adequate drainage of the pus. Much to the surprise of our surgical colleagues, it has been established that effective control of the suppuration can be gained with the use of the relatively narrow bore radiological drains that can be inserted under either ultrasound or computed tomographic guidance (16,17). In certain circumstances, it is now accepted (contrary to time honoured surgical teaching) that insertion of a drain can be omitted, and simple aspiration of pus in combination with systemic antibiotic therapy will suffice. This form of therapy has been particularly successful in the treatment of pyogenic liver abscesses (18), when there is a good blood supply to the tissues around the abscess, ensuring access of the antibiotics to the focus of infection (*Figure 13*). However, the majority of intra-abdominal abscesses do require drainage and a variety of specially designed catheters are now available for the purpose (19). Guide wires can be inserted through sheathed needles [e.g. 18-gauge Longdwell, (Becton Dickinson)], and either Pigtail catheters or larger bore double-lumen sump catheters can be passed over the guide wire into the cavity. Alternatively, trocar catheters with an inner stylet can be inserted directly into the abscess under imaging guidance. Subsequent management of the catheter depends upon the clinical course. All patients should be treated with appropriate antibiotics and many abscesses will cease to drain within 24−48 h, in which case the catheter can safely be removed without further ado. If continuing drainage

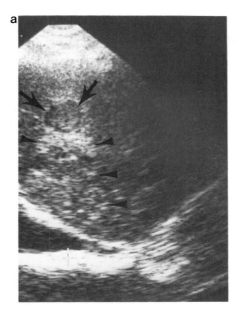

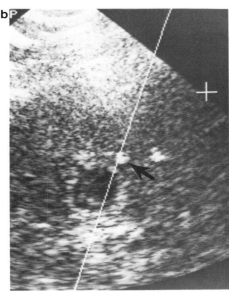

Figure 13. (**a**) Small liver abscess deep within the liver (note posterior enhancement). (**b**) Needle tip approaching abscess which was then aspirated.

takes place, then further imaging, by ultrasound, computed tomography or sinography/fistulography will be needed to plan further therapy (15).

4.4 *Drainage of viscera*

Interventional uro-radiology relies heavily upon the ability of ultrasound to guide needles, guide wires and catheters into the urinary tract. Antegrade pyelography in both the orthotopic and transplanted kidneys is performed by ultrasound-guided puncture of the renal collecting system with a fine needle followed by injection of radiographic contrast medium under fluoroscopic control. Not only is this a diagnostic procedure in its own right, but it is also the first stage of a percutaneous nephrostomy (20), when it is usually performed in the radiographic screening room using a mobile ultrasound unit, although the entire percutaneous nephrostomy procedure can be performed under ultrasound guidance alone. Ultrasound-guided suprapubic bladder catheterization is a simple procedure that is worthy of mention as it may be used to come to the rescue of the surgeons—especially in neonates (20).

There is much less dependence upon ultrasound for biliary tract catheterization procedures, and most radiologists are content to use radiographic screening. However, ultrasound-guided puncture of the gall bladder for drainage of empyema of the gall bladder can prove life-saving in elderly patients who are unfit for surgery, and percutaneous transhepatic gall bladder puncture is receiving attention for gallstone dissolution purposes.

Other fluid-containing anatomical structures that have been the target of ultrasound guided puncture include the portal venous system for transhepatic portal venography and venous sampling, the pancreatic duct for percutaneous pancreatography, the stomach for feeding gastrotomy and percutaneous pancreatic cysto-gastrotomy, and various systemic arteries and veins for placement of intravascular catheters.

5. Interventional ultrasound in obstetrics

Pre-natal diagnosis depends upon either the ultrasound demonstration of a structural abnormality of the fetus, or the pathological study of fetal tissue gathered using interventional ultrasound techniques—which also offer a route for certain pre-natal therapies.

5.1 *Amniocentesis*

This is probably the most widely practised of all interventional ultrasound procedures, and most authorities

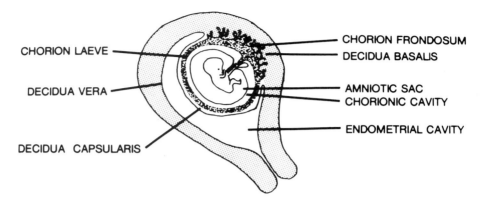

Figure 14. Anatomy of the uterus at around 9–10 weeks of gestation.

would agree that some form of ultrasound guidance for amniocentesis is absolutely mandatory. At its simplest, this guidance will involve preliminary ultrasound scanning for assessment of placental position and location of an accessible pool of liquor. The amniocentesis should follow the scanning with the minimum of delay and without the patient moving her position. The placenta and fetal parts are thus avoided and the needle is inserted to a predetermined depth. In the vast majority of cases, direct real-time visualization of the needle insertion is unnecessary, but it should be said that the performance of such simple procedures, using either a puncture attachment or indirect needle-guidance, is a valuable opportunity for the aspiring interventional ultrasonologist to gain practical training in needle-guidance methods. In a small proportion of cases, especially when there is oligohydramnios, access to liquor will be difficult and a needle guidance system is then essential.

5.2 Chorion villus biopsy

The problem with amniocentesis is that in order to obtain sufficient cells for genetic analysis, the procedure has to be delayed until the 16th week of gestation and the cells then have to be cultured for another 3–4 weeks. If indicated, termination of pregnancy will thus take place at around the 20th week of gestation. Chorion villus biopsy enables fetal tissue to be obtained as early as the 8th week of pregnancy, and adequate material can be harvested so that direct chromosome analysis can be performed, enabling a karyotype to be available within hours of the procedure. Termination of pregnancy in the first

Figure 15. The 18-gauge outer needle and 20-gauge inner cannula used for transabdominal chorion villus biopsy.

trimester runs much less risk of physical and emotional harm to the mother than termination in mid-second trimester.

A comprehensive review of the subject has recently been published (21). The aim is to obtain villi from the chorion frondosum (*Figure 14*). The transcervical route may be used for the passage of a cannula into the chorion frondosum under ultrasound visualization, or transabdominal chorion biopsy may be performed using a needle-guide system. This latter approach is preferred in the author's department (22,23). An 18-gauge needle (*Figure 15*) is inserted through the abdominal wall and myometrium into the edge of the placenta. An inner 20-gauge cannula is then passed through the outer needle into the placenta (*Figure 16*) and villi are aspirated. The inner cannula is withdrawn and the sample is inspected. If insufficient villi have been obtained then further aspirations can be performed by re-inserting the cannula through the outer

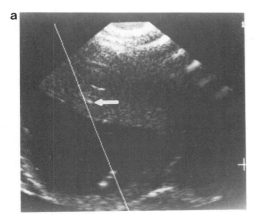

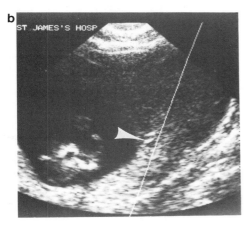

Figure 16. (a) Needle tip (arrow) is seen within an anterior placenta. (b) The inner cannula (arrowhead) is advanced beyond the outer needle tip into a posterior placenta.

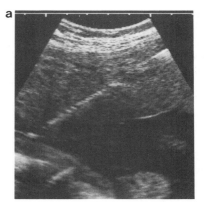

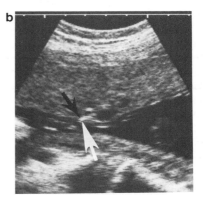

Figure 17. Percutaneous ultrasound-guided fetal blood sampling from an umbilical vessel. (a) Free-hand guidance to show the length of the needle traversing the placenta towards the cord insertion; (b) Tip of needle (arrows) within a cord vessel.

needle. Of the first 100 such procedures performed in this manner at St James' University Hospital, Leeds, there was only a single instance of a hugely obese patient from whom insufficient material was obtained for the intended karyotyping, gene probe analysis or enzyme analysis to be performed (23). The procedure is carried out using local anaesthetic without sedation, and has proven to be quick, safe, reliable and acceptable to the patients.

5.3 Fetal blood and other tissue sampling

The acquisition of a sample of blood from the fetus opens up a vista of possibilities for pre-natal diagnosis

way beyond the genetic abnormalities diagnosable via amniocentesis and chorion villus biopsy. A wide range of biochemical, serological, microbiological and other studies can be undertaken with such a sample.

Fetal blood may be obtained by puncture of an umbilical cord vessel under fetoscopic guidance (24) or, less invasively, by percutaneous ultrasound-guided umbilical cord puncture (*Figure 17*) (25), or by ultrasound-guided puncture of the fetal heart.

Other fetal tissue that is obtainable for analysis in appropriate clinical circumstances includes fetal skin and liver, each of which requires biopsy needle insertion under direct ultrasound guidance.

5.4 Fetal therapy

The most long established form of ultrasound-guided

fetal therapy is intra-uterine transfusion for rhesus haemolytic disease. Compatible blood is infused into the fetal peritoneal cavity via a fine catheter passed through an outer needle which has been inserted into the fetal abdomen under ultrasound visualization (*Figure 7*). An elegant refinement of this technique is to infuse the blood directly into the umbilical vein which has been punctured in its intrahepatic portion (26).

Other forms of ultrasound-guided fetal therapy include shunting of urine from the fetal bladder into the amniotic cavity by insertion of a double-pigtail stent, and similar shunting of dilated fetal ventricles for decompression of hydrocephalus, but both these manoeuvres, although technically feasible, are of controversial value. Finally, there is selective fetocide, in which a noxious substance can be injected into the fetal heart by ultrasound-guided puncture. Such procedures have been used in twin pregnancies where an unhealthy fetus is felt to be endangering a normal fetus, and in multiple pregnancies when sacrifice of some of the fetuses will increase the survival chances of the remainder.

6. Conclusion

Interventional ultrasound is a rapidly developing subject. Advances in technology have led to improved imaging and diagnosis, and better visualization of puncture needles. The portability of ultrasound equipment means that procedures can now be performed upon sick patients at the bedside and in intensive care units. Progress in the wider field of interventional radiology has resulted in improved needle and catheter design, and to the increasing acceptance of, and demand for, guided interventional procedures by clinicians.

Applications of interventional ultrasound that have (for reasons of brevity) had to be omitted from the foregoing, include needle-guidance by endoprobes (transrectal, transvaginal, transgastric), intra-operative ultrasound-guided punctures (of brain, pancreas and other abdominal viscera), ultrasound-guided coeliac plexus block, ultrasound-guided insertion of radio-active seeds (for tumour therapy), ultrasound-guided parathyroid ablation and many others. There is no doubt that the ingenuity and inventiveness of interventional ultrasonologists all over the world will ensure continuing expansion of this exciting and rewarding field.

7. References

1. Holm,H.H., Kristensen,J.K., Rasmussen,S.N., Northeved,A. and Barlebo,H. (1972) Ultrasound as a guide in percutaneous puncture technique. *Ultrasonics,* **10**, 83–86.
2. Goldberg,B.B. and Pollack,H.M. (1972) Ultrasonic aspiration transducer. *Radiology,* **102**, 187–189.
3. Otto,R.C. and Wellauer,J. (1986) *Ultrasound-guided Biopsy and Drainage.* Springer-Verlag, Berlin.
4. Aindow,J.D., Deogan,D.S., Robins,P. and Lesny,J. (1986) Fine-needle biopsy—enhanced needle visualisation using tip mounted miniature polymer transducers. Presented at *4th International Congress on Interventional Ultrasound,* Copenhagen.
5. McDicken,N., Anderson,T., McKenzie,W.E., Dickson,H. and Scrimeour,J.V. (1984) Ultrasonic identification of needle tips in amniocentesis. *Lancet,* **2**, 198–199.
6. Lindgren,P.G. (1982) Percutaneous needle biopsy. A new technique. *Acta Radiol. Diagn.,* **23**, 653–656.
7. Ubhi,C.S., Irving,H.C., Guillou,P.J. and Giles,G.R. (1987) A new technique for renal allograft biopsy. *Br. J. Radiol.,* **60**, 599–600.
8. Riley,S.A., Ellis,W.R., Irving,H.C., Lintott,D.J., Axon, A.T.R. and Losowsky,M.S. (1984) Percutaneous liver biopsy with plugging of needle track: a safe method for use in patients with impaired coagulation. *Lancet,* **2**, 436.
9. Solbiati,L., Livraghi,T., De Pra,L., Ierace,T., Masciadri,N. and Ravetto,C. (1985) Fine-needle biopsy of hepatic haemangioma with sonographic guidance. *Am. J. Radiol.,* **144**, 471–474.
10. Ferucci,J.T., Wittenberg,J., Margolies,M.N. and Carey,R.W. (1979) Malignant seeding of the tract after thin-needle aspiration biopsy. *Radiology,* **130**, 345–346.
11. Smith,F.P., Macdonald,J.S., Schein,P.S. and Ornitz,R.D. (1980) Cutaneous seeding of pancreatic cancer by skinny-needle aspiration biopsy. *Arch. Intern. Med.,* **140**, 855.
12. Saini,S., Mueller,P.R., Ferucci,J.T., Simeone,J.F., Wittenberg, J. and Butch,R.J. (1983) Percutaneous aspiration of hepatic cysts does not provide definitive therapy. *Am. J. Radiol.,* **141**, 559–560.
13. Bean,W.J. (1981) Renal cysts: treatment with alcohol. *Radiology,* **138**, 329–331.
14. Goldstein,H.M., Carlyle,D.R. and Nelson,R.S. (1976) Treatment of symptomatic hepatic cyst by percutaneous instillation of pantopaque. *Am. J. Radiol.,* **127**, 850–853.
15. Irving,H.C. and Robinson,P.J. (1988) The diagnosis and management of intra-abdominal abscesses and fluid collections. In Simpkins,K.C. (ed.) *A Textbook of Radiology, Volume 4—The Alimentary Tract.* H.K.Lewis, London.
16. Van Sonnenberg,E., Mueller,P.R. and Ferucci,J.T. (1984) Percutaneous drainage of 250 abdominal abscesses and fluid collections. Part 1: Results, failures and complications. *Radiology,* **151**, 337–341.
17. Gronvall,S., Gammelgaard,J., Haubek,A. and Holm,H.H. (1982) Drainage of abdominal abscesses guided by sonography. *Am. J. Radiol.,* **138**, 527–529.
18. Berger,L.A. and Osborne,D.R. (1982) Treatment of pyogenic liver abscesses by percutaneous needle aspiration. *Lancet,* **1**, 132–134.
19. Mueller,P.R., van Sonnenberg,E. and Ferrucci,J.T. (1984) Percutaneous drainage of 250 abdominal abscesses and fluid collections. Part 2: Current procedural concepts. *Radiology,* **151**, 343–347.

20. Irving,H.C., Arthur,R.J. and Thomas,D.F.M. (1987) Percutaneous nephrostomy in paediatrics. *Clin. Radiol.,* **38**, 245−249.

21. Liu,D.T.Y., Symonds,E.M. and Golbus,M.S. (eds) (1987) *Chorion Villus Sampling.* Chapman and Hall, London.

22. Lilford,R.J. (1986) Chorion villus biopsy. *Clin. Obstet. Gynaecol.,* **13**, 611−632.

23. Lilford,R.J., Linton,G., Irving,H.C., and Mason,M.K. (1987) Transabdominal chorion villus biopsy: 100 consecutive cases. *Lancet,* **1**, 1415−1417.

24. Rodeck,C.H. and Nicolaides,K.H. (1983) Ultrasound guided invasive procedures in obstetrics. *Clin. Obstet. Gynaecol.,* **10**, 515−539.

25. Daffos,F., Capella-Pavlovsky,M. and Forestier,F. (1983) A new procedure for pure fetal blood sampling *in utero. Prenatal Diagn.,* **3**, 271−277.

26. Bang,J., Bock,J.E. and Trolle,D. (1982) Ultrasound guided fetal intravenous transfusions for severe rhesus haemolytic disease. *Br. Med. J.,* **1**, 373−374.

Chapter 15

Safety of diagnostic ultrasound

Roland Blackwell

1. Introduction

The output level of some types of diagnostic ultrasound machine is sufficiently large for there to be a theoretical possibility of hazard. Whilst the existence of hazard has never been demonstrated experimentally, scanning equipment should never be used without thought and high standards of practice should be maintained.

In this chapter some of the important mechanisms for interaction between ultrasound and tissue are introduced and considered in relation to the acoustic output of commercial scanners. A brief resume is given of the experimental findings for the effects of ultrasound on humans, animals and other systems. Finally a check list for maximum safety is suggested.

2. Thresholds

To palpate a patient requires a pressure of about one pound and is considered beneficial. However, to run over the same patient with a steam roller weighing 10 tons is hazardous! Between the two limits there is a pressure beyond which benefit is lost. Similarly, with ultrasound there is an intensity threshold above which we expect a significant risk to be present. The range of intensities used in medicine is much the same as from palpation to a steam roller. The average intensity of a sound beam used in diagnosis is well below one hundredth of a Watt per square centimetre ($W\ cm^{-2}$) while the intensity of a beam used for ultrasound surgery may be as much as $100\ W\ cm^{-2}$.

It would seem straightforward to compare the output of the equipment we use with the threshold at which damage occurs and ensure that the threshold is never exceeded. Unfortunately the situation is complex. The living body is very robust and the threshold levels found for effects in the laboratory may be significantly higher for patients. Further, damage thresholds depend upon the way the output variables are combined.

3. Output variables

3.1 Energy, power and intensity

Forcing something to move involves *work* and gives *energy* to the object. Both work and energy are measured in *Joules*. Lifting a bag of sugar (1 kg) from the floor onto a table requires about 10 J of work and gives the bag 10 J of potential energy. It doesn't matter how long it takes to do the work, always the same energy is gained. Doing the work quickly uses more *power*. Power is the rate at which work is done. Lifting the bag in 1 s uses 10 W of power.

The energy in an ultrasound pulse moves the tissue structure, thus the beam has energy. There is more energy in the centre of the beam than at the edge and it is necessary to express this fact. The rate ($J\ s^{-1}$) at which the energy passes through a small area is commonly expressed in milliwatts per square centimetre ($mW\ cm^{-2}$). The value is called the *intensity* of the beam. Although the measurement is usually made over a very small area, perhaps $1\ mm^2$, the answer is expressed as if the same intensity had been spread with the same value over a complete square centimetre.

3.2 Average and peak values

Imagine you punch someone on the nose once every minute for an hour. You then become kind hearted and hold a handkerchief to their nose for a further hour to stop the bleeding. The pressure of the punches and the pressure of the handkerchief, both averaged over 1 h, are about the same, but the effect is very different!

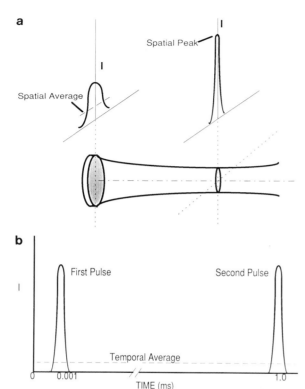

a

Spatial Peak

Spatial Average

b

First Pulse

Second Pulse

Temporal Average

0   0.001

TIME (ms)

1.0

Figure 1. (**a**) Plots of the sound intensity across the beam as it leaves the transducer, and at the position of the focus are shown. The intensity increases from the edge of the beam to the centre. The highest intensity anywhere in the beam, usually found on the axis at the focus, is called the spatial peak intensity. The spatial average intensity is the total power output divided by the area of the beam. This is often measured at the transducer when the area of the beam is the same as the area of the transducer, otherwise the edge of the beam is usually taken to be the point where the intensity has fallen to one tenth of the maximum. This is the value that would be obtained if the intensity were spread evenly across the beam. (**b**) The intensity at any one point in the beam changes with time. Each time a pulse passes, the intensity rises to a maximum. The intensity during the pulse may be very high. The peak value reached is the temporal peak intensity while the average value during the pulse is the pulse average intensity. The time between the pulses is about 1000 times longer than the pulse itself. The intensity, averaged over the time taken for the pulse plus the time between the pulses, is called the temporal average intensity and is very much lower than that occurring during the pulse.

In imaging we use repeated pulses, an event akin to the punches. Averaging the intensity over the time of the scan, the 'temporal average intensity', may give a false impression of the potential hazard.

The intensity may also be averaged over the area of the beam, to give the 'spatial average intensity'.

Intensity averaged both over the time of the scan and the area of the beam is called the 'spatial average temporal average' intensity and is designated I(SATA). This is the value usually quoted by manufacturers. However, the peak value is the maximum intensity found across the beam, the 'spatial peak intensity' and the maximum intensity during the pulse, the 'temporal peak intensity' and is designated I(SPTP) (see *Figure 1*).

In practice the instantaneous peak intensity is difficult to measure and the average intensity during the pulse is often used. The 'pulse average intensity' in the most intense part of the beam is designated by I(SPPA). Lastly, the spatial peak value averaged over the scan time is designated I(SPTA), the spatial peak temporal average.

3.3 *Pressure*

Output measurements are made with a device called a hydrophone. This measures the fluctuating pressure in a fluid as the sound wave causes the molecules to crowd into a volume and then out again. There is a formula relating intensity to the measured pressure and the frequency of the wave. The formula is inexact in practical conditions (1). Because pressure, the measured value, is reliable it is often quoted when considering safety. The instantaneous pressure may be high and is measured in *megapascal* (MPa). Atmospheric pressure is about one tenth of a MPa. If you sit on your hands they experience a pressure of about one fifth of a MPa.

4. Possible interactions associated with hazard

The biological effect of a pulse of sound depends upon the *dose* rate, the rate at which energy is deposited in the 'target'. The dose at any point in the tissues is, in practice, very difficult to estimate (2). It will depend upon the *exposure*, the total energy reaching the target, and many other variables. Because of these difficulties it is the exposure which is referred to in output measurements.

4.1 *Heating*

Only about one thousandth part of the sound energy entering the body returns into the transducer. The rest stays in the body and is converted to heat. Even if no heat was radiated away from the body the average temperature rise would be insignificant, typically less than 0.0001°C per min.

However, local heating in the beam can be significant. The local temperature rise depends upon the beam intensity, absorption coefficient of the tissues, the frequency of the sound pulse, the diameter of the beam and the time the beam dwells at the point under consideration. There are other factors, such as the thermal conductivity of the tissues and the cooling effect of the local blood supply. The considerable uncertainty in these factors makes temperature rise difficult to estimate. The initial temperature rise, before any cooling effects have time to take place, due to an I(SPTA) of 100 mW cm^{-2} and at a frequency of a few MHz is estimated to be at a rate of about 0.002°C per second. The temperature will rise exponentially over perhaps 10 min to a final value (3). The maximum possible local temperature rise due to diagnostic equipment is likely to be between 0.1 and 10°C.

An investigation using low output pulsed Doppler equipment [I(SPTA) 112 mW cm^{-2} at the skin surface] on 10 pregnant women, failed to detect any temperature change (accurate to 0.05°C) in fetal subcutaneous tissue over 10 min insonation at the site of measurement (4). The intensity of the beam could have been reduced by a factor of between 10 and 100 in passing through the abdominal wall.

It is known that a rise in temperature of the uterus of 2.4°C sustained over several hours may cause fetal damage in animals (5). It has been found (6) that there is a change in the uterine temperature of about 0.4°C when a woman moves from the lying to the standing position. It seems unlikely that under normal circumstances such temperature changes could be caused, even locally. However, special care should be taken when using endoprobes, where there is little attenuation between the probe and the target, and with the high output pulsed Doppler scanners.

Subtle effects of heating occur. For instance, the threshold voltage at which electric shocks may be felt was found to be lowered by ultrasound applied to the area. Any pulse regimes which caused an increase in skin temperature were effective (7). Using a physiotherapy machine in continuous wave and pulsed modes at an I(SPTA) of 250 mW cm^{-2} between 1 and 5 MHz the threshold sensitivity fell by about 20% according to the temperature rise induced. Interposition of a constant temperature bath between the probe and the skin caused the effect to vanish.

4.2 Cavitation

4.2.1 Stable cavitation

Within tissues there may be vapour-filled cavities of extremely small dimensions, some not much bigger than a red blood cell. These cavities, termed 'microbubbles', respond to sound waves by expanding and contracting with the low and high pressure regions of the sound wave. According to the original size of the cavity the bubble may grow in average size, or may contract.

Bubble growth is by a process called rectified diffusion (8). This is a complex procedure. In simple terms, as the cavity expands in response to the low pressure phase of the wave, dissolved vapour in the surrounding fluid diffuses into the cavity so that there is an increase in the vapour contained. The diffusion is through the enlarged surface area of the cavity. As the compression phase of the wave takes place some of the excess vapour is lost from the cavity, but less is lost than was gained. The net gain is related to the vapour concentration gradients around the cavity and the smaller surface area through which the vapour may diffuse out. The increased vapour in the cavity makes it grow. Growth continues until a resonant size is reached.

When resonance occurs, the bubble size increases and decreases with a relatively high amplitude and 'stirs' the surrounding fluids. This is a cause of 'microstreaming'. In the stirring process macromolecules, such as DNA, may be damaged. There is evidence that cell permeability may be altered with a consequent change in the charge distribution on the cell (9 – 11). This could have serious consequences, for instance during cell differentiation. There is a threshold below which this effect is not significant in tissues. At present it is believed to be about 200 mW cm^{-2} although the threshold increases with frequency. Cavitation occurs only when a long series of pressure fluctuations act on the cavity and is relevant only to continuous wave or to pulses of ultrasound containing many cycles. Cavitation is thought to be responsible, along with heating, for the beneficial effects of ultrasound physiotherapy (12).

4.2.2 Collapse (transient) cavitation

As the intensity of the pressure wave increases, a cavity may become unstable and, during the high pressure phase of the wave, the cavity may collapse

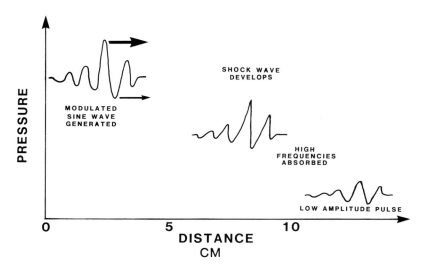

Figure 2. Pulses used in diagnostic work may have large amplitudes. The high pressure parts of the pulse travel slightly faster than the low pressure parts and, after several centimetres of travel, a shock wave builds up. The high frequency components of the shock wave are preferentially attenuated by tissue, causing localized heating. A low amplitude pulse remains.

entirely. This event is dramatic. The walls of the cavity, which becomes lenticular in shape, approach each other at supersonic speeds. The temperature in the entrapped vapour rises by several thousands of degrees centigrade and the electrons are stripped off the atoms of the vapour (13). Free radicals are produced which can modify cellular structures or result in cell death (14,15). When the walls impact bubble fragments result which may act as centres around which further cavitational events occur, enhancing the effect (16).

There is a threshold below which collapse cavitation will not occur. In aqueous solutions containing micro-bubbles of a critical size it is possible to achieve cavitation with single cycle pulses having pressure amplitudes less than produced by some diagnostic equipment (17,18). Whether cavities of critical size exist in the living body is unknown (19). The effects of transient cavitation on living tissues are uncertain but it is believed that the predominant effect would be cell death. Since many cells regularly die in normal tissues and are easily replaced, this may not have serious consequences.

4.3 Non-linear propagation

Ultrasound waves are usually described as a sine wave and drawn like ripples on a pond. This is appropriate only if the pressure fluctuations are small (20). When the fluctuations are large, as they are in pulse−echo diagnostic work, the wave is more like a large wave at the seaside which is turning into a breaker. The high pressure parts of the wave travel slightly faster than the low pressure parts and cause a shock wave to build up in which the pressure goes rapidly from a low value to a high one. The shock wave builds up over a distance of typically 7−10 cm, particularly through materials with a low attenuation coefficient, such as liquor amnii and urine (see *Figure 2*).

Because high frequencies are absorbed more strongly than low frequencies, the energy in the high frequencies associated with the shock wave will be absorbed rapidly, causing local heating to occur in a small volume (21). It is also possible that macro-molecules situated across the shock wave boundary could be damaged resulting in the potential for genetic changes. The associated forces at the boundary are termed Oseen forces.

4.4 Radiation force

When ultrasound passes through an attenuating medium a force is experienced by the medium. Small particles suspended in a tank of liquid can be seen moving across the tank when insonated by a high intensity pulsed Doppler beam. The resulting pressure on a perfect reflector is equal to twice the beam intensity divided by the speed of sound. The resulting

Table 1. Selected levels from surveys of output of commercial equipment.

Variable	Equipment type	Minimum	Median	Maximum
I(SPTA)	Mechanical sector	0.02	301	440
mW cm^{-2}	arrays	0.02	1.12	330
	Pulsed Doppler	40.00	213	4000
I(SPPA)	Mechanical sector	12 000	160 000	362 000
mW cm^{-2}	arrays	110	61 000	720 000
	Pulsed Doppler	280	25 600	150 000
Peak −ve	Mechanical sector	0.39	0.50	0.65
pressure	arrays	0.04	0.33	0.55
MPa	Pulsed Doppler	0.1	0.54	1.55

pressure on a reflecting plate can be used in a pressure balance to measure the total power of a beam. The force on a reflecting plate due to a 1 W beam is 0.135 g weight.

The movement of the medium due to radiation pressure can result in a small Bernoulli force (the force that lifts an aeroplane). The gradient across the beam results in radiation torque which causes small objects to spin in the beam.

All of these forces produce microstreaming. It is debatable whether these mechanisms are sufficiently powerful, at diagnostic intensities, to affect tissues (22).

quality images do not necessarily relate to the higher outputs.

The output of Duplex scanners is high, and in some cases extraordinarily high, with output levels towards the top of the physiotherapy range. The potential for hazard may be similar or even greater than with physiotherapy machines.

Peak negative pressures of 0.5 MPa are commonly found, that is about −5 atmospheres. After −1 atmosphere all further negative pressure must result in 'stretching' the molecular structure of the tissue. This level is certainly above the threshold for transient cavitation under ideal conditions.

5. Standards and typical outputs for commercial equipment

It is only recently that appropriate standards for measuring the output of ultrasound equipment have been available. The AIUM/NEMA standard (23) is probably the best known. Recently, the IEC have prepared a draft standard. Manufacturers should now quote the output of their equipment in a consistent form.

There have been a number of surveys of the acoustic output of modern ultrasound scanners. The AIUM publish figures provided by the manufacturers themselves for the output of commercial equipment (24). Independent surveys have also been published (25−29). The figures in *Table 1* give levels selected from these surveys.

It will be noted from *Table 1* that the range of output intensities is huge. Scanners producing low outputs are rather limited in their application. However, the better

6. Recommendations from professional bodies

Various professional bodies have studied the evidence available and have made recommendations for output levels below which it is believed that the possibility of hazard will be extremely small (30).

A widely quoted statement (31) prepared by the American Institute of Ultrasound in Medicine (AIUM) and reaffirmed in 1982 says:

'In the low megahertz frequency range there have been (as of this date) no independently confirmed significant biological effects in mammalian tissues exposed to intensities (SPTA as measured in a free field of water) below 100 mW cm^{-2}. Furthermore, for ultrasonic exposure times (i.e. total time: this includes off-time as well as on-time for a repeated pulse regime) less than 500 seconds and greater than one second, such effects have not been demonstrated even at higher intensities when the product of intensity and exposure time is less than 50 Joule cm^{-2}.'

Table 1 shows that there are many ultrasound scanners which produce pulses with an intensity in excess of 100 mW cm^{-2}. It may be inferred from the AIUM statement that such equipment could be used safely for less than 8 min provided that the total energy density is limited to 50 J cm^{-2}. For many pulsed Doppler devices, for instance, the energy density limit would be exceeded in less than 2 min use.

7. Studies on the effects of ultrasound

There are more than a thousand papers on the interaction of ultrasound with tissue. The vast majority have used conditions of exposure which are atypical of diagnostic use, which makes the results difficult to apply. The relevant data base is comparatively small. Excellent summaries of the data are available (30,32 − 34).

At diagnostic levels the only reported effects are increases in the rate of some naturally occurring feature such as low birth weight. To identify an effect of this kind, statistical methods have to be used. If a positive result is found the probability of it occurring by a random fluctuation in the numbers of the naturally ocurring event, that is by chance alone, is calculated. If less than one in 20 ($P < 0.05$), the effect is considered 'significant'. It follows that an experiment repeated 20 times will be expected to give one significant result by chance. Also, if a large number of different experiments are undertaken, significant results will also occur by chance once in 20 times. For this reason it is essential that positive results are confirmed independently before they are treated as reliable.

7.1 *Epidemiological studies*

It has been estimated that some 50×10^6 babies have been scanned *in utero* since diagnostic ultrasound was first used in pregnancy. In the UK about 80% of all pregnancies are scanned (35). Agencies such as the OPCS publish national statistics of variables such as birthweight, childhood cancers and notifiable problems. There has been no noticeable change in the percentage of low birthweight babies born since scanning became common place and the number of

cases of leukaemia and childhood cancer have fallen in recent years.

There have been a number of specific studies of human groups published recently.

Cartwright *et al.* (36) investigated all cases of childhood cancer reported in the Midlands between 1980 and 1983. There were 555 children with cancer. The ultrasound history of the mothers was compared with that of 1110 randomly selected mothers of children who did not develop cancer. There was no difference found in the incidence of ultrasound exposure in the two groups, neither was there a difference in the rate of malignancy in the exposed versus unexposed groups or any relationship with increasing number of scans.

Kinnier Wilson *et al.* (37) undertook a retrospective study of British children who had died of cancer between 1972 and 1981. There were 1731 cases. The mothers' ultrasound history was compared with 1731 matched controls. No association was found between ultrasound exposure during pregnancy and malignancy in the children up to 6 years of age.

Lyons *et al.* (38) compared 300 children who had been exposed *in utero*, with their siblings who had not. The pairs were followed up to 6 years of age and compared for height, weight, congenital malformations, neoplasms and developmental problems. No significant difference was found.

Moore *et al.* (39) studied the records of babies born in three hospitals in Denver between 1968 and 1972. They compared 1061 exposed children with 1074 controls and found a higher incidence of low birth weight (<2500 g) in the exposed group. There was no attempt to take account of reasons for the scan, such as poor obstetric history. Stark *et al.* (40) took the same data set and were able to match 425 of the exposed children with 381 controls. There was no significant difference between the two groups. Stark looked for a very wide range of physical, neurological and developmental effects and found more dyslexics in the exposed group, although the difference was not statistically significant.

In Japan Mukubo (41) compared the incidence of anomalies in babies born in hospitals before and after ultrasound was introduced. The incidence dropped from 2.54% to 2.13%. He also compared a group of 470 children whose mothers were scanned at different stages in pregnancy with 1000 children who were not insonated. Comparisons were made of a wide range of variables at birth and every 2 months up to 3 years of age. There was no significant difference in any measure at birth or in any of the subsequent paediatric assessments.

In a large detailed study Lyons and Coggrave-Toms (42) examined 2428 children but found no significant increase in the incidence of malformations, neoplasms, speech or hearing disorders or developmental problems.

The largest general study so far undertaken was by the Environmental Health Directorate of Canada (43) who in 1979 sent a questionnaire to all known clinical users of ultrasound in Canada requesting that all examinations and subsequent difficulties be reported for that year. More than 1.2×10^6 scans were reported involving 340 000 patients. Only two adverse cases were reported. One was a case of assault and the other unspecified as to its nature. This study is of sufficient size to have spotted any dramatic problem.

It is recognized that it is necessary to undertake large-scale prospective controlled trials to determine whether there are adverse affects of scanning, but preventing some patients access to a scan raises such serious issues that the trial is unlikely to be given ethical clearance in the Western world.

7.2 Animal studies

Animal studies can be carefully controlled and are likely to give helpful pointers to human effects. The studies have included investigation of fetal weight and the weight and condition of individual organs, fetal abnormalities and resorption, the nervous and immune systems and behavioral changes.

Using high intensity continuous wave (CW) or long pulses of ultrasound a number of positive results have been found, although even these have frequently not been confirmed when repeated in independent laboratories. At levels below I(SPTA) 100 mW cm^{-2} no independently confirmed positive result has, as yet, been repeated.

Of particular interest is the report of Takabayashi (44) who found an increase in abnormalities in fetal C3H mice after insonation with prolonged high intensity pulses. The length of the pulses was increased and the amplitude decreased so that the total power was constant. The effect did not persist. This suggests that the observed effects are related not just to heating but also to the peak intensity. It was further found that the effect could not be repeated even at very high intensities when the pulse length was less than 3 μs, which suggests stable cavitation as the cause.

Much of the work reported on animal behaviour has

to be discounted because the experimenters assessing the animals' response have known which animals had been insonated. This can unconsciously bias judgement. Reliable work must be performed 'blind'.

Child *et al.* (45) have shown that the larvae of fruit flies can be killed by diagnostic ultrasound. These larvae, at the appropriate stage of their development, have gas bubbles in their abdomens and it seems likely that a cavitational effect is responsible.

7.3 In vitro studies

Experiments with cell preparations are the easiest to perform and control. The results of *in vitro* work cannot be assumed to have direct relevance to the human, but do point to further investigations which should be made.

There are technical difficulties which should be understood.

(i) It is almost impossible to remove microbubbles from *in vitro* preparations, making cavitation almost inevitable.

(ii) Preparations may be susceptible to the direct heat conducted from the transducer which becomes warm due to the inefficient transduction process.

(iii) Standing waves may be set up in the container causing local high intensities.

Of the many *in vitro* experiments just two notable examples have been chosen for comment.

7.3.1 SCEs

An increase in sister chromatid exchanges (SCEs) is an event which often indicates the presence of a mutagenic agent. Just before a cell divides its complement of chromosomes is doubled, each chromosome producing a duplicate of itself. The identical chromosomes are termed 'sister chromatids'. By marking the material in the cell with a fluorescent tracer the duplicate chromatid material can be identified (46). It is found that occasionally sister chromatids break and then rejoin the opposite chromatid. There is no change in the base sequence so that the genetic code is unchanged. This event is a 'sister chromatid exchange'. It occurs naturally, but with an increased frequency in the presence of a mutagen. The increased rate of SCEs is not specific to mutagens but does occur much more frequently in their presence. There have been two reports of an increase in SCEs caused by diagnostic levels of ultrasound (47). The report causing most concern was by Leibeskind who insonated human lymphocytes. A number of attempts to repeat these results have proved negative (48,49).

7.3.2 Cell wall

Leibeskind (50) found that 3T3 mouse cells responded dramatically to ultrasound by reduced contact inhibition, increased mobility and in their surface features. The effects persisted over 10 generations of the cell. This work has not been independently repeated. It has been pointed out that 3T3 cells are difficult to work with because of their natural instability. Other workers have noted that cell motility is affected by high intensity CW ultrasound and have suggested that there is a change in the cell permeability and hence charge distribution on the cell surface (9−11). The charge distribution on the cell wall seems to be important in the process of cell differentiation.

8. Practical steps to minimize risk in scanning

In practice we have to make a risk/benefit judgement. In most cases there is no question that the value of the information we obtain will far outweigh any theoretical hazard. However, that should not give us the right to be careless in the use of ultrasound (51,3).

Next time we purchase a machine there will be the opportunity to ask what the output levels are so that we can reject equipment using extremely high intensities, but until then the following are suggested guidelines.

(i) Never scan unless there is a clear clinical objective, for example, do not scan models at exhibitions.

(ii) Make sure that your equipment is regularly checked so that it gives the best results and is accurate.

(iii) Discover if the sound output is disabled when 'Frame Freeze' is used. If not, remove the probe from the patient while making measurements.

(iv) Use the output attenuator to reduce the sound output to the lowest levels consistent with giving the quality of image you require.

(v) Discover the output of your equipment. If it exceeds an I(SPTA) of 100 mW cm^{-2} then calculate the time you can use the equipment before 50 J cm^{-2} is exceeded and stay within that time.

(vi) Take special care when scanning sensitive organs, for example the very early pregnancy, the eye and gonads; or scanning when microbubbles have been introduced into the sound field, such as in oocyte recovery.

(vii) Make sure that you keep up to date with scanning technique so that the patient gets maximum value from a scan with minimum exposure.

(viii) Ensure that trainee ultrasonographers do not spend an undue time on any one patient.

(ix) Keep abreast of experimental findings on the safety of ultrasound.

9. Conclusion

There are biological effects of ultrasound which occur at high intensities. The pulse regimes used for imaging and Doppler have not been widely used in investigation of bioeffects, and it is unknown whether there is any hazard from them. Epidemiological studies have had negative results, but have been neither prospective nor of large enough scale to show subtle effects. Mechanisms by which damage at diagnostic intensities could theoretically occur, given the right conditions, have been identified. Careful use of the modality should be encouraged.

10. References

1. Martin,K. (1986) Portable equipment and techniques for measurement of ultrasonic power and intensity. In *Physics in Medical Ultrasound*. Report 47. Institute of Physical Sciences in Medicine. 2 Low Ousegate, York, pp. 20−29.

2. Duck,F.A. (1987) The measurement of exposure to ultrasound and its application to estimates of ultrasound 'dose'. *Phys. Med. Biol.*, **32**, 303−326.

3. NCRP (1983) *Biological Effects of Ultrasound: Mechanisms and Clinical Implications*. NCRP Report No. 74.

4. Soothill,P.W., Bilardo,C.M. and Rodeck,C.H. (1987) The temperature of fetal tissues is not increased by scanning with pulsed Doppler ultrasound equipment. In *Obstetric and Neonatal Blood Flow*. Sheldon,C.D., Evans,D.H. and Salvage, J. (eds), Conference Proceedings Vol. 2, Biological Engineering Society, pp. 17−19.

5. Lele,P.P. (1979) Safety and potential hazards in the current applications of ultrasound in obstetrics and gynaecology. *Ultrasound Med. Biol.*, **5**, 307−320.

6. Southerland,I.A., Randall,N.J., Wertheim,D.F.P., Reginald, P.W. and Beard,R.W. (1987) Postural changes in pelvic blood flow. In *Obstetric and Neonatal Blood Flow*. Sheldon,C., Evans,D.H. and Salvage,J. (eds), Conference Proceedings Vol. 2, Biological Engineering Society, pp. 77−83.

7. Williams,A.R., McHale,J., Bowditch,M., Miller,D.L. and Reed, B. (1987) Effects of MHz ultrasound on electrical pain threshold perception in humans. *Ultrasound Med. Biol., 13*, 249–258.

8. Crum,L. and Hansen,G.M. (1982) Growth of air bubbles in tissue by rectified diffusion. *Phys. Med. Biol., 27*, 413–417.

9. Hughes,D.E., Chou,J.T.Y., Warwick,R. and Pond,J. (1963) The effect of focussed ultrasound on the permeability of frog muscle. *Biochim. Biophys. Acta, 75*, 137–139.

10. Chapman,I.V., MacNally,N.A. and Tucker,S. (1979) Ultrasound induced changes in rates of influx and eflux of potassium ions in rat thymocytes *in vitro*. *Ultrasound Med. Biol., 6*, 47–58.

11. Hrazdira,I. and Adler,J. (1981) Active interactions of ultrasound with the cell surface. In *4th European Congress on Ultrasound in Medicine*. International Congress Series 547, Excerpta Medica, Amsterdam, pp. 4.

12. Dyson,M. (1985) Therapeutic applications of ultrasound. In *Biological Effects of Ultrasound. Clinics in Diagnostic Ultrasound, Vol. 16*. Nyborg,W.L. and Ziskin,M.C. (eds), Churchill Livingstone, pp. 121–123.

13. Love,L.A. and Kremkau,F.W. (1980) Intracellular temperature distribution produced by ultrasound. *J. Acoustic Soc. Am., 67*, 1045.

14. Carmichael,A.J., Mossoba,M.M., Riesz,P. and Christman, C.L. (1986) Free radical production in aqueous solutions exposed to simulated ultrasonic diagnostic conditions. *IEEE Transactions on Ultrasonics, Ferroelectrics and Frequency Control*. UFFC-33, pp. 148–155.

15. Edmonds,P. and Sancier,K.M. (1983) Evidence for free radical production by ultrasonic cavitation in biological media. *Ultrasound Med. Biol., 9*, 635–639.

16. Flynn,H.G. and Church,C.C. (1984) A mechanism for the generation of cavitation maxima by pulsed ultrasound. *J. Acoust. Soc. Am., 76*, 505–512.

17. Flynn,H.G. (1982) Generation of transient cavities in liquids by microsecond pulses of ultrasound. *J. Acoustic Soc. Am., 72*, 1926–1932.

18. Crum,L.A., Daniels,S., Dyson,M., ter Haar,G.R. and Walters, A.J. (1986) Acoustic cavitation and medical ultrasound. *Proc. Inst. Acoustics, 8*, 137–146.

19. Gross,D.R., Miller,D.L. and Williams,A.R. (1985) A search for ultrasonic cavitation within the canine cardiovascular system. *Ultrasound Med. Biol., 11*, 85–97.

20. Bacon,D.R. (1984) Finite amplitude distortion of the pulsed field used in diagnostic ultrasound. *Ultrasound Med. Biol., 10*, 189–195.

21. Hynynen,K. (1987) Demonstration of enhanced temperature elevation due to nonlinear propagation of focussed ultrasound in dog's thigh *in vivo*. *Ultrasound Med. Biol., 13*, 85–91.

22. Nyborg,W.L. (1982) Biophysical mechanisms of ultrasound. In *Essentials of Medical Ultrasound*. Repacholi,M.H. and Benwell,D.A. (eds), The Humana Press, pp. 35–75.

23. AIUM/NEMA (1981) Standards Publication UL1-1981. Safety standard for diagnostic ultrasound equipment. *J. Ultrasound Med., 2*, S1–S4.

24. American Institute of Ultrasound in Medicine (1985) Acoustic data for diagnostic ultrasound equipment. Manufacturer's Commendation Panel. 4405 East–West Highway, Suit 504, Bethesda, MD 20814.

25. Carson,P.L., Fischella,P.A. and Oughton,T.V. (1978) Ultrasonic power and intensities produced by diagnostic ultrasound equipment. *Ultrasound Med. Biol., 3*, 341–350.

26. McHugh,D. (1986) IPSM survey of manufacturers data on acoustic output of diagnostic ultrasound equipment. In *Physics in Medical Ultrasound*. Report 47. Institute of Physical Sciences in Medicine. 2 Low Ousegate, York, pp. 36–46.

27. Duck,F.A., Starritt,H.C., Aindow,J.D., Perkins,M.A. and Hawkins,A.J. (1985) The output of pulse-echo ultrasound equipment. *Br. J. Radiol., 58*, 989–1001.

28. Duck,F.A., Starritt,H.C. and Anderson,S.P. (1987) A survey of the acoustic output of ultrasonic Doppler equipment. *Clin. Phys. Physiol. Measurement, 8*, 39–50.

29. NHS Procurement Directorate (1987) Joint UK Health Departments Ultrasound Equipment Evaluation Project. STD/87/18 (Available from: DHSS Leaflets, PO Box 21, Stanmore, Middlesex HA7 1AY.)

30. Wells,P.N.T. (ed.) (1987) The safety of diagnostic ultrasound. *Br. J. Radiol.*, Supplement No. 20.

31. AIUM (1984) Safety considerations for diagnostic ultrasound. AIUM Publications, 4405 East–West Highway, Suit 504, Bethesda, MD 20814.

32. Williams,A.R. (1983) *Ultrasound: Biological Effects and Potential Hazards*. Academic Press.

33. WHO (1982) Environmental Health Criteria 22. Ultrasound. World Health Organisation, Geneva.

34. Stewart,H.D., Stewart,H.F., Moore,R.M. and Garry,J. (1985) Compilation of reported biological effects data and ultrasound exposure levels. *J. Clin. Ultrasound, 13*, 167–186.

35. RCOG (1984) Report of the RCOG Working Party on Routine Ultrasound Examination in Pregnancy. 27 Sussex Place, London NW1 4RG.

36. Cartwright,R.A., McKinney,P.A., Hopton,P.A., Birch,J., Hartley,A., Mann,J., Waterhouse,J., Johnstone,H., Draper,G. and Stiller,C. (1984) Ultrasound examination in pregnancy and childhood cancer. *Lancet, 2*, 999–1000.

37. Kinnier Wilson,L.M. and Waterhouse,J.A.H. (1984) Obstetric ultrasound and childhood malignancies. *Lancet, 2*, 997–999.

38. Lyons,E.A., Coggrave-Toms,M. and Brown,R.E. (1980) Follow-up study of children exposed to ultrasound *in utero*. In *Proceedings of the 25th Annual Meeting of the American Institute of Ultrasound in Medicine*, New Orleans, pp. 49.

39. Moore,R.M.,Jr, Barrick,K.M. and Hamilton,T.M. (1982) Effect of sonic radiation on growth and development. *Proceeding of the Meeting of the Society for Epidemiological Research, Cincinnati, Ohio*, June 16–18.

40. Stark,C.R., Orleans,M., Haverkamp,A.D. and Murphy,J. (1984) Short and long term risks after exposure to diagnostic ultrasound *in utero*. *Obstet. Gynecol., 63*, 194–200.

41. Mukubo,M. (1986) Epidemiological study: safety of diagnostic ultrasound during pregnancy on fetus and child development. *Ultrasound Med. Biol., 12*, 691–693.

42. Lyons,E.A. (1986) Human epidemiological studies. *Ultrasound Med. Biol., 12*, 689–691.

43. Environmental Health Directorate. (1980) Canada Wide Survey of Non-ionising Radiation Emitting Devices. Part II. Ultrasound devices. Report 80-EHD-53 (Environmental Health Directorate, Health Protection Branch: Ottawa).

44. Takabayashi,T., Abe,T. and Sato,S. (1981) Study of pulsewave ultrasonic irradiation on mouse embryos. *Cho'onpa Igaku, 8*, 286.

45. Child,S.Z., Carstensen,E.L. and Smachlo,K. (1980) Effects

of ultrasound on *Drosophila. Ultrasound Med. Biol.*, **6**, 127−130.

46. Martin,A.O. (1984) Can ultrasound cause genetic damage? *J. Clin. Ultrasound*, **12**, 11−19.

47. Leibeskind,D., Bases,R., Mendez,F., Elequin,F. and Koenigsberg,M. (1979) Sister chromatid exchanges in human lymphocytes after exposure to diagnostic ultrasound. *Science*, **205**, 1273−1275.

48. Miller,M.W. (1985) Does ultrasound induce sister chromatid exchanges? *Ultrasound Med. Biol.*, **4**, 561−570.

49. Jacobson-Kram,D. (1984) The effects of diagnostic ultrasound on sister chromatid exchange frequencies: a review of the recent literature. *J. Clin. Ultrasound*, **12**, 5−10.

50. Leibeskind,D., Padawer,J., Wolley,R. and Bases,R. (1982) Diagnostic ultrasound. Time-lapse and transmission electron microscopic studies of cells insonated *in vitro. Br. J. Cancer*, **45**, Suppl. V, 176−186.

51. Blackwell,R. (1987) Practical issues in minimising ultrasound exposure. In *Obstetric and Neonatal Blood Flow*. Sheldon,C., Evans,D.H. and Salvage,J. (eds), Conference Proceedings, Vol. 2, Biological Engineering Society, pp. 8−11.

Index